DATE DUE

An Education in Psychology

An Education in Psychology
James McKeen Cattell's
Journal and Letters from
Germany and England,
1880–1888

Cattell, James McKeen

Selected and edited by

Michael M. Sokal

The MIT Press
Cambridge, Massachusetts
London, England

#6914784
DLC

5-21-81 JH

This book was set in VIP Janson and VIP Gill Sans by DEKR Corporation and
printed and bound by Halliday Lithograph in the United States of America.

Publication of this volume has been aided by a grant from the National Historical
Publications and Records Commission.

Library of Congress Cataloging in Publication Data

Cattell, James McKeen, 1860–1944.
  An education in psychology.

  Bibliography: p.
  Includes index.
  1. Cattell, James McKeen, 1860–1944.
2. Psychologists—United States—Biography.
3. Psychologists—United States—Correspondence.
I. Sokal, Michael M.   II. Title. [DNLM:
1. Psychology—Personal narratives.   WZ 100 C369e]
BF109.C37A32   150'.92'4   [B]   80-22505
ISBN 0-262-19185-7   80-25505

For Charlene

Contents

Preface

The publication of a large scholarly edition of previously unpublished documents, even a selective edition such as this, requires some justification. The material presented here brings together at least five major interconnected themes of interest to those concerned with the late nineteenth century, and it is expected that scholars working in a wide variety of fields—history of science, history of education, experimental psychology, American social history—will find much that is useful.

The most important theme is the personal, professional, and scientific development of a major figure in the history of psychology, education, and American science in general. Probably most commonly known today as the originator of the *American Men and Women of Science* directories, James McKeen Cattell has been viewed as a significant figure for the past hundred years. But the reasons why he has been considered important have changed through the years. In the mid-1880s Americans and Britons knew him as the one English-speaking scholar who could report with any precision the import and details of the work coming out of Wilhelm Wundt's new laboratory of physiological psychology at the University of Leipzig. By the 1890s he was seen by many as the prophet of mental testing, a new and scientific methodology for the evaluation of individual differences with many important implications for social and educational policy. Among historians of psychology Cattell has been best known as the first person to take the new German psychophysical techniques designed to study the generalized human mind and use them to investigate the differences between individuals. But at the beginning of the twentieth century Cattell's reputation was based upon the way he took control of *Science*, which had been founded by Thomas Edison and supported extensively by Alexander Graham Bell, and converted it from a magazine that could not meet expenses to the most important and profitable general scientific journal in the United States. By 1918 Cattell had become known as a martyr in the cause of academic freedom, which had been heavily attacked in America during World War I. In the 1920s he again became known for his role in expanding psychology's role in American life, this time through his Psychological Corporation.

Only Cattell's first work, in experimental psychology, is represented in the documents present here. (The rest of his life is outlined in some detail in the postscript.) Still, Cattell's career after his return to the United States in 1888 was continuous with and rooted in his European experiences and education. For example, the Cattell who publicly referred to the president of Columbia University as a liar cannot be understood without reference to the twenty-two-year-old student who alienated the Johns Hopkins faculty. Similarly, the scientific basis for the 1923 establishment of the Psychological Corporation can be seen clearly in Cattell's work in Germany in the 1880s. All of the early strands of Cattell's life and personality appear in the documents that make up this volume.

Almost as important a theme as Cattell's personal development is his experience as a student in Germany and England, which can be seen as a case study of the experiences of the 10,000 Americans who studied in Europe between 1865 and 1914. The general phenomenon of the search for advanced education by Americans in Germany is discussed in some detail in this volume's introduction. Many of the documents presented herein do much to show exactly how one American student experienced life in Germany and England and how he profited from the European universities. The point here is not so much that Cattell was typical or even representative of the flood of American students (though annotations to the documents stress when he followed the examples of those who preceded him or helped set the course for those who followed him). Some of the documents are quite routine, but they help present Cattell's experience as an example of an important phenomenon in American social and intellectual history.

Closely intertwined with this strand is the picture these documents give of just what it was like to do experimental science in the 1880s. Here again the documents stress the day-to-day routine of a laboratory scientist and give what has been called the "fine texture of the past." Modern history of science emphasizes the conceptual evolution of past theoretical constructs and their relation to important unifying themes, but equally important to a rounded view of past science is a concern for the way in which it was actually practiced.

The development of experimental psychology during this period, especially by Wundt and his students, is another topic on which the documents shed much light. As stated, Cattell's early reputation was based on his work in psychology, and his reports of just what was going on at the Leipzig laboratory in the 1880s provide the most detailed picture available of the way Wundt worked during this period. These documents make clear just how Wundt expected his students to carry out the process of introspection and how he used his instruments; they also shed light on the English reaction to Wundt's work and how that work did and did not relate to the concerns of such important thinkers as Henry Sidgwick and James Ward. Thus, this volume contributes to a comparison of two psychological traditions.

During the period covered by these documents Cattell studied at several major universities and reported details about three of them: Johns Hopkins, Leipzig, and Cambridge. His letters and journal entries about these institutions can be seen as a comparative study of late-nineteenth-century university education in America, Germany, and England. This theme is of course intertwined with all others, but it differs in that James Cattell's "audience"—his parents—helped determine the significance of his letters. William C. Cattell was an important figure in the development of American higher education and thus was sensitive in many ways to the issues his son raised in their correspondence; in addition, his ties with American educators and educational institutions enabled him to arrange for his son's first positions within this community. William Cattell's letters to his son do much to magnify the insights into American higher education gained from the son's letters.

In addition to these five major themes at least two others emerge from the Cattell documents. It is probable that documents explicitly excluded relate more closely to these themes than many of those selected. These two concerns are the internal history of the Cattell family and James Cattell's psychohistorical development. Both of these topics are important, and both are illuminated somewhat by documents included here. Readers of drafts suggested that the annotations on these points be expanded, but to do so well would require skills that the editor lacks. It is probably better to hint at these themes than to introduce amateurish interpretations at critical points. In the annotations to some of the documents that concern Cattell's personal development some mention is made of one recent psychological interpretation of the nature of the transition from adolescence to adulthood. However, no attempt is made to analyze Cattell's overall development and his relations with his parents in psychoanalytic terms, as one reader suggested. In the same way, although an analysis of the dynamics of the Cattell family would be of great interest to family historians (especially as its male members all achieved some repute in their chosen fields), the editor has no ability as a family historian, and a discussion of this topic would require the inclusion of many documents not closely related to other themes. Scholarship is, after all, based in part upon a division of labor, and the editor hopes that other historians will use the documents presented here to explore these matters in detail.

The focus of this volume, then, is on one man's personal development and on the role he played in a number of late-nineteenth-century scientific, educational, and intellectual currents. As a selective edition, it is not presented as a definitive interpretation of Cattell's life or of any of the developments with which he was involved. Like most such collections, this volume has been prepared with the hope that it will stimulate further scholarship in related areas.

Acknowledgments

I cannot acknowledge here all those who have made it possible for me to complete this volume. So many people have helped me in so many ways (some obvious and some not so obvious) that a simple list of their names would fill pages. The individuals I mention here are those who have played a direct role in helping me prepare this edition, and really represent the efforts of many other men and women.

In particular the help of the Cattell family in all of my work must be stressed. Dr. and Mrs. McKeen Cattell, Dr. Psyche Cattell, and Marcel and Elizabeth Cattell Kessel have all gone out of their way many times to support my study of their father and father-in-law. They have often disagreed with my views of his career, but our arguments have always been good-natured and substantial. Most important, these disagreements have always led me to a deeper understanding of Cattell's life, if not necessarily to one that his children would share.

Whatever skill I have as a historian rests upon the education I received in the Program in History of Science and Technology of Case Western Reserve University. Though all members of its faculty played large roles in my training, three were and are particularly important to me. Melvin Kranzberg established the program and led it through its most productive years, and Robert E. Schofield first excited me with a vision of what scholarship should be. I also owe a major debt to Robert C. Davis, who stimulated my interest in the history of psychology and who directed my doctoral dissertation on Cattell's career as a psychologist. In fact, any discussion of what I owe Bob Davis is bound to understate the debt manyfold, and it is a special pleasure to acknowledge this debt here.

Since leaving graduate school I have worked within two "invisible colleges" whose interests center around the history of American science and the history of psychology. Many members of each group have done much to help me with my work, and some I thank here for specific contributions to this edition. More generally, however, I want to thank Michele L. Aldrich, Maxine Benson, Wolfgang Bringmann, John C. Burnham, and William R. Woodwood for the very kind assistance and extremely helpful suggestions that they each provided while I was preparing this volume. Also important were suggestions made at a meeting of the informal Wellesley College Colloquium on the History of Psychology. The group's members, especially Arthur Blumenthal, Lorenz Finison, Laurel Furumoto, and Theo Kalikow gave me the benefit of their own research, and this volume is better for it. Similarly, the "invisible college" of historical editing and its visible counterpart, the National Historical Publications and Records Commission, have done much to help my work. A summer at the NHPRC-sponsored Institute for the Editing of Historical Documents (sometimes called Camp Edit) helped me develop insight into the editorial process, and two members of the NHPRC staff, Roger Bruns and George Vogt, have been especially helpful. In addition, the publication of this volume has been supported in part by an NHPRC subvention.

Of course, no collection of this sort can be prepared without the extensive assistance of librarians and archivists. Particularly helpful in this area have been Ronald E. Robbins of the David Bishop Skillman Library at Lafayette College and the staff of the Manuscript Division of the Library of Congress, especially Paul G. Sifton, Ronald Wilkinson, and Mary Wolfskill.

Worcester Polytechnic Institute, where I have been teaching since 1970, has contributed extensively to the completion of this volume. Ray E. Bolz, Vice-President and Dean of Faculty, supplied a small grant at a critical point that allowed "hard copy" to be made from what seemed like miles of microfilm. In addition, the Financial Aid Office made available, over a number of years, the help of several undergraduates under the Federal College Work Study Program. These students—Jeffrey Farash, Robert A. Fisher, Bradley W. Jarvis, Kim Brian Lemerise, David C. Patrick, and John W. Roche IV from WPI, and Margaret Brodmerkle, Judy Bourassa, and Diane Mercon from the Salter Secretarial School—proved invaluable. The Interlibrary Loan Section of WPI's George C. Gordon Library, staffed by Diana Johnson and Cosette Mailloux, sometimes made me forget that I lacked direct access to a major research library. And the secretaries of the WPI Department of Humanities, Carol Garofoli Pupecki and Loria Gourouses, were always pleased to help me as deadlines approached. Most important, my colleagues in Humanities at WPI often went out of their way to help me, and contributed greatly to the completion of this volume in many ways. Cattell's activities in Europe ranged far beyond experimental psychology, and my colleagues, who teach and write in a broad spectrum of humanistic disciplines, served as a corps of experts with whom I could readily consult. They also listened sympathetically to my ideas as they evolved, gently deflating some of the more extravagant ones and supporting those that deserved it. In some ways, parts of this volume are truly the product of a department.

During the 1973–1974 academic year, I was a visiting postdoctoral research fellow at the National Museum of History and Technology of the Smithsonian Institution. I did not work on this edition during that year, but the research I did while in Washington helped greatly to clarify my ideas about Cattell, the history of psychology, and the history of science in general. Conversations with and advice from several members of the Institution's staff (especially Audrey B. Davis, Brooke Hindle, Uta C. Merzbach, and Nathan Reingold) were particularly important in this respect, and I owe them all a great deal.

Most of the work that led directly to this edition was supported by a grant from the Program in History and Philosophy of Science of the National Science Foundation. I owe the program and especially its Director, Ronald J. Overmann, a great deal, as this grant gave me the time and resources I needed to complete this edition expeditiously. I am especially grateful for an extension that the program arranged when a major illness prevented me from completing this volume in accordance with my first schedule. In addition, the Johnson Fund of the American Philosophical Society provided for the microfilming of almost 4,000 documents at the Library of Congress.

A major benefit of the NSF grant was that it provided funds for the final typing of the manuscript and for a research assistant. Kathleen Kelly prepared several drafts as well as the final typescript, and did so in a way that made revising almost a joy. Bruce Zellers was the first research assistant to work on this edition. He was succeeded by Patrice Rafail, whose contributions to this volume can only be underestimated. Many of its strengths are the direct results of the skill and insight she brought to her work.

Drafts of large sections of this book were read by several distinguished scholars, and a number of them even read all 1,200 pages of the typescript. These individuals, Ross W. Beales, Jr., Ludy T. Benjamin, Jr., Hamilton Cravens, Solomon Diamond, John A. Popplestone, and Dorothy Ross, all suggested improvements, and I have incorporated many of their ideas. If I have not accepted all of their suggestions, it has been with many qualms, and at my own peril. But I feel that my acquaintance with Cattell's life and career enables me better than anyone else to balance the many facets of late-nineteenty-century life revealed in the documents. I have learned much from all of these scholars, and of course none of them are responsible for any errors of fact or interpretation that appear in this volume.

For several years before the publication of this edition I worked closely with William A. Bayless and Helene J. Jordan of Rockefeller University Press, who helped shape the material into something worth publishing. Similarly, Vicki P. Raeburn of Columbia University Press did much to encourage my efforts. However, what success this book will have will be based in large part on what the staff of the MIT Press did to the typescript I submitted. The skill with which they converted those 1,200 dog-eared and sometimes poorly written pages into a beautiful and readable volume must be acknowledged with thanks.

Many friends provided encouragement during the long months while I worked on this book, and did much to buoy my spirits when I doubted I would ever finish it, especially while I was ill. David Riesman gave me what only can be called inspiration, and Harold G. Reiss did much to support my self-confidence. Many other friends, including most of those whose names I've mentioned here in other contexts, took gratifying interest in my work; I especially thank Guy and Jo Anne Beales, Pat and Mike Burke, Bill and Joy Dalton, Helen F. Eckerson, Gayle Janes, and Ruth and Norm Wallace.

My children, Kate and Matt, are happy that Daddy's book is finally completed. At six and four years, they do not understand much about historical scholarship. But they do know how much this book means to me, and I appreciate their understanding greatly.

The dedication of this volume expresses my major debt, and I am pleased for many reasons to acknowledge all that I owe my wife, Charlene Key Sokal. I had thought that I would close these acknowledgments with a gushy tribute to her, but it perhaps would be more appropriate for me to paraphrase Cattell's first words to his parents about his fiancée, and keep the most important reasons for my love for her my own secret.

All manuscript material quoted in this volume is published with the permission of the literary executors of the author of each item. The Cattell family has most kindly dedicated the literary rights to the James McKeen Cattell papers to the public, and this action is an excellent example of the way they have helped the work of scholars. The following other executors and repositories are also due thanks: Mr. Alexander R. James, for permission to quote a letter from William James; the Department of Special Collections of the Milton S. Eisenhower Library at Johns Hopkins University, for permission to quote material in the Daniel Coit Gilman papers; the David Bishop Skillman Library at Lafayette College, for permission to quote material and reproduce photographs in its collection; the University of Pennsylvania Archives, for permission to quote a letter from Wilhelm Wundt found in the faculty application files for 1886; and the Presbyterian Historical Society, for permission to quote material in the William C. Cattell papers.

Notes on the Documents

The 455 documents that make up the bulk of this volume form only a small part of the extensive documentary record James McKeen Cattell left of his life and career. For example, in the Library of Congress is a collection of Cattell papers estimated at about 51,000 items,[1] and in the Division of Special Collections of the Columbia University Libraries in New York is a collection of about 500 documents entitled "James McKeen Cattell, 1860–1944, and Academic Freedom."[2] In addition, large collections of Cattell correspondence are included in the papers of such important American psychologists as Edwin G. Boring, Lewis M. Terman, and Robert M. Yerkes,[3] and in those of other important scientists such as Charles B. Davenport, Simon Flexner, and George Ellery Hale.[4]

The Cattell collection at the Library of Congress, the most important collection, was acquired by the library's Manuscript Division in three major stages. The first set of about 15,000 items was donated in 1957 by Jaques, Cattell's youngest child.[5] This collection was supplemented in 1966 by a

[1]*James McKeen Cattell: A Register of His Papers in the Library of Congress* (Washington, D.C.: Manuscript Division, Reference Department, Library of Congress, 1962), as supplemented by an additional unpublished register, available at the manuscript reading room.
[2]Described in catalog of Columbia University Libraries Division of Special Collections.
[3]Edwin G. Boring, "Psychologists' Letters and Papers," *Isis* 58 (1967): 103–107; Michael M. Sokal, comp., *A Guide to Manuscript Collections in the History of Psychology and Related Areas* (Wellesley, Mass.: Wellesley College Colloquium on the History of Psychology, 1977).

[4]Whitfield J. Bell, Jr., and Murphy D. Smith, comps., *Guide to the Archives and Manuscript Collections of the American Philosophical Society* (Philadelphia: American Philosophical Society [*Memoirs*, vol. 66], 1966); Daniel J. Kevles, ed., *Guide to the Microfilm Edition of the George Ellery Hale Papers, 1882–1937* (Washington, D.C.: Carnegie Institution of Washington, 1968).
[5]"Annual Report on Acquisitions: Manuscripts," *The Library of Congress Quarterly Journal of Current Acquisitions* 15 (1958): 179–196; Grover C. Batts, "The James McKeen Cattell Papers," ibid. 17 (1960): 170–174.

small set of papers given by Quinta Kessel, Cattell's third daughter and fifth child.[6] In 1971, primarily as a result of the editor's efforts, McKeen Cattell, the eldest son and second child, was able to begin the transfer of about 35,000 more items.[7] The original gift from Jaques Cattell, as supplemented by the donation from Quinta Kessel, has been cataloged by the library as the "James McKeen Cattell papers." The later gift from McKeen Cattell has been cataloged separately as the "Addition to the James McKeen Cattell papers."

The documents included in this volume are drawn primarily from two different series in both sections of the Cattell papers in Washington. The first series, "Journals," is part of the original set of papers and includes several informal diaries that Cattell kept as a boy. Most important, this series contains Cattell's record of his years in Europe, beginning with his departure for England in July 1880 (see 1.1) and closing in August 1887 with his appointment at the University of Pennsylvania and his engagement to Josephine Owen (see 6.102). The second series of documents, "Family Correspondence," is split between the original set of Cattell papers and the Addition. It includes items written between 1835 and 1944, and consists primarily of letters written by and to William and Elizabeth Cattell and their two sons, James and Henry. For the purposes of this edition the most important letters are those from James Cattell to his parents and his brother and their letters to him, but this correspondence is only one part of this series. In all it consists of about 4,000 documents, and includes extensive correspondence between William and Elizabeth Cattell, written primarily during his business trips on behalf of Lafayette College; correspondence between the senior Cattells and their brothers, sisters, nieces, and nephews; and letters to James McKeen Cattell from his aunts, uncles, cousins, and children. The "Family Correspondence" contains 1,403 letters directly relevant to the focus of this edition. All 194 entries in Cattell's journal for 1880–1887 are relevant.

In addition to the correspondence and the journal entries, two special documents were selected for inclusion in this edition: Cattell's "Journal of Experimental Work in Psychology" (3.12) and his "Statement Prepared at Prof. Wundt's Request" (5.17).

[6]Dorothy S. Eaton, "Recent Acquisitions of the Manuscript Division," *The Quarterly Journal of the Library of Congress* 23 (1966): 274–289.

[7]"Manuscript Division Acquisitions," ibid. 30 (1973): 333; 31 (1974): 273; 33 (1976): 376.

Notes On Style

The preparation of this volume involved three main editorial functions—selection, transcription, and annotation—which must be described.

It was obvious from the outset that this volume could not and should not include all 200 journal entries and 1,400 family letters representing the period of focus. However, it was not immediately clear how many items should be included, and a precise goal was never set. Instead implicit criteria were established and continually refined.

In general the criteria for selection revolved around the relevance of a document to one or more of the major developments of the 1880s (noted in the preface) with which the volume was to be concerned. Hence, a document was chosen for inclusion if it illustrated the way an American student lived and worked in Europe during this period; if it provided insight into the operations of the German, British, and American educational and scientific communities, especially by comparison; if it pertained directly to the development of experimental psychology, particularly by Wilhelm Wundt and his students; if it gave a good idea of what it was like to practice experimental science as a student; or, especially, if it helped trace Cattell's own scientific and personal development. In addition, several documents (such as 5.81 and 6.151) were selected purely on the basis of "human interest"; however, items pertaining solely to the Cattells' family life have been omitted.

Some may wish that no selection had been made, and that all the letters had been transcribed and annotated. Economic considerations made such a course impossible, but in any case the value of having all 1,600 documents in print is doubtful at best. Those who care to see the entire set of letters and the journal may consult the originals in Washington at the Library of Congress or the microfilm copies in Philadelphia at the Library of the American Philosophical Society. Scholars (especially psychologists and educators) with interests different from those of the editor might have wished for a different set of criteria. (For example, a historian of the American family might argue that the omission of items related solely to family life detracts from the value of the volume.) Still others would have applied the criteria in a different way. No selection can satisfy everyone; it is hoped that the criteria and their application represent a reasonable compromise.

Preliminary selection yielded about 1,000 items, a total too large for a single volume. The criteria were refined and made more explicit, and a second selection reduced the total to about 600 documents. These letters and journal entries were all transcribed and research for annotation was begun. It was still thought that 600 items were too many, and during the annotation process items that duplicated others or whose contents were outside the focus of the volume were culled. In this way 455 documents were finally selected for inclusion. The selection process did not involve many difficult or major decisions, once the criteria had been established;[1] those who have compared the documents included with those omitted have agreed with the choices made.

It was first thought that the entire contents of each letter or entry selected should be included. However, it was quickly realized that Elizabeth Cattell's long paragraphs about the health of her relatives and William Cattell's extensive reports on Presbyterian Church politics were too far from the major concerns of this edition; hence, such passages were omitted. For similar reasons it was decided to omit certain paragraphs of James Cattell's own letters—particularly his weather reports, advice on the furnishing of his parents' home, and dental complaints. All omissions are mentioned in the headnotes to the documents in question. Much of the omitted material can be described as gossip—"trivial, chatty talk or writing," to use the sense of the word given in the *American Heritage Dictionary*. Again, historians interested in American family life in the 1880s or in the operation of the Presbyterian Church during this period are referred to the Library of Congress and American Philosophical Society collections.

All documents in this volume were transcribed literally. Arguments in favor of expanded or modernized transcription seem to apply primarily to eighteenth-century items, or to documents written in an almost indecipherable hand or in code.[2] Cattell wrote very clearly, especially before he turned thirty; his mother's handwriting is also easily readable. His father's hand does present some problems, but the spelling and grammar they all used are close to those of the late twentieth century. Where the transcription of a name or any word is uncertain the fact is indicated in a note. Where one or

[1] For a similar conclusion about a selective edition of the letters of an important American political figure see Elting E. Morison, "Selecting and Editing the Letters of Theodore Roosevelt," *Harvard Alumni Bulletin* 60 (1958): 598–601.

[2] This point has been made well recently by G. Thomas Tanselle, "The Editing of Historical Documents," *Studies in Bibliography* 31 (1978): 1–56.

more words are totally indecipherable, because of wormholes or stains or (especially in the case of William Cattell) because of poor handwriting, the number of indecipherable words, if determinable, is indicated in the text in brackets. Where the writer made any but the most trivial changes in a document, the final version is presented and earlier readings are indicated in notes.

There is one important exception to the rule of literal transcription: Cattell, like many other students at German universities during the period, learned Pitman shorthand so that he would be able to record lectures verbatim.[3] Many of the short words (primarily prepositions) in his journal, especially in the entries written between July 1880 and January 1881, were written in shorthand. Although the transliteration of shorthand to English is not always straightforward, it usually can be accomplished with little uncertainty once the stenographer's idiosyncrasies are recognized.[4] The short words that Cattell wrote in shorthand in these early entries have been silently transliterated into English with the help of an expert stenographer.[5] The presence of shorthand in an entry is indicated in a headnote. More troublesome were several journal entries, all from 1881, written almost entirely in shorthand. Though most proper names, place names, and titles were written out, these entries were often difficult to transliterate, even with the editor and the stenographer working together. The published texts of these entries, therefore, are not as accurate as the others. When symbols were undecipherable their number (and hence the number of missing words) is indicated in brackets. For these entries no effort has been made to recover Cattell's original punctuation and spelling, especially as Pitman shorthand uses a phonetic system. However, both are implied by context, and every effort has been made to present Cattell's meaning, or at least the editor's understanding of it.

The heading for each letter and each journal entry indicates its author and addressee, where it was written, and the actual date on which it was written. In all cases James Cattell's name has been omitted; "Journal" refers to an entry in James Cattell's journal, "to parents" indicates a letter from James to William and Elizabeth Cattell, and "from Elizabeth Cattell" means a letter to James from his mother. The original place and date as indicated by the writer are usually omitted. James Cattell and his parents all regularly indicated from where they were writing, and the few exceptions are easily placed from internal evidence. Dating each letter is more complicated. James Cattell usually wrote only the day of the week and the date, without the year, at the head of his letters; for example, "Wed., Dec. 10th." His letters

[3] See R. H. Lotze, *Grundzüge der Psychologie* (Leipzig: S. Hirzel, 1881), which was the record of Lotze's last series of lectures as recorded by his son. See also document 2.63.
[4] A similar conclusion was reached by the editors of *The Papers of Woodrow Wilson*, vol. 24, *1912* (Princeton, N.J.: Princeton University Press, 1977); see pp. vii–xiii.
[5] Alice Clarken of Case Western Reserve University, Cleveland, Ohio.

have been dated through the use of a perpetual calendar. Elizabeth Cattell usually wrote the full date on each of her letters. William Cattell also did so at times, but on other occassions he gave only the day of the week ("Weds.," or "Satur. evening"). Both parents occasionally misdated their letters, sometimes by noting the incorrect year. The dates of these letters are determined from internal evidence. Also, as James Cattell sometimes noted (as in 5.29), it was not unusual for him to confuse dates. When the use of a perpetual calendar leads to a date that is obviously incorrect, internal evidence is used to date the letter correctly. When the dating of a letter has been corrected, the date written on the letter is indicated in a headnote.

James Cattell always addressed his parents as "Mama" and "Papa." His mother usually addressed her letters to "My own Dear Jim" or to "My own Dear Boy," while his father's letters usually began "My dear son." The regularity of these salutations has led to their omission here. Cattell usually closed his letters with "Your loving son Jim," and once complained to his parents that he ruined a business letter by closing it this way. His parents usually closed with "Your loving Mamma" and "Your loving Father." Except in unusual circumstances these closings have also been omitted here.

The annotation is extensive. The National Historical Publications and Records Commission's statement,[6] which stresses a "lean" approach to annotation, was used as a general guide, but that statement focuses on the annotation of very large collections of the papers of major political figures. For the sake of the wide audience that might be interested in this selective edition, whose subject is a relatively little-known figure who worked in a technical field, it was decided to annotate many seemingly peripheral items that indeed shed interesting sidelights on the several themes that run through the papers. Social historians might find the notes on the poems Cattell read an unnecessary waste of space, but they are asked to remember that all too many people with primarily technical education might need this information to help them understand the cultural context. Similarly, psychologists might argue that the identification of the tourist attractions Cattell visited in Europe and the ships on which he traveled is a distraction; they are asked to consider the needs of the social historians.

Probably more important than the notes identifying proper names are those that explain technical details, put the mention of a person into historical context, trace the origins of Cattell's ideas or use of words, or relate the contents of a document to published material by Cattell or others. Here, again, it was decided to try to establish the links that would help place Cattell's work into a larger context.

It became immediately clear that, with some exceptions noted below, each note should be kept as short as possible. This decision was based not only on the NHPRC guidelines, but also on concern about the ultimate cost and

[6]"Commission Policy Statement: On Annotation and Selectivity," *Annotation: The News-letter of the National Historical Publications and Records Commission* 4, no. 3 (1976): 2–3.

size of the volume and on the desire not to swamp the documents in annotation. Notes identifying most proper names and places are short and refer the reader to more detailed sources. Annotations have also been kept short through the use of short titles, which (except in the case of consecutive documents, where repeated cross-referencing seemed unnecessary) refer users to the document in which a given person or place was first identified. James McKeen Cattell's name is abbreviated to JMC in all notes.

Notes that do more than identify incidental people and places, however, are not abbreviated so extensively. In particular, notes about people or books important to Cattell's career or to one or more of the other major themes are intended to provide enough information about the subject to make its significance clear. These notes are kept as short as possible—in part, through the use of headnotes and editorial passages between documents. Short titles are sometimes used to provide cross references. However, this method is used more to establish links and provide additional context for the documents in question than for the sake of abbreviation. Such technical documents as Cattell's "Journal of Experimental Work in Psychology" and even his descriptions of experimental apparatus seemed to require extensive annotation for the sake of nonspecialists; even modern psychologists have been found to get lost without help. Historians of science specializing in late-nineteenth-century experimental psychology consider Cattell's procedures unique, and have found the annotation of these documents helpful.

Notes on Bibliographic Sources

The sources used most often are listed here, together with the abbreviation or short title to be used for each in the notes to the documents. Other sources used less frequently are cited in full in the notes, except when a second citation is shortened because it appears near the first mention of a given source. When no source is noted for a document, it is to be found in the Cattell papers at the Library of Congress.

Sources generally familiar to scholars are presented without full bibliographic data. Other extremely useful sources—including several important manuscript collections—are known primarily to specialists, and these are given below in full. Several of the listings are annotated in order to clarify how they were used in the preparation of this volume.

AMS: American Men of Science.

Baedeker, . . .: Most of the notes on geographical features mentioned in the documents are based upon one or another of the *Handbooks for Travellers* published by Karl Baedeker at Leipzig throughout the nineteenth century (see introduction to section 1). Among those that have been used most extensively are the following. *The Dominion of Canada*, 1st edition, 1894; *The Eastern Alps, Including the Tyrol* . . ., 4th edition, 1879; *Great Britain: England, Wales, and Scotland*, 1st edition, 1887; *Italy, From the Alps to Naples*, 2nd edition, 1909; *London and Its Environs*, 2nd edition, 1879 (for Section 1); 5th edition, 1885 (for Sections 5, 6); *Northern Germany*, 6th edition, 1877 (except as noted); *Norway, Sweden, and Denmark*, 8th edition, 1903; *Paris and Its Environs*, 10th edition, 1891; *The Rhine from Rotterdam to Constance*, 9th edition, 1884; *Southern Germany and Austria*, 7th edition, 1891; *Switzerland*, 14th edition, 1891; *The United States*, 1st edition, 1893.

Biographical Catalog of Lafayette College: John Franklin Stonecipher, comp. (Easton, Pa.: Chemical Publishing Co., 1913).

Biographical Sketch of William C. Cattell: Reprinted from *A Biographical Album of Prominent Pennsylvanians* (Philadelphia: American Biographical Publishing Company, 1888).

Brasch, *Leipzig Philosophen*: Moritz Brasch, *Leipzig Philosophen* (Leipzig: Adolf Weigel, 1894).

British Museum Catalogue, Compact Edition: Catalogue of Printed Books to 1955. (New York: Readex Microprint, 1967), 27 vols.

Cattell "Autobiography": Michael M. Sokal, "The Unpublished Autobiography of James McKeen Cattell," *American Psychologist* 26 (1971): 626–635.

Cattell collection: James McKeen Cattell collection, Division of Special Collections, Columbia University Libraries, New York.

Cattell papers: James McKeen Cattell papers, Manuscript Division, U.S. Library of Congress, Washington, D.C.

William C. Cattell papers: Presbyterian Historical Society, Philadelphia.

DAB: Dictionary of American Biography.

Diehl, *Americans and German Scholarship*: Carl C. Diehl, *Americans and German Scholarship, 1770–1870* (New Haven, Conn.: Yale University Press, 1978).

DNB: Dictionary of National Biography.

DSB: Dictionary of Scientific Biography, Charles C. Gillispie, ed. (New York: Scribner, 1970–1976), 14 vols.

Encyclopedia of the Social Sciences: Edwin R. A. Seligman, ed. (New York: Macmillan, 1937), 15 vols.

EoP: The Encyclopedia of Philosophy, Paul Edwards, ed. (New York: Macmillan and Free Press, 1967), 8 vols.

Galton papers: Francis Galton papers, D.M.S. Watson Library, University College, Gower Street, London.

Gilman papers: Daniel Coit Gilman papers, Johns Hopkins University Archives, Baltimore.

Grove's: Grove's Dictionary of Music and Musicians, 5th edition, Eric Blom, ed. (New York: St. Martin's, 1954), 9 vols.

Hart, *German Universities*: James Morgan Hart, *German Universities: A Narrative of Personal Experience* (New York: Putnam, 1874).

Howison papers: George Holmes Howison papers, Bancroft Library, University of California, Berkeley.

IESS: International Encyclopedia of the Social Sciences, David L. Sills, ed. (New York: Macmillan and Free Press, 1968), 17 vols.

International Cyclopedia of Music and Musicians: Oscar Thompson and Bruce Bohle, eds., 10th edition (New York: Dodd, Mead, 1975).

JHBS: Journal of the History of the Behavioral Sciences.

JHU Circulars: The Johns Hopkins University Circulars.

Königliche Conservatorium: Das königliche Conservatorium der Musik zu Leipzig (Leipzig: Das Conservatorium, 1893).

Man of Science: James McKeen Cattell, 1860–1944: Man of Science, A. T. Poffenberger, ed. (Lancaster, Pa.: Science Press, 1947), 2 vols. This source contains many of Cattell's most important papers, and translations (by Robert S. Woodworth) of several that had only been published previously in German. In general, the original version of each of Cattell's papers, as published in its original journal, is cited in preference to the versions published in this set. Exceptions occur, however; for example, the English translations in these volumes are cited along with the originals. In addition, where appropriate, notes on the details of Cattell's experimental work refer to this set.

McGraw-Hill Encyclopedia of World Drama: (New York: McGraw-Hill, 1972), 4 vols.

Men of Lafayette: The Men of Lafayette, 1826–1893: Lafayette College, Its History, Its Men, Their Record, Sheldon J. Coffin, ed. (Easton, Pa.: George W. West, 1891).

NAW: Notable American Women, 1607–1950: A Biographical Dictionary, Edward T. James, ed. (Cambridge, Mass.: Belknap Press of Harvard University Press, 1971), 3 vols.

NDB: Neue Deutsche Biographie.

OED: Oxford English Dictionary.

Oxford Classical Dictionary: 2nd edition, H. G. L. Hammond and H. H. Scullard, eds. (Oxford University Press, 1970).

Oxford Companion to . . .: Several of the excellent volumes of this series, published by the Oxford University Press, have been used extensively, including the following: . . . *American Literature*, 4th edition, James O. Hart, ed. (1965); . . . *Art*, 1st edition, Harold Osborne, ed. (1970); . . . *English Literature*, 4th edition, Paul Harvey, ed., revised by Dorothy Eagle (1967); . . . *French Literature*, 1st edition, Paul Harvey and Janet E. Heseltine, eds. (1959); . . . *Music*, 10th edition, Percy A. Scholes, ed., revised by John Owen Wood (1970); . . . *The Theatre*, 3rd edition, Phyllis Hartnoll, ed. (1967).

Pardee diary: "A Boy's Trip to Europe, 1869–1870" (typescript copy of diary kept by Israel Platt Pardee while accompanying the Cattell family on an inspection trip of European institutions of higher education). David Bishop Skillman Library, Lafayette College, Easton, Pa.

Paulsen, *German Universities*: Friedrich Paulsen, *The German Universities and University Study*, Frank Thilly and William W. Elwang, trs. (New York: Scribner, 1906).

Poggendorff: Johann Christian Poggendorff, *Biographisch-literarisches Handwörterbuch zur Geschichte der exacten Wissenschaften* (Leipzig: Barth, 1863–1904),

6 vols.; (Berlin: Akademie Verlag, 1955–   ), vols. 7–

Pre-1956 Imprints: The National Union Catalog: Pre-1956 Imprints (London: Mansell, 1968–), 685 vols. to date.

Ross, *Hall*: Dorothy Ross, *G. Stanley Hall: The Psychologist as Prophet* (Chicago: University of Chicago Press, 1972).

Skillman, *Biography of a College*: David Bishop Skillman, *The Biography of a College: Being a History of the First Century of Life of Lafayette College* (Easton, Pa.: Lafayette College, 1932), 2 vols.

Tanner, *Historical Register*: J. R. Tanner, *The Historical Register of the University of Cambridge* (Cambridge University Press, 1917).

Venn, *Alumni Cantabrigiensis*: Part 2, *1752–1900*, John Venn and John Archibald Venn, comps. (Cambridge University Press, 1940–1954), 6 vols.

Who Was Who [Br.]: *Who Was Who* (London: Black).

Who Was Who in America: (Chicago: Marquis).

World Who's Who in Science: A Biographical Dictionary of Notable Scientists from Antiquity to the Present, Allen G. Debus, ed. (Chicago: Marquis, 1968).

Introduction

American Students in Europe

Large numbers of young Americans interested in getting advanced educations in almost all subjects have always studied in Europe. This pattern has particularly held true in science and medicine—from the mid-eighteenth century, when many of the men who were to become leading colonial physicians earned their M.D.s at Edinburgh and Leiden, through the mid-twentieth century, when James D. Watson went to work at the Cavendish Laboratory at Cambridge.[1] In the humanities and the social sciences, also, many students and scholars have gone to Europe for their education, at least since the late eighteenth century,[2] though never in the same numbers as the scientists. When these men and women returned to the United States they often played major roles in the academic, scholarly, and scientific communities of their own country. For example, all the most important presidents of the new American universities founded in the late nineteenth century studied in Europe,[3] and the *American Journal of Philology*, that of chemistry, and that of psychology were all founded by Americans with European educations. During the last fifty years this pattern of European study has decreased in importance.[4] Nevertheless, the transit of ideas and methods from Europe to America did much to shape scholarship and science in this country.

The nineteenth century, when most Americans studying in Europe enrolled in German universities, was the most important period for this trans-

---

[1] Whitfield J. Bell, Jr., "Philadelphia Medical Students in Europe, 1750–1800," *Pennsylvania Magazine of History and Biography* 67 (1943): 1–29; James D. Watson, *The Double Helix: A Personal Account of the Discovery of the Structure of DNA* (New York: Atheneum, 1968).

[2] Diehl, *Americans and German Scholarship*.

[3] For example, Daniel Coit Gilman of Johns Hopkins, and Andrew Dickson White of Cornell.

[4] Stanley Coben, "The Scientific Establishment and the Transmission of Quantum Mechanics to the United States, 1919–32," *American Historical Review* 76 (1971): 442–466.

fer. The number of American students crossing the Atlantic during this period is known to have been large, though precise figures are unobtainable and estimates differ by at least one order of magnitude.[5] The University of Göttingen, for example, enrolled more than 1,100 American students between 1782 and 1910, of whom more than 60 earned their Ph.D. degrees.[6] These Americans studied a wide range of subjects and established by 1855 a quasiofficial "American colony, with its own rules, regulations, and rituals."[7] During this period a fairly large number of guides for Americans planning to study in Europe were published, each containing suggestions as to the easiest ways to learn the German language, sketches of the strengths and weaknesses of each university, general overviews of the German university system, and hints about financial matters and coping with German life.[8]

What these Americans were seeking in Germany varied from individual to individual, but some generalizations can be made. Some sought the social polish and the acquaintance with European art and music that only a grand tour could provide at the time.[9] Even those with less artistic interests made sure to take in as much art and music as possible, realizing that such cultural resources were then generally unavailable in America. An example of this attraction can be seen in the reaction of Americans (and Europeans) to Raphael's Sistine Madonna, which was brought to Dresden by the rulers of Saxony in 1793. It was hung in its own room in the Dresden Art Gallery, and there became a magnet for tourists and other visitors. Baedeker's, the

[5] Carl C. Diehl ("Innocents Abroad: American Students in German Universities, 1810–1870," History of Education Quarterly 16 [1976]: 321–354) estimates that about 640 Americans studied at German universities before 1870, sharply disagreeing with Charles Thwing (The Americans and the German University: One Hundred Years of History [New York: Macmillan, 1928], pp. 40–43), who calculates American enrollment during the period at about 1,200–1,500. Both Diehl and Jurgen Herbst (The German Historical School in American Scholarship: A Study in the Transfer of Ideas [Ithaca, N.Y.: Cornell University Press, 1965], p. 1) agree with Thwing's estimate of about 10,000 Americans for the entire period before 1920; however, Thomas N. Bonner (American Doctors and German Universities: A Chapter in International Intellectual Relations, 1879–1914 [Lincoln: University of Nebraska Press, 1963], pp. 39–40) estimates that as many as 15,000 Americans studied medicine at German-speaking universities between 1870 and 1914.

[6] Daniel B. Shumway, "The American Students of the University of Göttingen," Ger-

man-American Annals, new series, 8 (1910): 171–254; Paul G. Buchloh, ed., American Colony of Göttingen: Historical and Other Data Collected between the Years 1855 and 1888 (Göttingen: Vandenhoeck and Ruprecht, 1976).

[7] William W. Goodwin, "Remarks on the American Colony at the University of Göttingen," Proceedings of the Massachusetts Historical Society, 2nd series, 12 (1897–1899): 366–371.

[8] Some examples: William Howitt, The Student-Life of Germany (Philadelphia: Carey and Hart, 1842); James Morgan Hart, German Universities: A Narrative of Personal Experience (New York: Putnam, 1874); Henry Hun, A Guide to American Medical Students in Europe, (New York: William Wood, 1883).

[9] Ruth Ann Musselman, "Attitudes of American Travelers in Germany, 1815–1890: A Study in the Development of Some American Ideas" (Ph.D. diss., Department of History, Michigan State University, 1952; University Microfilms no. 04321). See also Herbert Cahoon, "The Grand Tour: Memorandum from J. Pierpont Morgan," New York Times, 22 April 1979, section 10, p. 5.

preeminent European guidebooks, awarded the painting two stars. George Eliot, who in 1858 visited it for three consecutive days, each day found it "harder and harder to leave," and one American tourist even claimed that "the enjoyment of this picture is well worth the double Atlantic voyage."[10] American students, no matter what their major interest, were equally impressed. Edward S. Joynes, later a teacher of languages at several Southern universities, called it "an inspired work, so unearthly, so heavenly" and "alone, unequalled, unapproached"; William Henry Welch, later Professor of Pathology at Johns Hopkins, thought it the "most beautiful and sublime of paintings"; and George Santayana, later to teach philosophy at Harvard, was just as strongly affected.[11]

The American students and professionals with specific interests also hoped to profit from their stays in Europe. The physicians, in particular, hoped to learn the latest techniques for diagnosis and treatment, particularly in such specialties as ophthalmology and otolaryngology, in which European practice was held to be far in advance of American methods.[12] Those interested in general internal medicine also felt that they would benefit from study in Europe, even when (in the early nineteenth century) they realized that the Europeans were as ignorant as they were about the etiology of most diseases.[13] These doctors knew that foreign study would increase their prestige in the eyes of colleagues and patients and would allow them to practice at a higher social level.

Practicing American lawyers, too, sought something at European universities that American law schools could not give them. Status, of course, was one goal, but many of those who went to study in Germany hoped to develop a grounding in Roman law that would enable them to see beyond the specific issues in any one case to the nature of the law itself. Of course, any background in Roman or German law would not be useful in the day-to-day practice of an American lawyer; Roman law was outmoded and German law was less fitted to America than to an empire. But at least some Americans looked beyond daily practice toward the philosophy of jurisprudence, and even perhaps toward the nature of justice itself. In the eyes

[10] Baedeker, *Northern Germany*, pp. 221–222; J. W. Cross, ed., *George Eliot's Life, As Related in Her Letters and Journals* (New York: Harper, 1903), vol. 2, p. 42; Andrew P. Peabody, *Reminiscences of European Travel* (New York: Hurd and Houghton, 1868), pp. 80–82.

[11] Edward S. Joynes, "Old Letters of a Student in Germany, 1856–57," *Bulletin of the University of South Carolina* no. 45, part 3 (April 1916), pp. 23, 55; Simon Flexner and James Thomas Flexner, *William Henry Welch and the Heroic Age of American Medicine* (New York: Viking, 1941), 90; George Santayana,

Persons and Places, vol. 2, *The Middle Span* (New York: Scribner, 1945), pp. 2–4.

[12] Bonner, *American Doctors*, pp. 30–31.

[13] Charles Rosenberg, "The Role of Learning in Nineteenth-Century American Medicine" (paper presented to the History of Science Society, Chicago, December 1970). For most of the nineteenth century, before the emergence of the germ theory, both European and American physicians realized how little they knew about infectious diseases, and were more concerned with the development of better methods of diagnosis and treatment than anything else.

of some observers, such concerns were not reflected in the curricula of most American law schools.[14]

But the most exciting developments at nineteenth-century German universities were taking place outside the professional faculties of law, medicine, and theology, in the "Philosophical Faculties." Here, where the arts and the sciences were being taught, a new type of education was emerging in the early 1800s. Large numbers of American students, both scientists and humanists, went to Germany for the explicit purpose of learning how to do science or, in general, any type of scholarship. The approach to scholarship that German universities tried to train into their students was called *Wissenschaft*. It was a research-oriented educational theory based on many sources.[15] *Wissenschaft* was institutionalized in 1809 when Wilhelm von Humboldt founded the University of Berlin for the Kingdom of Prussia. The immediate reason for the founding of this university was Prussia's loss of the University of Halle during the Napoleonic wars. More important, however, was that "Humboldt conceived the salvation of the German nation as coming from the combination of teaching and research" in *Wissenschaft*.[16] So successful was the University of Berlin, in all ways, that many new schools were modeled after it (for example, Bonn in 1818 and Munich in 1826) and most of the existing universities were reorganized in similar fashion.[17]

Several traditions supported the development of *Wissenschaft*. Two were the ideals of *Lehrfreiheit* and *Lernfreiheit*, the freedoms to teach and to study without interference from outside the university. Though often limited in fact by the structure of doctoral examinations, these traditions were usually well honored and were often especially well supported by the state or city ministries of education that had official control over the universities. Another important tradition was the freedom of professors and students to wander from school to school. Students often heard the best lecturer in each subject at different universities, and then took their degrees (though many of the Americans did not) at the school charging the lowest graduation fee. This mobility served to bolster both *Lehrfreiheit* and *Lernfreiheit*, and allowed both students and professors to get the most out of their education and research.[18]

Despite brilliant scholarship on the part of some American professors, American educational institutions in general lacked any such *Wissenschaft* tradition before 1876, when Johns Hopkins University was founded. American scholars in all fields therefore took advantage of the availability of such an education in Germany. In 1815 Edward Everett, just elected the first Eliot Professor of Greek literature at Harvard, "went immediately to Göt-

[14] Hart, *German Universities*, pp. 112–115.

[15] Diehl, *Americans and German Scholarship*; Roy Steven Turner, "The Prussian Universities and the Research Imperative, 1806 to 1848" (Ph.D. diss., Program in History of Science, Princeton University, 1973; University Microfilms no. 72-23, 225).

[16] Abraham Flexner, *Universities: American, English, German* (New York: Oxford, 1930), p. 312. See also Paulsen, *German Universities*, pp. 50–67.

[17] Flexner, *Universities*, p. 443.

[18] Ibid., pp. 317–327.

tingen to prepare himself for his new work."[19] Göttingen was one of the older German universities, founded in the 1730s, but it had quickly followed Berlin in adopting the practices of *Wissenschaft*. As it was located in the province of Hanover and had been founded by the Elector who also was King George II of England, its ties with the English-speaking world were strong. An American was enrolled as early as 1783. The German language spoken in the city of Göttingen was purer, clearer, and more grammatically correct than that spoken elsewhere, and Americans were advised to begin their studies there so as not to become infected by the ignorant and sloppy German of, for instance, Berlin.[20]

Everett, then, had many reasons to study at Göttingen, and in 1817 he became the first American to earn a Ph.D. from the university. Even before that date, he was joined by three other Harvard men, George Ticknor, Joseph Green Carswell, and George Bancroft. All were attracted to Göttingen by a vision of *Wissenschaft*, but it is clear that their understanding of the term was limited. For many Germans *Wissenschaft* implied a good deal more than a narrow focus on research methods. Instead, in the humanities at least, it included an empathetic and broadly based understanding of human culture within an idealistic philosophical framework known as *Neuhumanismus*. The four Harvard men were more interested in research techniques, and the way scholarly oriented higher education developed in America reflects their slant.[21]

Everett and his friends laid the foundation for the development of the American colony at Göttingen. More than one hundred other Americans were to follow them before the Civil War. By the 1840s the colony was thriving, and its members formally organized themselves in 1855.[22] Another practice dating from their period of study in Germany was that of American universities paying for the European education of their professors. Though this was never widespread, the precedent established by Harvard in paying Everett's, Ticknor's, and Bancroft's fees and other expenses is one to which other scholars and universities later appealed.[23]

Once the tradition of study in Germany had been established, Americans crossed the Atlantic in large numbers, seeking advanced research-oriented education in all fields. In the 1840s and 1850s, for example, many Americans

[19]Goodwin, "American Colony," p. 366. More generally, see Cushing Strout, *The American Image of the Old World* (New York: Harper and Row, 1963), pp. 62–73.

[20]Ibid. See also Shumway, "American Students," pp. 171–172; Hart, *German Universities*, pp. 7–8; and some of the documents presented in Philip Rahv, ed., *Discovery of Europe: The Story of American Experience in the Old World* (Garden City, N.Y.: Doubleday, 1960).

[21]Diehl, *Americans and German Scholarship*, pp. 70–100. Later in the century, other Ameri-

cans studying in Germany ignored their teachers' philosophical subtleties to focus on the technical methodology they used. For the way historians misunderstood Ranke's idealist concern for universality within individuals, see George G. Iggers, "The Image of Ranke in American and German Historical Thought," *History and Theory* 2 (1962): 17–33. For the way psychologists ignored Wundt's philosophical subtleties, see below.

[22]Shumway, "American Students," pp. 252–254; Buchloh, *American Colony of Göttingen*.

[23]Diehl, "Innocents Abroad," p. 333.

studied chemistry with Friedrich Wöhler at Göttingen and with Justus von Liebig in Giessen; Liebig's American students did much to shape the practice and the profession of chemistry in nineteenth-century America.[24] Similarly, the Americans who studied history and related fields played an important role in the development of American academic social science.[25] Examples can be drawn from other areas of scholarship, and, as noted, the most important American university presidents of the late nineteenth century studied in Europe. Although "American universities were not constructed from simple blueprints shipped over [from Europe], the fact is that American universities were conceived and staffed largely by people who had studied in Germany."[26]

The Origins of Experimental Psychology

The German university system was especially important to the development of the new discipline of psychology and to the establishment American experimental psychology. To be sure, psychological questions had been asked by philosophers from the time of Plato, and America did have a substantial tradition of mental philosophy.[27] But nineteenth-century Germany, with its research-oriented universities, proved to be the perfect setting for the emergence of experimental psychology. The causes for this development were many, and both social and intellectual factors were important. Positivistic and scientistic notions of the authority of experiment, and science in general, led those philosophers interested in psychological questions to the work of the neurophysiologists, who had themselves been able to develop their science in the suitable environment the university provided. And philosophy-oriented physiologists were able to apply their work to psychological problems and to teach courses in "medical psychology," "inductive philosophy," and "nervous physiology" in an atmosphere enriched with a tradition of *Lehrfreiheit*.[28] By the 1880s the new discipline had emerged out of philosophy, though most Germans working in the "new psychology"

[24]Margaret W. Rossiter, *The Emergence of Agricultural Science: Justus Liebig and the Americans, 1840–1880* (New Haven, Conn.: Yale University Press, 1975). See also Charles Rosenberg, "Science and Social Values in Nineteenth-Century America: A Case Study in the Growth of Scientific Institutions," in Arnold Thackray and Everett Mendelsohn, eds., *Science and Values* (Atlantic Highlands, N.J.: Humanities Press, 1974), pp. 21–42.
[25]Herbst, *The German Historical School*.

[26]Diehl, "Innocents Abroad," p. 321.
[27]On the American tradition in mental philosophy, see below and A. A. Roback, *A History of American Psychology*, revised edition (New York: Collier, 1964).
[28]Wallace A. Russell, "A Note on Lotze's Teaching of Psychology, 1842–1881," *JHBS*, 2 (1966), 74–75; Wolfgang G. Bringmann, William D. G. Balance, and Rand B. Evans, "Wilhelm Wundt, 1832–1920: A Brief Biographical Sketch," *JHBS* 11 (1975): 287–297.

continued to call themselves philosophers, as psychology was usually only part of their interest and as philosophy was still the most prestigious subject.[29] Much has been made of "role hybridization" in explaining the establishment of new positions in psychology in the otherwise mature and saturated German university system.[30] There is some value in this idea, particularly when the German situation of the 1880s is compared with the expanding American university system of that time or with the supersaturated American system of the 1970s. But probably more important is the related concept of "idea hybridization," which stresses the way philosophical and physiological ideas combined to yield the intellectual basis for experimental psychology.[31]

Among the individuals who developed the new psychology in Germany were two who are important to the subjects considered in this volume. The first was Rudolph Hermann Lotze (1817–1881), whose career represents an almost perfect hybrid. He graduated from the University of Leipzig in 1838 with degrees in both medicine and philosophy. By 1841 he had published a *Metaphysik*, which expanded his more philosophically oriented ideas, and 1842 saw the appearance of his *Allgemeine Pathologie*.[32] In 1844 he was appointed professor of philosophy at Göttingen, and there he continued to work in both physiology and philosophy and also in the area where the two fields intersected. His next two books were most directly related to this intersection; both his *Allgemeine Physiologie* and his *Medicinische Psychologie* stressed the importance of physiology to the understanding of psychological questions.[33] In fact, the first part of the second book was primarily concerned with the "fundamental concepts of physiological psychology." But after the mid-1850s Lotze devoted most of his attention to more narrowly focused philosophical problems, though his work was still highly informed by physiology and stressed an antivitalist approach. His most important work was *Mikrokosmus: Ideen zur Geschichte und Naturgeschichte der Menschheit*,[34] which despite its metaphysical goals was based at least in part on his physiological understanding. His last two books, a second *Metaphysik* (1879) and a *Logik* (1881), were parts of a projected three-volume system of philosophy left unfinished by his death.[35] As might be expected, Lotze's philosophy stressed

[29]Paulsen, *German Universities*, pp. 55–56.
[30]Joseph Ben-David and Randall Collins, "Social Factors in the Origins of a New Science: The Case of Psychology," *American Sociological Review* 31 (1966): 451–465.
[31]Dorothy Ross, "On the Origins of Psychology," *American Sociological Review* 32 (1967): 466–469.
[32]Lotze, *Metaphysik: Drei Bücher der Ontologie, Kosmologie und Psychologie* (Leipzig: Weidmann, 1841); *Allgemeine Pathologie und Therapie als mechanische Naturwissenschaften* (Leipzig: Weidmann, 1842).
[33]Lotze, *Allgemeine Physiologie des körperlichen*

Lebens (Leipzig: Weidmann, 1851). *Medicinische Psychologie: Oder Physiologie der Seele* (Leipzig: Weidmann, 1852).
[34]Lotze, *Mikrokosmus: Ideen zur Naturgeschichte und Geschichte der Menschheit: Versuch einer Anthropologie*, 3 vols. (Leipzig: Hirzel, 1856–1864).
[35]William R. Woodward, "The Medical Realism of R. Hermann Lotze" (Ph.D. diss., Department of History of Science and Medicine, Yale University, 1975; University Microfilms no. 76-14, 576). See also *EoP* 5: 87–89; *DSB* 8: 513–516.

a reconciliation of idealism and realism, and he "saw his goal as uniting in a consistent fashion the results of scientific research with an ethical and religious world view."[36] From his point of view, the realms of fact, law, and value were "in harmony" and all just different aspects of the same reality. Hence, for Lotze, an empirical psychology was a way to approach the broader philosophical questions of ethics and metaphysics and to investigate such narrower problems as the nature of instinct and the perception of space. In all, the details of Lotze's "medical realism" were not as influential as his metaphysics, which impressed Josiah Royce and (for a time) George Santayana. Though Lotze died before he could attract many American students, his emphasis on the possibility of an experimental psychology proved to be important to such Americans as George T. Ladd and (as will be seen) James McKeen Cattell.[37]

Wilhelm Wundt (1832–1920) also was educated in medicine, but unlike Lotze, he apparently did not develop more philosophical interests until after he had started his medical career.[38] Wundt earned his M.D. in 1855 from Heidelberg and almost immediately started publishing the results of his physiological investigations, including books on the theory of sense perception. From 1857 through 1874 he was at the University of Heidelberg, first as a *Docent* and then as an *ausserordentlicher Professor*. During most of this time, Hermann von Helmholtz (1821–1894) was professor of physiology at Heidelberg, and the details of Wundt's relations with him during this period have been the subject of much recent disputation. Whatever these relations were, while at Heidelberg Wundt published his major books on physiological psychology, *Vorlesungen über die Menschen- und Thierseele* and *Grundzüge der physiologischen Psychologie*.[39] By 1874 Wundt's work had led to his appointment as professor of inductive philosophy at Zurich, and in the following year he moved to Leipzig as professor of philosophy. Wundt's new title was impor-

[36]*DSB* 8: 516.

[37]Woodward, "Lotze, the Self, and American Psychology," in R. W. Rieber and Kurt Salzinger, eds., *The Roots of American Psychology: Historical Influences and Implications for the Future* (New York: New York Academy of Sciences [*Annals*, vol. 291], 1977), pp. 168–177.

[38]The importance of Wundt and the details of his career have been the subject of much controversy and differing interpretations recently. The discussion presented here is an attempt to balance the conflicting views presented in the following: Arthur L. Blumenthal, "A Reappraisal of Wilhelm Wundt," *American Psychologist* 30 (1975): 1081–1088; "Wilhelm Wundt and Early American Psychology: A Clash of Cultures," in Rieber and Salzinger, eds., *The Roots of American*

Psychology, pp. 13–20: "The Founding Father We Never Knew," *Contemporary Psychology* 24 (1979): 547–550; Bringmann et al., "Wundt, A Brief Biographical Sketch"; Solomon Diamond, "Wilhelm Wundt," *DSB* 14: 526–529; Diamond, *The Roots of Psychology: A Sourcebook in the History of Ideas* (New York: Basic Books, 1974), pp. 695–700, 750–752; Theodore Mischel, "Wundt and the Conceptual Foundation of Psychology," *Philosophy and Phenomenological Research* 31 (1970): 1–26. See also Diamond, "Wundt's Early Career: 1856–1873" (paper presented to American Psychological Association, Toronto, 29 August 1978).

[39]Wundt, *Vorlesungen über die Menschen- und Thierseele* (Leipzig: Voss, 1863); *Grundzüge der physiologischen Psychologie* (Leipzig: Engelmann, 1874).

tant, not only because philosophy had always been the parent discipline of psychology but also because it was the most prestigious subject. Psychology had arrived.

At Leipzig Wundt established a number of institutional foundations for his work which within a few years after his appointment made Leipzig the place to study experimental psychology. By 1879 Wundt had established a laboratory for experimentation in psychology that had some connection with the university and was listed in the university *Verzeichniss* as a "Seminar für experimentelle Psychologie," designated "Privatissime aber gratis." By 1881 he also was sponsoring the publication of a journal, *Philosophische Studien*, to publish the results of the work he and his students were doing in the laboratory. Meanwhile, Wundt began what eventually was to become a series of revisions of his major psychological books, issuing the second edition of *Grundzüge der physiologischen Psychologie* in 1880, the third edition in 1887, and the second edition of *Vorlesungen* in 1892.[40]

Students from all over Germany and the rest of the world flocked to Leipzig to study the new experimental psychology with Wundt. Lotze had never really experimented as had Wundt, and had died in 1881. Wundt had established institutions to promote the new hybrid science, and young men attracted by visions of a scientific psychology within the tradition of *Wissenschaft* came to work with him. By 1885, more than thirty-five students had earned their Ph.D.s with him, many for dissertations in experimental psychology.[41]

By the 1880s Wundt was clearly more interested in broader philosophical issues than in psychological ones, hoping to expand the focus of his work into the wider concerns and higher status of philosophy.[42] He published a *Logik*, an *Ethik*, and a *System der Philosophie*.[43] But the flood of students interested in laboratory techniques prevented him from concentrating on these broader interests. Wundt tried to ease some of the burden of teaching by appointing (from 1885 on) trusted students as laboratory assistants to direct most of the work done by his other students. Whether directed by

[40] Many other editions of these books were published. The fourth, fifth, and sixth editions of the *Physiologischen Psychologie* appeared in 1893, 1902–1903, and 1908–1911, respectively; the third through sixth editions of the *Vorlesungen* were published in 1897, 1906, 1911, and 1919.

[41] Miles A. Tinker, "Wundt's Doctoral Students and Their Theses, 1875–1920," *American Journal of Psychology* 44 (1932): 630–637.

[42] Much of the recent literature reinterpreting Wundt and his work has stressed the range of his philosophical interests and the fact that experimental psychology represented only part of his philosophical system. This revisionist literature often decries Wundt's early American students, who presented their teacher solely as an experimental psychologist working primarily in the laboratory, and who deemphasized his more philosophical concerns. The distortions by these early psychologists, however, were no more serious than those of Everett and his Harvard colleagues, or those of the historians who studied with Ranke. See above.

[43] Wundt, *Logik*, 2 vols. (Stuttgart: Enke, 1880–1883); *Ethik* (Stuttgart: Enke, 1886); *System der Philosophie* (Leipzig: Engelmann, 1889).

Wundt or by one of his assistants, most of the work done under his auspices in the 1880s was based on a unified set of assumptions and made use of a technique in which he had much faith.

Wundt's experimental methods involved the use of physiological and other methods to study not the reaction of the human body to specific stimuli in well-defined conditions, but the immediate experience of the human consciousness within a given setting.[44] Wundt argued that he was experimenting on the human mind, and expected his students to learn to observe their own minds under controlled conditions by a technique known as introspection, which involved careful concentration and the gradually learned ability to distinguish between (mental) sensation and (physical) stimulus or between immediate and mediate experience.[45] Observers were asked to distinguish between those ideas and sensations that were perceived in the *Blickfeld* (field of attention) from those that were apperceived in the *Blickpunkt* (focus of attention). This stage of the process of introspection itself defined apperception, and for a while this perspective played an important role in Wundt's thought.[46] It was later extended by a number of Wundt's disciples,[47] criticized by Cattell (document 3.12), and attacked severely by (among others) the American behaviorists. It did, however, form the basis for experimental psychology.

Among the large number of students attracted to Leipzig and Wundt were many Americans. For them, the traditional American mental philosophy, based upon the "faculty" ideas of the Scottish realists, held little or no attraction in an era influenced by Darwinism and positivism. During the first twenty-five years of the American Psychological Association (1892–1917) at least twenty Americans earned German Ph.D. degrees in psychology—almost all from Wundt—and at least twice that number studied with him for one period or another.[48] At least two universities harked back to Everett's experience and sent members of their faculties to Wundt for their Ph.D.s, despite the fact that Ph.D.s in psychology had been offered in America since 1886. One young American who wanted an American Ph.D. was forced by his mentor to study for his doctorate with Wundt, despite

[44]This discussion is based primarily on Mischel, "Wundt and the Conceptual Foundation," pp. 11–16.

[45]The effort to reinterpret Wundt's work has led to the extreme statement that he never advocated the practice of introspection as an experimental method. See "Discussion: European Influence Upon American Psychology," in Rieber and Salzinger, eds., *The Roots of American Psychology*, pp. 66–73. The documents presented in the present volume show just how extreme this statement is.

[46]Wilhelm Wundt, *Grundzüge der physiologischen Psychologie*, 2nd edition (Leipzig: Engelmann, 1880), vol. 2, pp. 205–212.

[47]For example, Ludwig Lange, "Neue Experimente über den Vorgang der einfachen Reaction auf Sinneseindrücke," *Philosophische Studien* 4 (1888): 479–510.

[48]"In Memory of Wilhelm Wundt, by His American Students," *Psychological Review* 28 (1921): 152–158.

the student's view that he would learn more in America.[49] The student may have been correct; very few of the Americans who studied in Leipzig practiced Wundtian psychology after returning to the United States (see postscript). It can be argued that, by the mid-1890s at least, Americans worked with Wundt more for the prestige than for any great attraction to his psychological ideas. American psychological institutions matured greatly in the late 1880s and the 1890s, and by 1900 American psychology was clearly different from German psychology in many ways.[50] But the initial strength of American psychology came from the German training of many of its early leaders.

Conventional wisdom holds that the first American to earn a Ph.D. under Wundt was James McKeen Cattell.[51] Like most things that "everybody knows," this statement contains some truth, but only a bit. The first American to study with Wundt was apparently G. Stanley Hall, but he did not earn a Ph.D. from Leipzig. Wundt's first American Ph.D. (in 1885) was James Thompson Bixby (1843–1921), a Unitarian minister who studied philosophy in Germany and wrote a dissertation on Herbert Spencer's ethics. In the same year, Hugo Münsterberg earned a Ph.D. under Wundt; though not an American, he was to play an important role in the development of psychology in the United States.[52] Despite these facts, Cattell was clearly the first American to work extensively in Wundt's laboratory and the first American (in 1886) to earn a Ph.D. under Wundt for a dissertation in experimental psychology. He also was Wundt's first assistant, though the details of his appointment to this position have been misunderstood (document 5.111). Cattell was to prove one of Wundt's more important American students. He established psychology at the University of Pennsylvania and Columbia University, was a charter member of the American Psychological Association, and developed the concept of, and the first real examples of, psychological tests (see postscript). His experience with Wundt, who later wrote of him as "colleague" and "associate" rather than as a student, paved the way for other Americans to come to Leipzig.[53]

[49]Edward A. Pace and George M. Stratton earned their Ph.D.s with Wundt at the expense of the Catholic University of America and the University of California, respectively. Lightner Witmer left the University of Pennsylvania in 1890 and earned a Ph.D. at Leipzig in 1892 at JMC's insistence.

[50]Hamilton Cravens, *The Triumph of Evolution: American Scientists and the Heredity-Environment Controversy, 1900–1941* (Philadelphia: University of Pennsylvania Press, 1978), pp. 56–86.

[51]See, for example, Edwin G. Boring, *A History of Experimental Psychology*, 2nd edition (New York: Appleton-Century-Crofts, 1950), p. 347.

[52]*DAB* 2: 306–307; Tinker, "Wundt's Doctoral Students," p. 632.

[53]Wilhelm Wundt, *Erlebtes und Erkanntes* (Stuttgart: Alfred Kröner, 1921), pp. 311–312.

James McKeen Cattell was born on 25 May 1860. From his fourth year until he went to Europe in 1880 he was closely associated with Lafayette College in Easton, Pennsylvania, where his father was president from 1863 through 1883. As the German universities of the nineteenth century represented *Wissenschaft*, so the American colleges of the same period represented a concern for "discipline and piety," and for the development of the "mental faculties" of their students within a Protestant religious framework. This concept of education was supported by the Scottish realist "faculty" psychology (see below), which by the 1880s was losing its hold and by 1910 had lost most of the respect it had commanded in the mid-nineteenth century.[54] Nevertheless, the small American colleges of the period had much to recommend them, and Lafayette, under the leadership of the elder Cattell, was one of the best outside New England.

William C. Cattell (1827–1898) had been educated as a Presbyterian minister at the Princeton Theological Seminary, after having graduated from the College of New Jersey in 1848. Unlike most ministers, he continued his studies in what was called "oriental antiquities" through college and seminary, and rather than take a church he decided to teach—first at a preparatory school, and then at Lafayette, where he was professor of ancient languages from 1855 to 1860. He did accept a pastorate in Harrisburg, Pennsylvania, between 1860 and 1863, but this position did not satisfy him as did his work at the College. In 1863, he returned to Lafayette as president.[55]

The elder Cattell's term as president of Lafayette was a success by any standard. When he assumed the office, the college was almost bankrupt as a result of the Civil War, the faculty consisted of nine men, there were only thirty-nine students, and the libraries had only about 2,500 volumes. By the time of his retirement, Lafayette was still smaller than other Eastern schools, but its endowment was about $800,000, twenty-six faculty members taught 289 students, and almost 25,000 volumes were available in several libraries.[56] More important, the program reflected a growing sophistication and rigor, and the college had developed an excellent and well-deserved reputation in English language and literature, in Anglo-Saxon studies, and (through its Pardee Scientific School) in Civil Engineering and Chemistry.[57] William C. Cattell was clearly responsible for much of this change. He had attracted to

[54] Laurence R. Veysey, *The Emergence of the American University* (Chicago: University of Chicago Press, 1965), pp. 21–56. See also Frederick Rudolph, *The American College and University: A History* (New York: Knopf, 1962).

[55] *Memoir of William C. Cattell* (Philadelphia:

Lippincott, 1899); *A Biographical Sketch of William C. Cattell* (Philadelphia: American Biographical Publishing Co., 1888).

[56] *Men of Lafayette*, pp. 33–34, 36–38, 48–51, 57–58, 195.

[57] Donald G. Mitchell, "Lafayette College," *Scribner's Monthly* 13 (1876): 184–197.

Lafayette such well known scholars as James H. Coffin, the distinguished meteorologist, and the noted philologist Francis A. March (1825–1911).[58] In addition, William Cattell cultivated Ario Pardee, a wealthy local coal operator, and convinced him to fund the scientific school. As James McKeen Cattell later noted about his father, "he had wondrous winning ways, and these gave him a host of friends and made him the idol of many generations of college students."[59]

James McKeen Cattell was named for his maternal grandfather, James McKeen (1803–1871), a prosperous Easton merchant and manufacturer. McKeen supported his son-in-law's college, and his wealth helped make the William Cattell family more financially independent than the families of most ministers and college presidents. Cattell grew up in a home in which both secular learning and orthodox religious devotion were well respected. He and his brother Henry Ware Cattell (1862–1936), later to become a distinguished physician and medical editor, were schooled by their father in his home library. They had their father's career as a model, as they saw him work daily and at times accompanied him on his trips on college business. The most extensive of these trips was a thirteen-month tour of Europe in 1869–1870, during which William Cattell visited many English and Continental technical schools, making plans for establishing the Pardee Scientific School. The elder Cattell was accompanied by his wife and their sons and two children of Ario Pardee, and the party traveled through Britain, France, Scandinavia, Germany, and Russia. The four children even attended school for a while in Paris, and it was an experience none of them ever forgot.[60] James McKeen Cattell noted ten years later when his path crossed that of his previous travel.

By the age of fourteen, James Cattell was auditing classes at Lafayette, and four months after his sixteenth birthday he matriculated there. He enjoyed many of the social aspects of college life, joining a fraternity (as did most of his classmates), substituting on the varsity football team, and rowing for his fraternity's boat club.[61] His record as a student was excellent. In his third year, he won prizes for proficiency in mathematics and in English and an award for work in the history of the English language. He graduated with honors and was elected by his fellow students to deliver the class's honorary philosophical oration. Cattell's student years at Lafayette were highly successful in many ways.[62]

In all of his academic achievements, Cattell exhibited the influence of Francis A. March, the outstanding scholar on the faculty, whom he later

[58] A. Guyot, "Memoir of James Henry Coffin, 1806–1873," *National Academy of Sciences Biographical Memoirs* 1 (1874): 257–264; James Wilson Bright, "Address in Commemoration of Francis Andrew March, 1825–1911," *Publications of the Modern Language Association* 29 (1914): 1–24.

[59] Cattell "Autobiography," pp. 626–635.

[60] Pardee diary.

[61] Francis A. March, Jr., *Athletics at Lafayette College: Recollections and Opinions* (Easton, Pa.: Lafayette College, 1926), pp. 183–197 (see Figure 1).

[62] *Men of Lafayette*, p. 230.

wrote of as "the great teacher, the great scholar and the great man." Before coming to Lafayette, March had studied with Noah Webster, and worked extensively in philology, making his reputation in 1869 with his great work, *A Comparative Grammar of the Anglo-Saxon Language: In Which Its Forms are Illustrated by Those of the Sanskrit, Greek, Latin, Gothic, Old Saxon, Old Norse, and Old High German.*[63] He was an authority on the history of English and the major American contributor to the *Oxford English Dictionary*. At Lafayette, he taught not only English and philology but also the required senior course in mental philosophy. This course, the capstone of a liberal education in a late-nineteenth-century American college, was taught at most colleges by the president; however, as William Cattell was busy with fund raising, it devolved to March, the faculty's distinguished scholar. March's teaching of English and philology led Cattell to two of his prizes, but in many ways his philosophical teaching proved more important.

Philosophical ideas were not just discussed in March's formal course in mental philosophy. March was one of the many mid-nineteenth century Americans to have become enamored of a version of Francis Bacon's ideas. These intellectuals reduced these ideas to simple empiricism with a stress on the collection of large amounts of elemental data, and to what has been called "vulgar utilitarianism."[64] This understanding of Baconianism undoubtedly oversimplified Bacon's work, but it greatly influenced American thought during this period.[65] March made extensive use of this philosophy; for example, one of his earliest articles was entitled "The Relation of the Study of Jurisprudence to the Origin and Progress of the Baconian Philosophy," and his great *Comparative Grammar of the Anglo-Saxon Language* was described by one of his most distinguished students as being built of "thousands of interrelated details."[66] In his various lectures, he regularly honored Bacon as "the prophet of inductive science" and stressed Baconian empiricism as the best method of science. He always argued for the useful

[63]Francis A. March, *A Comparative Grammar* (New York: Harper, 1869); Cattell, "The American College," *Science* 26 (1907): 368–373.

[64]George H. Daniels, *American Science in the Age of Jackson* (New York: Columbia University Press, 1968), pp. 63–85, 102–117; Theodore Dwight Bozeman, *Protestants in an Age of Science: The Baconian Ideal and Antebellum American Religious Thought* (Chapel Hill: University of North Carolina Press, 1977); Paoli Rossi, "Baconianism," in Philip P. Wiener, ed.-in-chief, *Dictionary of the History of Ideas: Studies of Selected Pivotal Ideas* (New York: Scribner, 1973), vol. 1, pp. 172–179.

[65]For some reason, perhaps related to George Daniels's personal life after the publication of *American Science in the Age of Jackson*, the influence of Baconian ideas on nineteenth-century America has been deemphasized by other scholars. See, for example, Stanley M. Guralnick, *Science in the Ante-Bellum American College* (Philadelphia: American Philosophical Society [*Memoirs*, vol. 109], 1975), p. 122: "It was only the name of Bacon that so appealed to the American public. . . . [Few] scientists were really as Baconian in their practice" as Daniels claims. However, it can be argued that many Americans working in "peripheral" fields such as philology practiced what they thought to be Baconian science.

[66]Francis A. March, "The Study of Jurisprudence," *The New Englander* 5 (1848): 543–548; Bright, "Address in Commemoration of Francis Andrew March," p. 17.

goals of scholarship, stressing that "we should seek the truth, says Bacon, for generation, for fruit, and comfort."[67] Each year he took the Juniors of Lafayette through Bacon's *Essays*, providing exegesis and commentary.[68] Cattell throughout his later life was something of March's kind of a Baconian, always taking an empirical approach to scientific problems, collecting bits of data wherever he could, and usually stressing the utility of his work.

The more formal philosophy that March taught at Lafayette was the mental philosophy of the orthodox "common-sense" school of Scottish realism, which had developed out of the work of Thomas Reid and which had close ties with the Presbyterian theology of Lafayette College's founders. In Scotland, these ideas were developed subtly, often in conjugation with a sophisticated understanding of Baconianism,[69] but in America in the second half of the nineteenth century they had deteriorated into a series of dogmas. Particularly lifeless during this period was the "faculty" psychology that had developed out of the "common-sense" school. This psychology held that the mind had within it various active powers (faculties), and that these could explain why and how individuals were grateful, pitiful, devout, and the like. By around 1860, most of the scholarship being done in the field was mere scholastic disputation with overly precise terminologies applied to trivial points; for example, the classification of human desires within the categories of hunger, thirst, and sex.[70]

Faculty psychology did not satisfy Cattell, any more than it did any of the many other Americans influenced by Darwinian and positivist ideas; they sought a new philosophy in Europe. Despite March's excellent teaching and the well-respected text by Joseph Haven,[71] the papers Cattell wrote for the course were full of criticism of the Scottish school. March had his students prepare essays on such topics as "Is it possible to recall by direct effort of the will?" and "Is there any such thing as immediate perception?" Most students probably wrote essays reflecting their study of Haven, but Cattell answered the second question by stating, "I do not know, and I doubt if anyone else knows."[72]

Cattell's other formal introduction to philosophy was a course in moral philosophy taught at Lafayette by Alexander Ballard, professor of rhetoric.

[67] Francis A. March, "The Buildings and Apparatus of the Modern College," *Addresses at the Re-opening of Pardee Hall* (Easton, Pa.: Lafayette College, 1881), pp. 3–17; "The Future of Philology," *The Presbyterian Quarterly and Princeton Review*, new series, 33 (1894): 698–714.

[68] Coffin, *The Men of Lafayette*, pp. 63–68.

[69] J. C. Robertson, "A Bacon-Facing Generation: Scottish Philosophy in the Early Nineteenth Century," *Journal of the History of Philosophy* 14 (1976): 37–50.

[70] Frank M. Albrecht, Jr., "A Reappraisal of Faculty Psychology," *JHBS* 6 (1970): 36–38; Herbert W. Schneider, *A History of American Philosophy*, 2nd edition (New York: Columbia University Press, 1963), pp. 195–196, 202–209.

[71] Joseph Haven, *Mental Philosophy: Including the Intellect, Sensibilities and Will* (Boston: Gould and Lincoln, 1857, 1860, 1862, 1872, 1876, 1883). The numerous editions and reprintings of the book show how well-respected it was. See Schneider, *American Philosophy*, pp. 211, 550.

[72] JMC, manuscript undergraduate essays.

The text was Mark Hopkins's *Outline Study of Man*, an attempt to discuss "the mind and body in one system" that was closely related to Scottish realist ideas and featured a large foldout chart purporting to represent all human attributes.[73] Cattell questioned Hopkins's classification of human desires, and had Professor Ballard inform Hopkins himself of his objections. Hopkins replied: "Young Cattell is sharp, and the President is to be congratulated on having such a son. I agree with the suggestion he makes."[74]

The details of Cattell's dissatisfaction with Scottish realism are not hard to understand, but their intellectual sources are only vaguely apparent. Darwinian and positivistic ideas were being discussed in the popular journals received by the Lafayette libraries throughout this period.[75] In any event, by 1880 Cattell was involved in a deep study of the ideas of Auguste Comte, the founder of positivism as a philosophical system. This work resulted in Cattell's honorary philosophical oration on "The Ethics of Positivism."[76] A copy of this talk has not been found, but it is known that Cattell concentrated on Comte's ethical ideas, which were based on the concept of altruism, with the sacrifice of the mother in childbirth as a model.[77] Despite the ethical focus of his talk, Cattell could not have studied positivism without becoming aware of Comte's philosophy of science, which was the most influential part of his ideas in nineteenth-century America.[78] Comte stressed the authority of scientific (or positive) ideas over religious or metaphysical forms of thought. Furthermore, for Comte the best type of science was quantitative science, and hence for his disciples a mathematical approach to the study of the world was to be preferred.[79] Though positivism did not stress empirical methods as such, an empirical bias was clearly present in most of Comte's writing. This empiricism meshed well with Baconianism, on which Cattell's

[73] Mark Hopkins, *An Outline Study of Man: Or the Mind and Body in One System, with Illustrative Diagrams, and a Method for Blackboard Teaching* (New York: Scribner, Armstrong and Co., 1873, 1878; Charles Scribner's Sons, 1886, 1898). See Schneider, *American Philosophy*, pp. 213–215.

[74] "Extracts from Testimonials of the Faculty of Lafayette College to accompany the thesis of James M. Cattell, an applicant for a Fellowship in Philosophy in Johns Hopkins University, May 1882." (See document 1.17.)

[75] Lloyd F. Wagner, "A Descriptive History of the Library Facilities of Lafayette College, Easton, Pennsylvania, 1824–1941" (M.S.L.S. thesis, School of Library Science, Catholic University of America, Washington, D.C., 1931).

[76] *Men of Lafayette*, p. 9.

[77] Giacomo Barzellotti, *The Ethics of Positivism: A Critical Study* (New York: Charles P. Somerby, 1878).

[78] There is no satisfactory discussion of the reception of positivistic ideas in the United States, though the biographies of most important American thinkers with any relation to science or philosophy during the nineteenth century discuss their subjects' study of, or flirtation with, Comte's ideas. For example, see *The Education of Henry Adams* (Boston: Massachusetts Historical Society, 1918), pp. 62, 225, 479, 495; Donald Fleming, *John William Draper and the Religion of Science* (Philadelphia: University of Pennsylvania Press, 1950), pp. 43, 46, 49, 58–59, 140–144, 160–161; John Spencer Clarke, *The Life and Letters of John Fiske* (Boston: Houghton Mifflin, 1917), vol. 1, pp. 113–125, 136–141, 232–233, 346, 350; Ralph Barton Perry, *Thought and Character of William James* (Cambridge, Mass.: Harvard University Press, 1935), vol. 1, pp. 151, 464, 471, 520–525, Ross, *Hall*, pp. 42–44.

[79] *EoP* 2: 173–177; 6: 414–419.

attraction to positivism appears to have been partly based.

In 1880, when Cattell graduated from Lafayette College, he was approaching a philosophical position that combined aspects of Baconian and Comtean ideas. This perspective stressed empiricism and the usefulness of scholarship and science, but at the same time appreciated the power of mathematics. This point of view led Cattell to seek to collect facts, as Bacon would have urged, and to be certain that the facts that he collected were mathematical, as Comte's ideas would have him do. In all, this philosophy was to lead Cattell to his most significant scientific achievements.

However, Cattell's study of Comtean ideas seems to have had a more immediate effect. By 1880 he had begun to feel the tension between the orthodox Christianity of his home and his college and the antireligious biases of positivism. Many other Americans, such as John Fiske, William James, and G. Stanley Hall, had felt similar tensions,[80] but Cattell had another reason for feeling ambivalent toward the orthodoxy his father represented. There is no doubt that he loved his father and respected his achievements, but the fact that his father was the president of the college he attended was a problem. Cattell enjoyed his years at Lafayette, and got along well with his classmates, but he was two years younger than most of them, and he lived with his parents. His friends were always a bit wary of him and, as one later noted, although "he had the advantage of knowing from childhood some of his classmates, . . . at college [he] suffered the disadvantage of living apart from the more intimate undergraduate life."[81] Many young men have felt ambivalence towards their parents for far less than this. Furthermore, it has recently been suggested that Francis A. March and William C. Cattell provided the younger Cattell with two different models of achievement: the scholar and the administrator.[82] James Cattell's deep attachment to these two different men probably "helped shape the duality of his [later] career and his views of university administration."[83]

James McKeen Cattell in 1880 was, then, a bright young man with an excellent philosophical education and with at least some acquaintance with scholarship. One friend who had known him as a boy wrote that Cattell had "improved considerably, as he has grown older."[84] But still he was

[80]Clarke, *John Fiske*, vol. 1, pp. 113–125; Leon Edel, "Portrait of Alice James," in Leon Edel, ed., *The Diary of Alice James* (New York: Dodd, Mead, 1964), p. 2; Ross, *Hall*, pp. 42–44.

[81]Lafayette College, *Vigintennial Reunion, Class of 1880* (Easton, Pa.: Lafayette College, 1900), 18.

[82]Dorothy Ross, "James McKeen Cattell," *DAB* Supplement 3: 148–151.

[83]Ross, "Cattell," p. 151. Erik H. Erikson, *Childhood and Society*, 2nd edition (New York: Norton, 1973), pp. 261–263, describes the tension of late adolescence as one of "identity vs. role confusion," and writes of the "inability to settle on an occupational identity" as a major problem for many young people. "To keep themselves together they temporarily identify" with one or another hero. For the twentieth century, Erikson mentions "the heroes of cliques and crowds," but one could well imagine a serious nineteen-year-old in 1880 attaching himself to a distinguished scholar.

[84]Pardee diary, p. 273.

under much personal tension, and really did not know what to do with his life. The popular journals of the previous decade had often published articles about the attractions of the German universities, with their opportunities for scholarship, and Cattell even had access to a copy of William Howitt's *The Student Life of Germany* at the Easton Library Company.[85] His father, of course, knew what the German universities offered, so Cattell and his parents decided that he should spend at least one year at Göttingen, though with no definite aim in mind. He would not be a financial burden on his family, as one of James McKeen's financial ventures, the Warren Foundry in Warren County, New Jersey, was manufacturing water and gas pipes for the growing American cities of the period and was declaring high dividends regularly.[86] James Cattell had found an opportunity to expand his horizons, and he was looking forward to it.

But things were not all well within him. In 1880, Cattell began to experience the periods of deep depression that were to become very common the next year in Europe. His religious doubts and his ambivalence about his father were somewhat to blame for these spells. Also, as he had never been apart from his parents before, he may have been suffering the "separation anxiety" common in adolescents.[87] In any event, his uneasiness revealed itself in his contributions to the "Statistics of the Class of 1880." Here he answered such simple questions as "Favorite Pastime?" and "Besetting Sin?" and "Beau Ideal of Happiness?" with either "yes" or "no," listed his future occupation as "work," and explained his choice in one word: "money." One of Cattell's few serious responses to this questionnaire listed his "favorite department of study" as "philosophy."[88]

James Cattell graduated from Lafayette College on 29 June 1880. He did not deliver his honorary philosophical oration—all the speakers were excused because of the length of the program.[89] He had turned twenty a month earlier. Two weeks after graduation, he set off for Europe.

[85] Wagner, "Descriptive History"; *Catalogue of Books Belonging to the Easton Library Company* (Phillipsburg, Pa.: Cooley and Wise, 1855), p. 60; "Hart's German Universities," *The Nation* 19 (17 December 1874): 400–401.

[86] Cattell "Autobiography," p. 633; James P. Snell, *History of Sussex and Warren Counties, New Jersey* (Philadelphia: Everts and Peck, 1881), p. 561.

[87] Peter Blos, *On Adolescence* (New York: Free Press, 1962), p. 99.

[88] *Vigintennial Reunion*, p. 79.

[89] *Men of Lafayette*, p. 99.

I

The Student as Tourist and Scholar:
England and Germany, July 1880–September 1882

By 1880, steamship improvements had made the journey from America to the Old World easier and cheaper than ever before, and in Europe the new railroads eased travel between cities. Many Americans took advantage of these conveniences, and by the mid-1880s the annual flow of Americans to Europe was twice that of any year before the Civil War.[1] Few of these tourists were attracted to Germany; most preferred the historic associations of Britain and the scenery of France and Italy. However, on Thanksgiving Day 1885, about 250 Americans— most of them probably students—sat down to dinner together in Berlin.[2]

Beginning in 1836, when the first of *Murray's Infallible Handbooks* was published, American tourists had formal guides on which to rely, and at least one American studying in Germany in the mid-1850s is known to have made extensive use of these guides.[3] More important and more widely used than Murray's guides was Harper's *Handbook for Travelers in Europe and the East*, first published in 1860.[4] But even more popular were the Baedeker handbooks. First published by Karl Baedeker of Leipzig in 1839, these guides had an enviable and well-deserved reputation for reliability and accuracy that led quickly to their worldwide use. The first English editions of these handbooks appeared in 1861. A system of one or more stars, which had been introduced in 1844, directed travelers to noteworthy and outstand-

[1]Diehl, *Americans and German Scholarship*, pp. 110, 186; Foster Rhea Dulles, *Americans Abroad: Two Centuries of European Travel* (Ann Arbor: University of Michigan Press, 1964), pp. 102–103.
[2]Dulles, *Americans Abroad*, p. 56.
[3]Dulles, *Americans Abroad*, p. 56. These handbooks were issued by John Murray, a

London publisher whose fortune they made. See Elizabeth A. Osbourne, ed. *From the Letter-Files of S. W. Johnson* (New Haven, Conn.: Yale University Press, 1913), p. 41.
[4]This handbook, edited by William Pembroke Fetridge, was later joined by many others. Eventually the series was entitled "The American Travellers' Guides."

ing hotels, sights, and restaurants.[5] Traveling students used the Baedeker guides as much as anyone.[6]

During his first two months in Europe, James Cattell purchased at least three Baedeker guides: those for London, Holland and Belgium, and Northern Germany.[7] He made extensive use of them, staying at starred hotels, visiting starred sights, and eating at starred restaurants. He later noted that he "looked at all sights marked with a star in Baedeker with the utmost conscientiousness," while complaining that he had forgotten most of what he saw.[8] In this, Cattell fell into the pattern often complained about by commentators on American culture. Cattell's descriptions of what he saw sometimes correspond almost word-for-word with the descriptions in whichever Baedeker guide he was using at the time. Some of the closest of these correspondences are those between entries in Cattell's journal (for example, document 1.2) and the descriptions in the Baedeker guide to Great Britain. This similarity seems remarkable, as the journal entry was written in 1880 and Baedeker's *Great Britain* was not published until 1887. The seemingly strange correspondence can be explained by the fact that Cattell, during his Leipzig years (see 5.24 and 5.143), became friendly with James Fullerton Muirhead, who was to compile and edit this guide for Baedeker.[9] When Cattell was at Cambridge, he helped Muirhead prepare the description of that university for the 1887 guide to Great Britain (see 6.63).[10] The possibility that Cattell might have lent Muirhead the record of his earlier travels is reinforced by the guide itself (p. v), which notes that Muirhead had personally visited "the greater part of the districts described." The compilers of most Baedeker guides were proud to have seen all that they wrote about.

James McKeen Cattell set out for Europe in the summer of 1880 with great expectations, and with some sadness. The early entries in his journal present no evidence that he was later to emerge as a distinguished scientist, but their concerns—particularly the interest in all aspects of European culture—reflect those of most Americans who studied in Europe during this period, at least during the first months abroad.

[5] Herbert Warren Wind, "The House of Baedeker," *The New Yorker*, 22 September 1975, pp. 42ff.
[6] John W. Burgess, *Reminiscences of an American Scholar: The Beginnings of Columbia University* (New York: Columbia University Press, 1934), pp. 87, 113

[7] JMC's account book for 1880–1881 (Cattell papers), entries for 9 August, 15 August, and 8 September 1881.
[8] JMC to parents, 11 April 1886.
[9] *Who Was Who* [Br.], 1929–1940, p. 980.
[10] See also Muirhead to JMC, 19 April 1887 and 3 May 1887 (Cattell papers).

1.1 Journal, aboard the S.S. *Indiana* at sea, 13 July 1880

(Many short words have been transliterated from Pitman shorthand.)

Many people begin to keep a journal on New Years & on their outward bound voyage. Few keep it up until the next New Years or until their return voyage. Whether I will be one of the faithful ones remains to be seen. I have not decided what to write yet—whether only what I see or also what I think. Perhaps I had better compromise by writing what I do.[1]

I am on the steamship Indiana of the American line.[2] I sailed from Phila last Wed. at 10 o'clock. Harry, Eddie, Uncle Lige and Mr Barrett came down to see me off.[3] They introduced me to the captain & several of the passengers—Dr Gross, Mr. Dittrick & Mr Ashmead.[4] It was hard to say good bye and I felt very sad when I thought about those at home, who, I fear, shall find my absence hard to bear. I feel my separation from them all the more as they are the only ones whom I especially mind not seeing for a year. I had a pleasant sail down the bay. I soon got acquainted with most of the passangers and now I think I speak to every one—the ladies always excepted.[5] Cousin Jo sent me down a very hansome basket of flowers.[6] As I reached the mouth of the bay I saw Cape May on our side and the breakwater on the other. In the evening I became a little sea-sick but since then I have been perfectly well. The sea has been as calm as a lake & all the passangers keep on the deck, very few of them being at all sick. Yesterday afternoon the wind blew a little & the sea was pretty well covered with white-caps. Some of the passangers were at the dinner table. My state room is number P. It is the third room from the stern on the left hand side. My roommate is a Mr Stevenson of Pitsburg who was at Lewisburg

[1]It is interesting to note that, in beginning his journal, the young man later to be known as one of the precursors of behaviorism (see John C. Burnham, "On the Origins of Behaviorism," *JHBS* 4 [1968]: 143–151) considered noting only his behavior.

[2]A 3,100-ton iron screw steamship that had been in service since the mid-1870s (Warren Tute, *Atlantic Conquest: The Men and Ships of the Glorious Age of Steam* [Boston: Little, Brown, 1962], pp. 110–111. JMC's passage cost him $75.00 (account book, 1880–1881, 7 July 1880).

[3]Henry Ware Cattell (1862–1936), JMC's brother; Edward James Cattell (1856–1936), JMC's cousin, who many years later became a popular author (under the pseudonym Francis H. Hardy) and the editor of several commercial publications (*Who Was Who in America*, 1897–1942, p. 204); and Elijah Gilmore Cattell (1818–1899), JMC's uncle and Edward's father.

[4]Samuel David Gross was a prominent Philadelphia physician and (*DAB* 8:18–20) who in 1880 was professor of surgery at Jefferson Medical College. He had taught at Lafayette in the 1830s. See also *Autobiography of Samuel D. Gross, M.D., with Sketches of His Contemporaries, Edited by His Sons*, reprint edition (New York: Arno, 1972). William Harris Ashmead was an employee of the Philadelphia publisher J. B. Lippincott and later a distinguished entomologist (*DAB* 1:392–393).

[5]See 1.3: "the first time I ever talked ten minutes to a young lady in my life!"

[6]Josephine Cattell Fithian, daughter of Joseph Fithian and Esther Gilmore Cattell Fithian (William Cattell's sister and James's Aunt Hetty). Jo Fithian later married Edward Wheeler Hitchcock, who held pastorates in New York and Paris. (Mrs. Edward Hitchcock, *The Genealogy of the Hitchcock Family* [Amherst, Mass.: Carpenter and Morehouse, 1894], p. 400.)

College through Soph year a year ago.[7] *He is an S.X.*[8] *He makes a good enough roommate, but I must confess I don't like him very much. He is what I would call at College soft, fresh, green. I wondered how on earth his Mamma & sisters let him come to Europe by himself until I found that as he had been safely put on board at Phila he would be received by an uncle at Liverpool who would take charge of him, and take him to Paris if he had to and then send him back home. There are a good many queer people on the boat, typical characters, though sometimes I doubt if there is such a thing or rather if every one is not a typical character. Perhaps the man I like best is a Mr Sulzburger a Philadelphia lawyer of age about 30 years.*[9] *He is a Jew, but what I call an advanced thinker—an apostle of modern culture and development; he is remarkable well read and can talk well, is perfectly at home on any topic of conversation and imparts a great deal of valuable information. I would like to describe all my fellow passengers, but it is very hard to write here. The saloon is close and the ship rocks—then it is hard to do anything that approaches to mental work on shipboard.*

[7]James B. Stevenson, a member of the Bucknell University class of 1882 who did not graduate and later entered the Pittsburgh business world (*Bucknell University Alumni Catalog, 1856–1926*, p. 91).

[8]A member of Sigma Chi fraternity. JMC was a member of Delta Kappa Epsilon at Lafayette (see figure 1).

[9]Meyer Sulzberger, later a distinguished jurist in Philadelphia (*DAB* 18: 205–206).

Cattell landed in Liverpool, and from there began a tour of the English lake district, Scotland, and central England. He made sure to see all of the recommended sights, stayed at Temperance hotels, and paid special attention to sites with literary associations. For example, at Keswick in the lake district he attended the church to which the romantic essayist Robert Southey had belonged (Journal, 27 July 1880; Baedeker, *Great Britain*, p. 401). Of course, he visited Stratford-on-Avon (Journal, 6 August 1880; Baedeker, *Great Britain*, pp. 245–249).

Cattell's descriptions in his very long entry on his trip through the lake district (Journal, 27 July 1880) are those that most closely parallel the description in Baedeker's *Great Britain*; for example, Furness Abbey: "one of the largest and most picturesque ruins in England" (Journal); "The ruins . . . are among the most extensive and picturesque in England" (Baedeker, p. 384); Derwentwater: "perhaps the most beautiful of all the lakes" (Journal); "perhaps the loveliest of the English lakes" (Baedeker, p. 401); an excursion at Buttermere: "said to be the finest drive in England" (Journal); "perhaps the finest drive in the Kingdom and should on no account be omitted" (Baedeker, p. 403).

The tour continued.

(Many short words have been transliterated from Pitman shorthand. Brackets represent spaces Cattell left to be completed later; the appropriate words have been filled in where possible.)

Yesterday I worked pretty hard. I got up at 6.30 & took the train to Oxford, arriving there about nine. I found Oxford exceeding interesting & only wish I could have spent more time there. I walked up from the station & first visisted the [Bodleian] library.[1] It contains some [460,000] volumes, some of them very rare. Also some interesting relics—autograph manuscript of Milton Pope Burns &c. Next I went into the [Sheldonian] theatre where degrees are conferred.[2] From the cupola I had a fine view. I next went through the [Ashmolean] museum of antiquities—very interesting.[3] Then I visited [   ] College. The [   ] chapel is a magnificent building. Thence I went to the [Christ Church] College—the most interesting of them all. The library contains a fine collection of paintings. The cathedral is very old & curious. The dining room is one of the finest halls in the world.[4] Then I visited St. Mag[dalene] College.[5] From this I went to the museum. I also visited during the day various other Colleges & Churches. My body and mind were both completely tired out when I took the train at 6 o'clock to London. I hope to visit Oxford again before I go home. I would enjoy spending a week there. I arrived in London about 8 & went to the Waverley hotel in Lawrence lane Cheapside.[6] It is cheap enough here, but I have a miserable little room under the roof & the public rooms are too small. In short it is a 2nd or 3rd class hotel. Today I went to hear Dr Parker preach, but he is away & I had the pleasure of hearing some nameless Methodist brother.[7] I also went to St. Pauls & heard part of the service there.[8] This afternoon I walked & rode about the city & looked up a hotel. I am going to the Westminster tomorrow.[9] It is one of the best & most fashionable hotels in the city. It seems very extravagant for me to go there but I only pay 3½ guineas per week & this includes everything. If I would stay where I am it would be possible for me to get along on 2½ guineas per week, but I would probably spend three. Then one feels so much more like a gentlemen at a first class hotel and you have fine reading rooms &c. I think too it will be better for my health to take substantial & regular meals and at the Westminister. I will have a 3/6 breakfast, meat, lunch & 5/6 table or lite dinner. I also have a good room facing the street on the 3rd floor.

[1]Starred by Baedeker (*Great Britain*, p. 229).
[2]Starred by Baedeker (*Great Britain*, p. 230), which stressed the view from the cupola.
[3]Surprisingly, not starred by Baedeker (*Great Britain*, p. 230).
[4]"One of the largest and most fashionable colleges . . . . The Hall . . . contains numerous good portraits . . . . The Cathedral . . . serves . . . as the Chapel of Christ Church" (Baedeker, *Great Britain*, pp. 225–226).
[5]"Perhaps . . . the most beautiful in Oxford"

(Baedeker, *Great Britain*, pp. 222–223).
[6]Not listed by Baedeker.
[7]Joseph Parker, the minister of the City Temple (a Congregational church) from 1869 through 1901, was noted by Baedeker (*London*, p. 45) as a distinguished preacher (see also *Who Was Who* [Br.], 1897–1915, p. 547).
[8]"London's most prominent building" (Baedeker, *London*, p. 77).
[9]The Westminster Palace Hotel was included by Baedeker (*London*, p. 67) among those "very handsomely fitted up."

From 8 August Cattell spent almost two weeks in London, visiting just about all the sights that have long attracted Americans (Journal, 15 August 1880). On 18 August he took an hour's trip out of the city.

1.3 Journal, London, 22 August 1880

(Many short words have been transliterated from Pitman shorthand.)

Wednesday I went to Cambridge and besides seeing the University buildings, was present at the conferring of degrees.[1] The Drs wore red gowns & the others black. The undergraduates crowded the galleries and kept cheering and hissing & shouting out rude jokes all through the ceremony & latin speaches.[2] Mr Gross introduced me to Mrs & Miss McCallister & I took the latter to the ceremony, the first time I ever talked ten minutes to a young lady in my life![3] Thursday I spent the morning at the British Museum and the afternoon at the Zoological Gardens. I met Mr Stevenson there.[4] The Crystal Palace occupied all of Friday. Sat. I devoted to the Houses of Parliament, Westminister Abbey and the National Gallery. Sunday I heard Spurgeon in the morning, and Canon Liddon at St Pauls in the afternoon.[5] They both preached good sermons & had large audiences. The latter preached a whole hour! I spent most of Monday at South Kensington Museum. The objects which have been bought all have the price attached. The Schliemann collection is at S. Kensington. I also walked through the National Portrait Gallery and visited the Albert Hall and Memorial and Hyde Park. I forgot to mention that on Wed. I went to the American

[1]On 11 August twelve distinguished British and foreign physicians, including Samuel D. Gross (see 1.1) were honored with degrees of Doctor of Laws as part of the forty-eighth meeting of the British Medical Association. (*Autobiography of Samuel D. Gross*, vol. 2, p. 124.) Among those also honored was F. C. Donders, a man whose work was later to be extremely important to JMC (see 3.12).

[2]The major uproar occurred when an honorary degree was conferred upon William Gull, a physician universally unpopular because of his public attacks on the competence of his colleagues (*Autobiography of Samuel D. Gross*, vol. 2, p. 124; *DNB* 8: 776–777).

[3]"Mr. Gross" is Albert Haller Gross, a distinguished attorney who accompanied his father on this trip (*Who Was Who in America*, 1897–1942, p. 491). "Mrs and Miss McCallister" are probably Elizabeth Stewart Macalister and Edith Florence Boyle Macalister, the wife and daughter of Alexander Macalister, the Irish-born anatomist then teaching at the University of Dublin, who

was attending the meeting of the association, and, from 1883, Professor of Anatomy at Cambridge (*Who Was Who* [Br.], 1916–1928, p. 656). Edith Macalister later married her distant cousin Donald MacAlister, who was later to be one of Cattell's closest friends (see 6.62). JMC's lack of contact with young women his own age at a time when he was maturing physiologically may have contributed to his tension.

[4]See 1.1.

[5]John Spurgeon, the distinguished Congregational minister (*Who Was Who* [Br.], 1897–1915, p. 668), had been praised extravagantly by other American visitors (see Andrew P. Peabody, *Reminiscences of European Travel* [New York: Hurd & Houghton, 1868], pp. 20–23). In 1869, William Cattell had heard Spurgeon and had been impressed with his ability (Pardee diary, 25 July 1869 and 1 August 1869). Henry Parry Liddon (*DNB* 11: 1102–1107), Canon of St. Paul's from 1870 to his death, in 1890, was noted by Baedeker (*London*, p. 44) as one of the three most eminent preachers in London.

Exchange & found Prof Moore, Markle, Reading & Wilbur had registered there.[6] *I*
called on them all. Markle had left the city a few hours before, the others two days.
I have not seen a single person since I left Phila. that I had ever met before!
Tuesday I visited Westminister Abbey including Jerusalem Chamber in the morning &
the National Gallery in the afternoon. In the evening I saw "The World" at the
Drury Lane Theatre. Wed. I visited the Temple with its church and Lincoln's Inn also
Smithfield markets. In the afternoon I visited the British Museum. In the evening I
went to the Aquarium with a gentleman from the hotel. Four or five of the gentle-
men who live at the hotel have gotten acquainted with me & been very kind.
Thursday I went to Dulwich where there is a fine collection of paintings especially
rich in the Dutch masters. In the evening I went to the Alexandra Palace, with about
50,000 other people. I also ordered a suit of clothers & overcoat at Holson's 6
Grace Church St for £ 5-5 each.[7] *Friday I spent the morning at S. Kensington &*
Indian Museums. In the afternoon & evening I attended the session of the House of
Commons. A gentleman from the hotel get me a pass for two to the Speakers
Gallery. I heard the leaders of both parties speak, including Hartington Harcourt &
Northcote.[8] *Sat. I visited Hampton Court & Richmond. Got lunch at the Star &*
Garter. This morning I heard Dr. Newman Hall preach & this afternoon went to
Kew.[9] *I am sorry that I have not been able to write more than this mere synopsis,*
still it is better than nothing, & I will have more time when I get to Göttingen.

[6]The American Exchange and Reading Room and the American Traveller's Reading Room were both recommended by Baedeker (*London*, p. 51). James W. Moore was Professor of Mechanics and Experimental Philosophy at Lafayette College from 1872 to 1909 (*Who Was Who in America*, 1897–1942, p. 860). John Markle was a classmate of JMC at Lafayette and later a major coal executive (ibid., p. 277).
[7]Not among those tailors recommended by Baedeker (*London*, p. 21).
[8]The Marquis of Hartington, born Spencer

Compton Cavendish, was later the eighth Duke of Devonshire (*DNB*, 1901–1911: 323–329). He was an MP from 1857 to 1891 and a major Liberal figure. William George Harcourt (*DNB*, 1901–1911: 198–212) was another distinguished Liberal and served as an MP from 1868 to 1904. Stafford Henry Northcote (*DNB* 14: 639–644) was from 1885 an important Conservative MP.
[9]Newman Hall, an eminent Evangelical Congregational minister well known as a preacher (*Who Was Who* [Br.], 1897–1915, p. 306).

Cattell left England on Wednesday, 25 August, and traveled to Antwerp, where he was greatly impressed by the paintings of Peter Paul Rubens. From there he went to Brussels, visited Waterloo, and viewed much Flemish art. On 1 September he left for Cologne, where he "immediately visited the grand cathedral finished just two weeks before but commenced in 1248. I climbed the towers (475 ft.) on scaffolding & without permission the ladder to the very top of the spire, something that will be impossible to [do] when the scaffolding is removed." (Journal, 14 December 1880.) (Baedeker [*Rhine*, pp. 26–31] double-starred the cathedral, presented a long history of its construction, and noted that it had officially opened on 15 October 1880.)

From Cologne, Cattell sailed down the Rhine, stopping at Bonn, Koblenz, Mainz, and Frankfurt, where he caught a train to Göttingen, stopping at Kassel. He arrived at Göttingen on 9 September.

1.4 Journal, Göttingen, 14 September 1880

(A description of the trip up the Rhine has been omitted.)

Thursday at 2½ I arrived in Göttingen and went to the Crown Hotel.[1] As yet I have not seen very much of Göttingen & nothing of the University. My first impressions of it are very pleasant. G. is an old German town situated in a pleasant region with mountains in the distance. Thurs (the 9th Sept) I bought a list of the lectures & catalogue of the professors & students.[2] I found the lectures did not begin till Oct. 15 & probably not till a week later. I thought they would begin a month earlier, so have about five weeks on my hands, still it does not make much difference since I can visit Berlin & Dresden & will have time to study up a little German. I called on one or two of the Profs. & was directed to Prof John the Prorector.[3] He received me very kindly, told me I would not be enrolled for about a month & recommended me to Frl. Bartling.[4] I went to see her Friday afternoon & was very much pleased. I moved from the hotel here Sat. morning. I have 2 room facing the south on the third floor. Adjoining is the plan for the rooms.[5] There are some engravings & photos on the wall. Only one piece of carpet. I have breakfast at 8 or later if I wish. Dinner at 1½. Coffee & rusk at 4 & Supper at 8. I only take in the evening a glass of milk & roll & butter, which I have at 6. The bread is excellent. Very good bread, butter & milk. For breakfast I have cold meat or eggs, oat-meal rolls & butter, honey & milk. For dinner I have soup, meat with vegetables & desert. They have a nice hot supper. The family is composed of 3 sisters (though I think one of them is only on a visit home) & their aunt. The former are 30–35 years old & the daughters of a deceased

[1] Baedeker (*Northern Germany*, p. 99) simply noted Göttingen as a "pleasant town with 17,000 inhabitants, . . . famous for its University." The only thing starred is one hotel—not the one frequented by Americans. For many years the Krone hotel was the unofficial headquarters of the American colony in Göttingen (Diehl, *Americans and German Scholarship*, p. 139; see also George W. Magee, ed., *An American Student Abroad: From the Letters of James Francis Magee* [Philadelphia: Magee, 1932], pp. 47, 55; Burgess, *Reminiscences of an American Scholar*, p. 94).

[2] Hart (*German Universities*, pp. 1–18) recommended just such a purchase.

[3] Richard Eduard John, a professor in the Göttingen faculty of jurisprudence (*Allgemeine Deutsche Biographie* [Leipzig: Duncker, 1875–1912], vol. 50, pp. 688–690).

[4] The daughter of Friedrich Georg Bartling, longtime professor of botany and director of the botanical garden at the university (*NDB* 1: 611–612). In 1871, another American student, John William Burgess—later a political scientist at Columbia University and Berlin—lived with the Bartlings at Göttingen. He later wrote of them as "the most cultivated family in town and the leading family of the university in the society of the place," and described the three young daughters (then about sixteen, eighteen, and twenty) as "charming and accomplished [and speaking] the most perfect Hannoverian German" (Burgess, *Reminiscences of an American Scholar*, pp. 97–98).

[5] See figure 4.

professor of the University.[6] *The ladies are kind, well educated, & cultured. They speak excellent English & so far we have been speaking mostly English, though sometimes they address me in German. When I return from my excursion to Berlin we shall speak German altogether. There are at present two other boarders here an Englishman Dr Marsden who has been studying chemistry at the University for 4 months but who knows no German & a Swede Mr. Kjellberg who has learned to speak German very nicely in the last 2½ month.*[7] *They both expect to leave soon.*

[6]Two of these women later married men who had boarded at their house while studying at the University. Mary married the English physiologist Allan MacFadyen in 1890, and Jenny married the American theologian Thomas Hall. See 6.39.

[7]Robert Sydney Marsden was later a chemist and physician (*Who Was Who* [Br.], 1916–1928, p. 703). Anders Lennart Kjellberg, from Uppsala, became an archaeologist (*Pre-1956 Imprints* 298: 354–355).

1.5 Journal, Göttingen, 19 September 1880

I have spent the last week quietly at Göttingen working about 6 hours a day and taking frequent & sometimes long walks in the vicinity. I can try to find time when I return to describe G. & the country around about & can only say now that the scenery is beautiful but the town itself is not very prepossessing. In my work I studied Whitney's grammer for one or two hours per day spent about 8 hours at reading 100 pages of Grim's Kinder u. Hausmärchen and several hours on the German testament.[1] *I also practised short-hand for about an hour a day & did some writing. Thursday I walked 12–14 miles through the woods alone & only passed two houses. I found a letter from Mamma & one each from Papa & Harry awaiting me when I arrived here & I have since received 2 from Mamma & one each from Joe, Fred & Brown.*[2]

[1]William Dwight Whitney, *A Compendious German Grammar* (New York: Holt, 1869). Whitney had been in the 1850s an important figure in the American colony at Göttingen; see Diehl, *Americans and German Scholarship*, pp. 120–130, 141–143. Jacob Ludwig Grimm and Wilhelm Karl Grimm, *Kinder- und Haus-Märchen* (Berlin: Realschul Buchhandlung, 1812–1815; often reprinted) was

one of the German philological school's most significant documents.
[2]"Fred" is Frederick Green, a classmate of JMC at Lafayette and later a lawyer in Easton (*Men of Lafayette*, pp. 231), and "Brown" is William Findlay Brown, another of JMC's Lafayette classmates, later a lawyer, businessman, and public servant in Philadelphia (*Men of Lafayette*, p. 231).

The day after he made the preceding entry Cattell embarked on an extended trip that took him to Hanover, Bremen, Hamburg, Copenhagen, and Berlin, where he arrived on Thursday, 30 September. Throughout this trip he spent much time at museums and art collections, viewing the paintings, sculpture and historical artifacts that represented European culture to the young American. He was especially impressed by the Thorvaldsen Museum in Copenhagen (double-starred by Baedeker; *Northern Germany*, pp. 149–151) and the many neoclassical sculptures of Bartel Thorvaldsen found in that city. In this admiration and in his choice of hotels and other attractions,

Cattell accepted the recommendation of Baedeker. He exhibited the conventional taste of many late-nineteenth-century Americans and Europeans who apotheosized Thorvaldsen and other artists whose work, from today's perspective, is overshadowed by the French impressionists of the same period. (The impressionists' work was minimized and even ignored by the contemporary Baedeker guides to France).

At Berlin, Cattell began attending musical events regularly and, like many other Americans, spent much time with an English acquaintance. The following entry, though written after his return to Göttingen, is a typical account of his experiences during this visit.

1.6 Journal, Göttingen, 20 October 1880

(Many short words have been transliterated from Pitman shorthand.)

On the morning of Tuesday the 5ᵗʰ Dr Marsden called on me & in the evening we went together to the Opera—Tanhäuser with Herr Niemann as hero.[1] After Dr. M. left I visited the Ravené col. of paintings (C. Hübner, Meissonier, Preyer &c) & then went to the museum which I can never see enough of. Wed I visited the National Gallery & museum together & in the afternoon visited Charlottenburg with its Flora & the Mausoleum in which King Fred William III & his wife Louise are burried. The marbles by Rauch are celebrated. In the evening I went to Opera—Margarethe.[2] Thurs. I spent at Potsdam. I had a charming day (tho it rained in the evening as it did every day I was in Berlin) & saw the beautiful park San-souci to great advantage. Of course I visited the dif. palaces, orangeries &c. I also had the good fortune to see the Prince imperial with his wife & youngest children. In the evening I went with Dr M. to the Bilsesche concert.[3] It was very crowded. I remembered distinctly the time [four symbols unclear] was only 11 years ago.[4] Friday morning I visited the Kunstgewerke museum & in the afternoon wrote home. I had written to Harry the day before on his Birthday. At sundown I attended service in the synagogue & in the evening visited the panopticum (wax-works). I spent Sat in the Nat. Gallery, Kunst-Ausstellung den Akademie & museum. At 5 in the afternoon I left for Dresden where I arrived about 8. Hotel Goldener Engel. Sunday I went to the English church it was crowded. In the afternoon I visited the gallery of paintings. I need not describe them, for I can not forget them, if I wanted to. The vision of the Sistine Madonna

[1]Albert Niemann, tenor at the Berlin Court Opera, 1866–1888 (*Grove's* 6: 89).
[2]Margarethe is the heroine of Gounod's *Faust*. The German version of this opera was sometimes performed under the title *Margarethe* (*Grove's* 3: 729–732; *Pre-1956 Imprints* 96: 509). See also 2.41 and 5.5.

[3]Benjamin Bilse regularly conducted well-attended concerts in Berlin from 1868 through 1884.
[4]JMC had attended a *Bilsesche* concert with his family in 1869 (Pardee diary, 14 December 1869).

still floats before me & the memory of the other paintings can not soon be effased.[5]
In the evening I went to the Opera—Don Carlos. Mon. morning I spent in the
Museum Johanneum (weapons armor & porcelain); in the afternoon I visited the
park of Alterthümer & Rietschel museum. In the evening [unclear] *I have been*
attending the [unclear] *quite often. I go largely for the practice. It gives me an*
understanding of German.

[5]This painting had special meaning for JMC because he had recently studied the ethics of positivism, which stressed the altruism of childbirth and motherhood. He later spent 600 marks—a large sum, even for a wealthy young man—on a copy of the picture, which is still in family hands (account book for 1880–1881).

From Dresden, Cattell went to Leipzig, where he was impressed with the great annual *Messe*, or Michaelmas Fair, "crowded with booths in which every imaginable variety of . . . articles were for sale" (Journal, 1 November 1880). He also took advantage of this visit to the center of German publishing to purchase a large number of books, including sets of the works of Johann Wolfgang von Goethe, Gotthold Ephraim Lessing, Johann Gottfried von Herder, and Friedrich Gottlieb Klopstock. From Leipzig he returned to Göttingen; he arrived on 16 October and was "right glad . . . to get back to my temporary home" (Journal, 1 November 1880).

On 18 October 1880 Cattell paid his matriculation fee of 20 marks (about $15), and on 15 November he paid another 50 marks for lecture fees, including 17 marks for the privilege of hearing "Prof. Lotz" lecture on psychology (account book for 1880-1881). On 20 November he bought Lotze's *Logik* and *Metaphysik*.

Little is known of Cattell's next four months in Europe, but it was during this time that he was drawn to Lotze and to a career in philosophy. His studies with March had shown him the possibility of such a career. And though he apparently had not known of Lotze before his arrival in Göttingen—as evidenced by the misspelling of Lotze's name in his account book—Cattell had been primed by his education at Lafayette to appreciate a philosophy that reconciled science and "the arts." But the details of Cattell's "conversion" to philosophy are not clear. It is clear, however, that philosophy for Lotze was grounded in empirical science, not in speculation. Perhaps Cattell underwent a change similar to that experienced by another American who had studied with Lotze earlier. Elihu Root had gone to Germany in 1871 planning to study theology, and, like Cattell, went first to Göttingen to learn German. After a chance encounter with Lotze, he spent the next few years studying physics in Berlin (earning a Ph.D. in 1876) and returned to America to teach the subject at Amherst College. As Burgess remembered it, Root described his conversion as follows:

I have, as a result of my study with Lotze, finally made up my mind to devote my life to the study of physics instead of theology. I am convinced that I can discover more of truth in the investigation of nature than in reading the speculations of theologians. (Burgess, *Reminiscences of an American Scholar*, p. 102; see also *Biographical Record of Amherst College, 1821–1939*, p. 156.)

Cattell studied hard at Göttingen, read widely, and like many others took verbatim notes of Lotze's lectures in shorthand. But there is no record of his daily life. His social activity during this period is also only sketchily recorded, though a letter from a friend mentions Cattell's having written that "life moves slowly along day by day in Göttingen" (William Woods, Jr., to Cattell, 19 February 1881). He did participate, like many who preceded him, in the affairs of the American colony (Diehl, *Americans and German Scholarship*, pp. 132–139), but, as one member of the group noted in August 1883, "the Colony was very small in numbers and . . . [was] little more than a name" (Paul G. Buchloh, ed., *American Colony of Göttingen: Historical and Other Data Collected Between the Years 1855 and 1888* [Göttingen: Vandenhoeck and Ruprecht, 1976], p. 83). Still, Cattell became friends with several of the Americans studying at the university at this time—notably Gwilym George Davis, a physician from Philadelphia; Lucien Ira Blake, a graduate of Amherst later to become a distinguished electrical engineer; and James Lewis Howe, an Amherst classmate of Blake's who was to be a longtime professor of chemistry at Washington and Lee University (see *Who Was Who in America*, 1897–1942, pp. 104, 300; 1951–1960, p. 422). Cattell and Howe were even chosen to design an official American colony cap "of dark blue and a white rim at the top, with a red and white band" (Buchloh, *American Colony*, p. 83). The colony got permission from the university's prorector for its members to wear the cap "should anyone choose to do so." Cattell purchased a fencing foil (account book for 1880–1881), but he probably got more use from the ice skates he bought at the same time.

Cattell visited Hanover, Brunswick, and Hildesheim during a four-day vacation in January. Meanwhile, he continued working. To develop further his acquaintance with Lotze's ideas, he purchased for 23 marks a copy of Lotze's great three-volume metaphysical treatise, *Mikrokosmus: Ideen zur Naturgeschichte und Geschichte der Menschheit: Versuch einer Anthropologie* (3 vols.; Leipzig: S. Hirzel, 1856–1864). When the term ended at Göttingen early in March 1881, Cattell left for a tour of southern Germany. He planned to travel alone until May, at which time he would meet his parents in England, where they would begin a long vacation. He would travel with them, but he would be able to continue his reading and studying during the planned long stays in France and Switzerland.

Cattell visited Dresden and Prague before arriving in Vienna, where he spent eleven days. As usual, he viewed as much art and attended as many operas as possible. From Vienna he traveled to Salzburg, Munich, Nuremburg, and Heidelberg, and from there he followed the Rhine to Holland, where he spent a week. He would later remember this period as one of the most depressed that he had ever spent (see 2.4). Arriving in London in the middle of April, Cattell attended concerts, visited the British Museum, and made preparations for his parents' arrival, as recorded in the following entry written later at Ventnor, a resort spa on the Isle of Wight.

1.7 Journal, Ventnor, Isle of Wight, 6 May 1881

(Transliterated from Pitman shorthand except for proper names, titles, and place names, which were originally written in longhand.)

Wednesday I heard Irving and Ellen Terry play Tennyson's "The Bells." [1] *Thursday afternoon I met Prof. Owen & Joe Dickson.* [2] *We took dinner & we talked with the professor until 12 o'clock. Friday I took dinner and went to the theatre (Criterion) with Dr. Moritz.* [3] *Saturday afternoon I went to Liverpool in order to be in time for the arrival of the Britannic and its precious cargo. Sunday I waited in suspense: Telegrams were received in Liverpool on Sunday. Monday I went down to the wharf hourly, the tender started at 8. In an hour I was again with my father, mother & brother, after a separation of 9½ months. My joy was great and I believe it is still greater. Harry has changed but my parents less than I had expected. Papa is very much broken and is at times very low spirited.* [4] *He had come around to regain his self by rest. Then went to Chester in the afternoon.* [5] *Tuesday we drove to Eaton Hall, Harwanden.* [6] *Wednesday we went to Warwick where they stayed until Monday. We drove to Stratford-on-Avon Thursday. I went there again Saturday and saw Shakespeare's "Twelfth Night" acted in the Memorial Theatre.*

[1] JMC mistakenly attributed this English adaptation by Leopold Lewis of *Le Juif polonau* to Tennyson, whose poetry he was reading during this period (see 1.16, 1.19).

[2] William Baxter Owen, longtime professor of Latin and Greek at Lafayette College (*Who Was Who in America*, 1897–1942, p. 925), and Joseph Benjamin Dickson, who had been a student at Lafayette from 1879 through 1881 (*Men of Lafayette*, p. 245).

[3] The Criterion was noted for operettas (Baedeker, *London*, p. 37). Edward Ralph Moritz was a chemist later to become known for his work in fermentation and brewing (*Pre-1956 Imprints* 395: 367).

[4] William Cattell had just led Lafayette College through a major crisis: the reconstruction of its main building on campus after a fire. He had always been prone to overwork and depression. (Skillman, *Biography of a College*, vol. 2, pp. 3–26.)

[5] "Strangers arriving in Liverpool should unquestionably devote a day to this most interesting city" (Baedeker, *Great Britain*, pp. 271–275). See also Journal, 27 July 1880.

[6] Eaton Hall was "an example of an English aristocratic manor," and Harwarden was the home of William Ewart Gladstone, then prime minister (Baedeker, *Great Britain*, pp. 278–279).

From central England the Cattells had traveled to Ventnor, a spot "much frequented . . . by persons suffering from complaints of the chest" (Baedeker, *London*, p. 235). William Cattell looked to recover his health, but soon he grew bored. By the middle of May, the family had moved on to London. James Cattell enjoyed his stay in the capital; he attended the theater often and even sat in on several sessions of Parliament (Journal, 22 May 1881).

Though James rejoiced in his reunion with his parents and his brother, all was not well with him. While he was traveling with the family the tension and depression he had experienced at Lafayette began to recur. He had been lonely the previous winter (Journal, 20 January 1881, 28 January 1881), and like other Americans (Diehl, *Americans and German Scholarship*, pp. 126–127) had suffered from depression while studying in Europe. The reunion with his father, especially after James's five months of studying a philosopher whose ideas did not mesh with orthodox Presbyterianism, led to the reemergence of the problems that had bothered him before.

1.8 Journal, London, 24 May 1881

(Transliterated from Pitman shorthand.)

Tomorrow I will be 21 years old. I seem on some sort of peak from which I can look back on the ground that I have travelled the foreward to a land that lies before me. It is true much of the land is covered by much of the forgotten, the unknown, but above it there may rise summits illumined by the light of truth. If there be such I think that it well to mark them mine. When I look back on the past, I almost doubt continued personality. Am I the same as child of 15, a boy of 10, and a student of 5 years ago? I can scarcely believe it. If life means anything the past is of awful moment. For the past of yesterday makes the present of today. The future of tomorrow and this year a link of the unbroken chain that led from the cradle to the grave. Life is a mystery of all mysteries the strangest is ones own life. What has made me what I am? Am I a machine turned by the irresistible force for the unknown end? And wither am I being led? For the moment let me rise above the doubts that the labors of my life will never subdue. For the moment let me look upon the perfect image of truth. And there do & see written by very noblest [five symbols unclear] my life be in so far blessed as it is sacrificed with ardor of love and when I have passed thru struggles, trials, I will with peace & joy enter into a blessed oblivion of death.

At the end of May 1881 the Cattells left London for Paris, where James marked his birthday with an ominous journal entry but soon found much to occupy his interests and apparently allay his depression.

1.9 Journal, Paris, 6 June 1881

(Transliterated from Pitman shorthand except for the last line, which was written in Greek.)

I am not superstitious but on my last birthday I yielded to the old superstition. After 12 o'clock as the 21st year of my life was dawning I picked up the Bible, opened it at random to a verse that first met my eye, lines thus in Greek:

$$\pi\acute{\alpha}\nu\tau\alpha\ \delta\grave{\epsilon}\ \tau\alpha\hat{\upsilon}\tau\alpha\ \dot{\alpha}\rho\chi\grave{\eta}\ \dot{\omega}\delta\acute{\iota}\nu\omega\nu.\ ^{1}$$

[1]Matthew 24:8. In the King James Version this verse is translated as "All these *are* the beginning of sorrows."

1.10 Journal, Paris, 7 June 1881

(Transliterated from Pitman shorthand except for proper names, titles, and place names, which were originally written in longhand.)

The evening before leaving London I heard Patti sing in Rosinni's Semiramidis.[1] Wednesday the 25ᵗʰ we came over via Dover & Cal.[2] We went to the Hotel de la Couronne and Monday moved up to the Hotel Beaujon, Rue Balzac. I have been very busy ever since we arrived in Paris. Today I seem to have accomplished so much. We have breakfast at noon. As I have a French lesson in the morning, could never go out until after breakfast. The distances are so great that one spends half his time getting from place to place. Still it is really pleasant to ride thru these beautiful constructed streets. Though I have by no means thoroughly examined them, I have visited most of the important public buildings. Collections—Louvre the Salon Luxembourg, Notre Dame, Madeleine, St. Roch, Pantheon St. Sulpice the Sorbonne the Library Trokadero &c. But all these must be revisited. I have attended Comedie Francais twice. Moliere's "Medecin malgre lui," Dumas "fils" La Princesse de Bagdad [five symbols unclear] Corneille's Horace and Le Menteur. I have attended several lectures at College de France, Renan, on Guizot and Mommsen.[3] Saturday I

[1]Adelina Patti was a world-renowned coloratura soprano (*Grove's* 6: 593–594).

[2]"Cal.": Calais.

[3]Baedeker (*Paris*, p. 236) noted that "the lectures are intended for the benefit of adults, and are of popular character. The public are admitted gratis, ladies included." Joseph Ernest Renan was a critic and historian, best known for his *Vie de Jesus* (1863), who rejected both the "supernatural content of religion" and the simple-minded view of science of some of Comte's followers (*EoP* 7: 179–180). Francois Guizot was a French historian and statesman who worked to "reconcile the interests and ideology inherited from the *ancien regime* with the growing forces of democracy" (*Encyclopedia of the Social Sciences* 4: 225–226). Theodor Mommsen, a German classical historian, was a professor at Berlin from 1858 to 1885. His work on Roman legal history was important, but most influential were his techniques of historical analysis. (*Encyclopedia of the Social Sciences* 5: 576–577.) Earlier, in Leipzig, JMC had purchased Mommsen's *Römische Geschichte* (3 vols.; Berlin: Weidmann, 1854–1856) (Journal, 1 November 1880).

attended the funeral of Littré.[4] It was a strange sight to see this unworthy opponent to Christianity buried with all the ceremonies of the church and to see Ernst Renan sprinkle his coffin with holy water. There was really a disgraceful scene at the grave, some positivists protesting against this form of burial.[5] In the afternoon they attended session in the Chamber of Deputies. We had seats in the [two symbols unclear] but the proceedings were uninteresting. Gambretta was not present.[6] We made a [one symbol unclear] of S. S. Cox.[7] I called upon him that Sunday eve. He is a very pleasant man. For example, he had been to Versailles to see [one symbol unclear] play: he said that as it was Sunday he was glad that they were not working.[8]

[4]Émile Littré, a positivist philosopher, was for a time Comte's "principal disciple" (*EoP* 4: 487). He broke with Comte in 1852 for both personal and political reasons, and after that his followers and Comte's often clashed.
[5]According to his wife and daughter, Littré accepted baptism and confessed on his deathbed, and was thus given a Roman Catholic burial. His followers did not accept his family's account, and for some months after his death positivists and Catholics accused each other of fraud. There is, apparently, no record of Renan's participation in Littré's funeral. (Stanislas Aquarone, *The Life and Works of Émile Littré* [Leiden: A. W. Synthoff, 1958], pp. 138–140, 161–163.)

[6]Leon Michele Gambetta was a leading French Republican during the Third Empire and the period of the Commune. He was premier of France from November 1881 through January 1882. (*Encyclopedia of the Social Sciences* 6: 555.)
[7]Samuel Sullivan Cox (*DAB* 4: 482–483) was a longtime Democratic congressman from Ohio and (later) New York, best known for his travel narratives, including *A Buckeye Abroad* (New York: Putnam, 1852) and *Arctic Sunbeams* and *Orient Sunbeams* (New York: Putnam, 1882).
[8]The palace was starred by Baedeker (*Paris*, pp. 287–307), as were various of its rooms, galleries, and gardens.

1.11 Journal, Paris, 18 June 1881

(Transliterated from the Pitman shorthand except for proper names, titles, and place names, which were originally written in longhand.)

Things are moving along quietly. My days seem to be pretty full. I do not accomplish very much. I take a French lesson every morning, but as I have no time to study I only try to get pronunciation. They had to discharge our French teacher, as he came in late, drank and was good for nothing. They have a lady they like very much. I have been at the Salon and Louvre several times, would never tire of them. Have also been at collections in the Library. Thursday I was at Versailles. Have been at the theatre. At the Chatelet, I saw Michael Strogoff, [ten symbols unclear] at the Gymnase, I saw Madame de Chamblay by Dumas. Last night I was at the Grand Opera and saw Les Huguenots. I had sat (3 f)[1] in the fourth gallery but at the end of the fourth act went downstairs to take a seat in the orchestra (17 f) no one objecting. I have been reading some French. Have finished two of G. Sand's novels

[1]The price of the seat, in francs.

Elle et Lui & *M. Sylvester.*[2] *I enjoyed them much. They show best thoughts that are troubling our age. Sometimes I feel strong & ready to wrestle with these questions of such awful moment, ready to push into intricies of philosophy and master humanity in the search for truth that I know will elude me, ready with the fervent trust to cast myself before the juggernaut of advancing humanity eludes us. Glad to be stone on which man shall step to climb to heights of which I do not even dream. But at other times I feel discouraged and sick at heart and [one symbol unclear] as a paradise I long to enter.*

[2]George Sand (Amantine Aurore Lucile Dupin) published *Elle et Lui* in 1859 and *Monsieur Sylvestre* in 1865. The former was widely read and concerned her open affair with Alfred de Musset, a gifted romantic who squandered his talent through drink, drugs, and his reckless life with Sand. (*Oxford Companion to French Literature*, pp. 501–502, 659–660.)

1.12 Journal, Geneva, 5 July 1881

(Transliterated from Pitman shorthand except for proper names, titles, and place names, which were originally written in longhand.)

I came here last Saturday, 2ⁿᵈ. I stayed at Paris in order to do some work in the National Library. I am studying philosophy with Prof. Lotze in order to write the essay, through which I may perhaps obtain fellowship at Johns Hopkins.[1] *I was at the Library every day for a week and half. I took a good many notes, but will not write an essay until next Winter. During the 5 weeks I stayed there I saw most of the sights of Paris—It is unnecessary to go over them here. I saw quite a number of plays at the Comedie Francaise. Britannicus, Le Monde ou l'on s'ennuie, Le Marriage de Figaro, Le Bataille des Dames*[2] *&c. I read papers good deal. Mme. Dardier [six symbols unclear] six of whom are home. There is daughter of 18, son of 17, and 4 younger children. I am much pleased with the whole family—it's an excellent place to learn French.*[3]

[1]As an active American college president, William Cattell was well aware of the exciting new university in Baltimore, of its stress on scholarly activity, and of the great honor attached to a fellowship from Johns Hopkins. In addition, he longed to have his son home in America, and probably urged him to apply for the fellowship, which carried a $500 stipend as well as an exemption from the university's $80 tuition. In any event, the elder Cattell had corresponded about fellowships with officials at Johns Hopkins as early as 1880 (William Cattell to Daniel C. Gilman, 6 April 1882, Gilman papers; *JHU Circulars*, 1882–1883, pp. 28–30, 44; see also document 1.17 below and Laurence R. Veysey, *The Emergence of the American University* [Chicago: University of Chicago Press, 1965], p. 318).

[2]*La Guerre des Femmes* by Alexandre Dumas père (*McGraw-Hill Encyclopedia of World Drama* 1: 502–506).

[3]The entire Cattell family grew to be good friends with the otherwise unidentified Dardier family. See documents 1.9 and 4.13.

While the Cattells were at Geneva, James again experienced periods of depression, alternating between "feeling happy and miserable" (Journal, 26 July 1881). Minor family crises sometimes took on major proportions for him, but he did experience some enjoyable events:

By good luck I did [something] rather remarkable—today blindfolded I won a game of chess playing against both Harry and Albert Dardier. (Journal, 12 August 1881.)

However, James's tendency towards depression was soon to be reinforced.

1.13 Journal, Geneva, 17 August 1881

(Transliterated from Pitman shorthand except for proper names, which were originally written in longhand.)

I have received a terrible blow. Prof. Lotze is dead.[1] This ruins important plans—I was going to study under Prof. Lotze this winter at Berlin, and hoped to come back again in several years to take my Ph.D. under him. I was going to write the thesis on his philosophy which I hoped would win me the fellowship to Johns Hopkins but this is not to be. I had [two symbols unclear] and Prof. Lotze's life, the philosophy has the possible solution of the duties and difficulties that surround me. I said this in a letter to W. Smith adding "It is well to have a light from a [one symbol unclear] of hope, although neither may be in the [one symbol unclear] of despairing."[2] Now even this is gone. [Two symbols unclear] of truth "these things are the beginning of sorrows."[3]

[1]Lotze died after one semester at Berlin, "from homesickness, according to Göttingen opinion," as JMC ("Autobiography," p. 631) later noted. However, in 1871 Burgess had noted that Lotze had "worked himself almost to death and was practically an invalid" (Burgess, *Reminiscences of an American Scholar,* p. 99).

[2]"W. Smith" is probably William Benjamin Smith, a graduate of the University of Ken-

tucky who received a Ph.D. at Göttingen in 1879. Though primarily educated in the sciences (he later was a professor of mathematics at the University of Missouri and Tulane University), Smith was also greatly interested in philosophy, and professed that subject at Tulane from 1906 through 1915. (*Who Was Who in America,* 1897–1942, p. 1150.)

[3]See document 1.9.

For the rest of August 1881 the Cattells stayed at Geneva, but in September they resumed their travels, visiting Berlin and Switzerland and gradually moving south to Italy. There they visited several cities, while James made plans to return to Germany for the winter.

1.14 Journal, Leipzig, 22 October 1881

(Transliterated from Pitman shorthand except for place names and proper names, which were originally written in longhand.)

Uncle Alec[1] told me at Geneva that he stopped keeping the journal because he found that he devoted too much time to it. I must consider this. Bellagio is one of the most beautiful places in the world[2]—we stayed there until Wednesday and then went to Venice. Venice is "the fairy city of the world," and we have much to interest and attract us. Saturday afternoon we went to Utine,[3] where we spent Sunday, and Monday we went to Vienna—a 14 hour ride. We met at Vienna Mr. Hatch, Mr. Stewart of Brooklyn. Pappa left Friday for Bohemia. He is delegate from our general assembly for the religious celebration they have been having at Prague.[4] Mamma and I followed Monday and left him again Wednesday, going to Dresden. On Saturday the 13th we came here to Leipzig—my home for the winter. This week I have spent in finding a family to board with, getting matriculated, next week the lectures begin.[5]

[1]Alexander Gilmore Cattell, the oldest of William Cattell's brothers, who had served as U.S. senator from New Jersey in 1866–1871 (*Who Was Who in America, 1607–1896*, p. 99.)
[2]Bellagio, a resort on Lake Como, north of Milan, was described by Baedeker (*Italy*, p. 21) as "perhaps the most delightful spot" in northern Italy.
[3]A town on the border between Italy and Austria, less than one hundred miles from Venice (Baedeker, *Italy*, pp. 67–68).
[4]William Cattell had been appointed commissioner to the Reformed Church in Bohemia, and played a major role in the development of Bohemian sunday schools (Biographical Sketch of William C. Cattell, p. 9).
[5]JMC eventually found rooms at Inselstrasse 11 (JMC to parents, 24 May 1885).

It is not totally clear why James Cattell chose to study at Leipzig in 1881. He had apparently not known of Lotze when he first went to Göttingen, and though he had decided to devote himself primarily to philosophy, the Leipzig philosophers were not well known to Americans. In any case, Leipzig was definitely a second choice necessitated by Lotze's death. Perhaps Cattell was attracted by Leipzig's generally high reputation. Only seven years earlier, Hart had written of Leipzig as "beyond question, the leading German university at the present day," stressing that "Berlin has been outstripped in the last ten years" (*German Universities*, p. 373). Or perhaps one of his fellow students recommended the university to him; many Americans went on to Leipzig from Göttingen (Buchloh, *American Colony of Göttingen*, p. 81).

Despite his hopes for Leipzig, Cattell's depressions only deepened and increased in frequency during the fall and winter of 1881. His parents had returned to the United States, and he was left alone again, at a university without an American colony like the one that had buoyed so many students at Göttingen.

1.15 Journal, Leipzig, 14 December 1881

(Transliterated from Pitman shorthand.)

If I would have my photograph taken every day it would show [two symbols unclear] and some combination of many very interesting features; but suppose I could take the photograph of my true self every night at 12! Who would recognize it! Sometimes it would show a heart full of despair and desire for death and thoughts of suicide. Sometimes the unquiet, restless heart ready for action and ambition sometimes tired—longing for rest and love, sometimes a sad stern heart that beats but would show sometimes a bitter mocking heart that scorns and despises, sometimes a strong brave heart ready to work and wait [three symbols unclear] very sad mixture of these. I often ask myself, who am I, where will I be in 10 years from now? Will I be dead, will I be a cynical skeptic, will I be a good true man? Will I be [three symbols unclear] humility and divinity, a shadow, an echo of [two symbols unclear] Torn and beat from this endless chain of brooding thoughts, what am I doing here in Leipzig? Will [seven symbols unclear] hear 17 lectures a week, 8 of which I never miss, 5 very seldom and four pretty often. And lectures on other subjects—Wundt, psychology; Heinze, logic; Baur, Shakespeare, Goethe, Schiller; Seydel, history of philosophy; Hermann, modern systems of philosophy; Teliner, Berkeley.[1] Then I am trying to study by Lotze's philosophy in order to write that Johns Hopkins essay.[2] Now I feel cheerful, my reading buries the tension; sometimes I read Shakespeare and once in a while a novel. I train my mind for passionate study by reading all the reviews in papers. I frequent the theatre and opera with diligence to escape being bored, [two symbols unclear] in which I am but moderately successful. I take flute, dancing and fencing lessons[3]—these to help pass present and future time. I must die sometime after 30 I know—so the Insurance Co. says,[4] and I fear from this that it is even longer, for they are still no more fond of cheating than the rest of us. As I am a member of the [one symbol unclear], I wear a little red

[1]Max Heinze (*NDB* 8:447) was professor of the history of philosophy at Leipzig from 1875 until his death. He was best known for his work on the pre-Socratic Greek philosophers, and later became a good friend of JMC (see 5.125). Gustave Adolf Ludwig Baur was a professor of theology at Leipzig and the author of many books. The course JMC mentions discussed the three poets "in their relation to religion" (*Pre-1956 Imprints* 40: 171–172; *Verzeichniss . . . auf der Universität Leipzig,* 1881–1882, pp. 3–6). Rudolf Seydel was a "speculative theologist" and the author of works on the relations between Buddhist and Christian thought and on science and religion (Brasch, *Leipzig Philosophen,* pp. 109–125). Konrad Hermann was best known for

his work in esthetics and the philosophy of history (Brasch, *Leipzig Philosophen,* pp. 295–312).
[2]See 1.12.
[3]By 1881 dueling had declined in its importance to German students, as well as in its ferocity and in the extent of the injuries it caused. Still, it was an important institution in that it continued a tradition long identified with German university life. (Hart, *German Universities,* pp. 65–83, 309–312; see also John Ross Brown, *An American Family in Germany* [New York: Harper, 1866], pp. 154–157.)
[4]This point is unclear, but JMC reiterated it elsewhere in the Journal. See, for example, 6.11.

cap with green and orange, a band of the same rainbow colors over my bosom.[5] *I exercise in the gymnasium twice a week and six evening kniep, a very stupid German institution.*[6] *I must be in a constant state of readiness to fight duels and drink beer. I suppose I can sleep some 7 to 8 hours, which is a grain of comfort any how.*

[5]These colors identified JMC's group as the Misnia, a student *Burschenband* that had been founded at Leipzig in 1837 and was to be suspended in 1893 (Werner Meissner and Fritz Nachreiner, *Handbuch der Deutschen Corpstudenten* [Frankfurt: Verlag der Deutschen Corpszeitung, 1927], p. 295; see also Joseph M. Leiper to JMC, 24 April 1882: "Your picture, or what I supposed was yours, came a few days ago. Scarcely knew you with the little cap & badge, you look German enough to have never seen America"). By 1881, many of the German student societies—known as *Verbindungen* (unions), or *Corps*, or *Burschenschaften* (boys' clubs)—had dropped many of the traits so often criticized by American visitors (specifically, the heavy drinking and dangerous dueling) to become similar in many ways to the Ameri-

can social fraternities. Their members continued to fence and drink, but to a much lesser degree than in the earlier part of the century. And the newer *Burschenschaften* regularly placed much emphasis on "systematic physical exercise," which would have attracted JMC, while continuing to require their members to wear their colors (Harry D. Sheldon, *Student Life and Customs* [New York: D. Appleton, 1901], pp. 10–36).
[6]*Kneipe*: literally, "pincers," but figuratively, "the word *Kneipe* has a double meaning. It denotes the place where the drinking is done . . . or it denotes the drinking itself, the carouse" (Hart, *German Universities*, pp. 138–139). See also P. H. Fridenberg, "At a German Kneipe," *Columbia Spectator* 25, no. 5 (19 December 1889): 57–58.

Cattell celebrated Christmas by visiting Dresden (and probably the Sistine Madonna) with Lucien Blake, an American friend from Göttingen. But New Year's Day found him alone again at Leipzig.

1.16 Journal, Leipzig, 31 December 1881

(Transliterated from Pitman shorthand.)

"A year is going; let him go."[1] *It has brought no happiness to me, this I know— dying I have found no comfort, no peace—darkness of duty, despair is around me. My pistol is on me—I wish with all my heart that bullet was in my brain but dare not put it there.*[2] *Nameless unceasing dread of ending all. I must try another way it may be the last. It can be made different from the past one. I have been drinking— I can, I must stop. I know I waver—I am the unflying bird, in the dark and cold ground, with a broken wing. I must start even though it be toward the light and rocky shore, but wither toward the light. What is life, what is duty, what is truth? I am the snowflakes falling into the snow, the speck flying into the darkness, 1882 "To thine own self be true."*[3]

[1]From Tennyson's *In Memoriam* (stanza 106, line 7).
[2]See other entries: "I have been holding my pistol in my hand the past ten months" (3 December 1881); "Sometimes [my] heart [is]

full of despair and desire for death and thoughts of suicide" (1.15); "This time last year I wanted to kill myself" (2.12).
[3]From *Hamlet* (act 1, scene 3).

Cattell recovered from this depression, and continued to study and to work on his thesis on Lotze for his application to Johns Hopkins (see 1.12). His parents were continually interested in his progress, and their letters to him were full of queries about the essay. And William Cattell had no qualms about using his position in the American academic community to further his son's chances.

1.17 From William Cattell, Baltimore, 8 February 1882

(Two paragraphs of gossip have been omitted.)

I came to Balt. yesterday—spent two hours with Mr. Garrett[1] last evening and today have seen Prof. Gildersleeve (my old college friend) Mr. King one of the Trustees whom I met in Geneva & Pres^t Gilman who has just ret^d after an absence of nearly 4 months—sick.[2] Nothing could have been more pleasant than these interviews & I certainly made some good points with Mr Garrett & Mr. King—Mr. G. asked me to give him copies of yr recommendations;[3] and both he & Mr. K. said they would take suitable opportunities to impress upon the Fac^y their personal interest in yr success. At the same time I have learned that personal influence—such as secures scholarships here, or the causa honoris degrees with us—will not secure Fellowships. There is an impressive no. of applicants—many of them College Professors; some even from Europe. So that the fac^y have adopted stringent rules and, for their own protection, stick to them. Still your chances are favorable and I am greatly encouraged by my visit.

And now as to the business part of it. Yr paper ought to be here by the middle of May, I shall file yr letters of recommendation at once & you better concentrate yrself on yr paper.

Gilman told me that a brief paper of a dozen pages—making one sharp point—was just as good as a labored essay of fifty pages. I told him that you had proposed

[1]John Work Garrett (*DAB* 7: 163–164) had been a student at Lafayette College for two years in the 1840s. From 1858 he was president of the Baltimore and Ohio Railroad. A good friend of Johns Hopkins, he had been one of the original trustees.

[2]Basil Lanneau Gildersleeve (*DAB* 7:278–282), a distinguished philologist, was a member of the Johns Hopkins faculty from its beginning. He had graduated from Princeton, where he had known William Cattell, and had been one of the first Americans to earn a Ph.D. (in 1853) from Göttingen. Francis T. King was a banker and merchant who had graduated from Haverford College, outside of Philadelphia, and who was generally involved with Baltimore philanthropies. He had been one of the original trustees of Johns Hopkins. (John C. French, *A History*

of the University Founded by Johns Hopkins [Baltimore: Johns Hopkins Press, 1946], p. 18.) Daniel Coit Gilman was president of Johns Hopkins from its founding in 1876 and a major figure in the transmission of the ideal of *Wissenschaft* to the United States (see introduction). He was a friend, or at least an acquaintance, of William Cattell (see 1.21) and later was to play an important role in JMC's life (see section 2). (*DAB* 7: 299–303; Fabian Franklin, *The Life of Daniel Coit Gilman* [New York: Dodd, Mead, 1910].)

[3]William Cattell later had printed a sheet entitled "Extracts of Testimonials of the Faculty of Lafayette College, to accompany the thesis of JAMES M. CATTELL, an applicant for a Fellowship in Philosophy in Johns Hopkins University, May 1882" (Cattell papers).

to write upon Lotze and of the remark in one of yr recent letters that you were somewhat discouraged about undertaking it since you had found that Jos. Cook had really not understood him.[4] *He said that just that one thing w^d be a capital point to make: & he urged me to write & advise you to discuss it if possible, with Prof. Wundt—and Dr. Gregory[5]—and then write the critique: starting out with the statement that as a Pupil of Lotze you felt constrained to differ from Mr. Cooke as to &c &c &c &c—selecting the most important point for a brief discussion but referring if you thought proper to others that might be discussed. The Dr. said you must not hesitate to get all the help you can from such talks or discussions with Wundt, Gregory or others. This was expected:—and you w^d be supposed to have done this in any event. Could you not show yr essay—or the draft of it to Gregory? I am sure he w^d be pleased to be of any service to you that is in his power.*

[4]Flavius Joseph Cook (*DAB* 4: 371–372) was a popular speaker, whose Monday lectures in Boston often oversimplified scientific and philosophical issues. His antievolution views were attacked by Asa Gray in the *New Englander* (32 [1878]: 100–113), and his misunderstanding of Lotze's ideas was criticized as "theological charlatanism" by John Fiske in the *North American Review* (132 [1881]: 287–295).

[5]Casper Rene Gregory (*DAB* 7: 601–602) was a graduate of the University of Pennsylvania (1864) with a Ph.D. from Leipzig (1876). He had studied at the Princeton Theological Seminary—William Cattell's alma mater—and had served as Pastor of the American chapel in Leipzig in the late 1870s. In 1882, he was involved in the editing of the *Theologische Literaturzeitung* at Leipzig. From 1884 to 1889, he was a *Privatdocent* at the University of Leipzig, and from 1889 he was professor of theology there.

The elder Cattells continued to express their concern in their every letter, reminding their son that the thesis was due in Baltimore by the middle of May and urging him not to worry if he did not get the fellowship: "if our business affairs continue as prosperous as they are now, we can afford the expense of sending you to Baltimore" (Elizabeth Cattell to JMC, 10 February 1882). William Cattell even repeated the substance of his letter of 8 February in a later letter (26 February): "suppose just that mail should be lost." And other members of the Lafayette faculty were called upon to support James Cattell's application.

1.18 From Elizabeth Cattell, Easton, 7 April 1882

(Several paragraphs of gossip have been omitted.)

Prof March saw Dr Gillman last week, he said the D^r had said your essay had not arrived yet, but the Prof said he threw in a good word for you, but dear Jim I have just been looking over the annual report of the University, it is wonderful how many of the Fellows have studied at Leipsic, of course you know your essay must reach the University before May 12th. I hope you may succeed, it will be a great honor but on the other hand I hardly see how it is possible, you are so young, and many of the applicants have been Professors.

Meanwhile, Cattell continued his studies at Leipzig, not totally happy, but not subject to the extreme fits of depression that had troubled his earlier.

1.19 Journal, Geneva, 9 April 1882

(Many short words have been transliterated from Pitman shorthand.)

I wish that I had some account of my winter at Leipzig. What isn't writ, isn't writ, would that it were. I change so from day to day that I can scarcely believe there to be the continuation of one personality. My thoughts, feelings and actions at different times would make up a curious collection of characters—all of little enough interest to any one else, but if I could photograph them the album would be interesting to just one, me. What I was last winter and what I did would fill many pages—what I thought and saw still more. But these I have all emptied into the bottomless ocean the past. Whether all the monotony of constant change, the battles with the shadows, the never ending see-saws of hope and despair, of calm and [one word unclear] have exhausted me or made me strong, I do not know. This is life and I accept it. Its opposite is [one word unclear], and this too I accept. Death and life are alike be it known, but I live on—sometimes acting, sometimes dreaming. What did I dream about last night?—dream after dream, only remembered by vague images, unconnected, illogical, unmeaning. It was all true and real then. Now it is all faded into nonsense. What did I do last winter? What difference does it make? It is all gone, all unreal, all untrue. Perhaps I suffered. Even so it gives me no pain now: perhaps some hours were happy, even so they can give me no pleasure now. Thoughts of past (Nice Apr 14th) misery do not make me sad now nor thoughts of past pleasure make me happy—"A sorrows crown of sorrow is remembering happier things"[1]—this all one—all the same—'tis an unreasonable illogical thing, this life of ours. Well, I lived through the winter somehow. I heard Wundt and Heinze lecture. I wrote an essay on Lotze—not yet finished—this involved some study. I usually studied in the morning, but did not accomplish very much. "Work without hope draws nectar in a sieve. And hope without an object cannot live."[2] I belonged to the Turnverein and saw a good deal of student life.[3] We had "Convent" "Kneipe" and "Spielabend" each once a week, "Fuchsstudien" twice, "Frühkneipe" Sunday morning and "Brummen" in the afternoon—all these compulsory.[4] Then there was really "Kneipe" every evening and Frühschoppin every morning.[5] We had fencing six

[1]From Tennyson's *Locksley Hall* (line 76).
[2]From Coleridge's *Work Without Hope* (lines 13–14).
[3]*Turnverein:* gymnastics club. Many students at German universities used the facilities of the *Turnvereine* (Paulsen, *The German Universities and University Study*, p. 364).
[4]*Convent Kneipe*: drinking meeting. *Spielabend*: evening of play. *Fuchsstudien*: literally, "fox studies," but as "Fuchs" was the term applied to first year students, "Fuchsstudien" probably meant some sort of meeting for first-year students, perhaps to practice their drinking (see Hart, *German Universities*, p. 71). *Frühkneipe*: early drinking. *Brummen*: literally, "to growl." A *Brummen* was a drinking bout in which two students competed at draining a mug of beer quickly. The term is derived, perhaps, from the sound of the beer pouring down the gullets (see Hart, *German Universities*, p. 71).
[5]*Frühschöppeln*: early tippling.

times a week—four times compulsory. We fought twenty-seven duels in less than two months. I might write a good deal about all this, but perhaps I can remember it for thirty years, and after that I can't take my journal to heaven with me. I was at Göttingen for a couple of days Christmas vacation. I saw more of Rolfe Δ.K.E. Amherst '80 and Robinson, Princeton '81 than of any of the other Americans at Leipzig.[6] I might also mention McCoy, Strong & Zinkeisen.[7] John Porter came a few days before I left. I saw lots of operas and plays. It was rather a pleasant winter— I look back with real pleasure on the time I spent sleeping, and part of the time I spent eating.[8]

[6]Henry Winchester Rolfe (*Who Was Who in America*, 1961–1968, p. 807) was a classical scholar who studied at Leipzig from 1881 to 1883. He later taught at Cornell, Swarthmore, the Universities of Pennsylvania and Chicago, and Stanford, and was one of JMC's best friends. William Andrew Robinson later taught Latin and Greek at Marietta College and at Bucknell and Lehigh Universities (*After Fifty Years: The Record of the Class of 1881* [Princeton, N.J.: Princeton, 1931], p. 267).

[7]George Alexander Strong was a graduate of Amherst (1880) who later attended the Episcopal Theological Seminary in Cambridge and then served as pastor at several important Episcopal churches (*Biographical Record of Amherst College, 1821–1939*, p. 234). Max Zinkeisen, probably the son of Herman Zinkeisen, a Milwaukee merchant, had registered at Leipzig as a student of law (*Personal Verzeichniss der Universität Leipzig*, winter 1881–82, no. 100, p. 131; *Milwaukee City Directory for 1874–1875*, p. 400).

[8]JMC had written ironically before (see introduction).

Meanwhile, the elder Cattells' concern about James's essay and about his chances of getting the Johns Hopkins fellowship grew.

1.20 From Elizabeth Cattell, Easton, 10 April 1882

(Several paragraphs of gossip have been omitted.)

I have just received your letter from Leipsic, and drop a line at once fearing you will not have your essay in Baltimore in time. What troubles me is, your last letter is dated the 19th of March, and it did not reach Easton until the 10th of April taking 22 days. I think I wrote in my last letter you must have your essay in Baltimore by the 12th of May. Your Papa is in Baltimore today.

1.21 From William Cattell, Baltimore, 10 April 1882

(Several paragraphs of gossip have been omitted.)

The enclosed slip shows why I am here; but you can readily imagine that I had another object in view.[1] I brought down yr letters of recommendation & left them

[1]The slip has not been found. However, it is known that on 10 April 1882 William Cattell attended a reunion of Lafayette alumni in Baltimore (William Cattell to Daniel C. Gilman, 6 April 1882, Gilman papers).

with Dr. Gilman. Prof. March was here last week and spoke to them all very kindly of you—and I suppose yr application is in good shape. Mr Garrett assured me yesterday that he w^d express his interest in yr success & so did Mr. King, whom I met in Geneva. This will do no harm;—and I also had a talk with Prof. Sylvester[2] this morning, in his room (he lives in the Mt Vernon) and saw Prof. Gildersleeve:— of course the "Fellowship" came in "incidentally"—but all these things are of weight, and you start, at least, fair; the great difficulty is in the no. of applicants— with only two vacancies;—possibly only one; and a very great pressure of personal influence (as in the present case!)—So you mustn't be disappointed if you dont succeed.

It is my peculiar misfortune now to worry over things that may not happen;— among these is the failure of just the one mail that carries yr thesis. Let us know when you send it—so that I may find out that it has reached its destination. If it does not, of course, that ends it: while if it does, that begins it only. I will let you know as soon as the matter as decided.

[2]James Joseph Sylvester, English born and educated and professor of mathematics at Johns Hopkins from 1876 to 1883. From 1883 until his death, he was Savilian Professor of Geometry at Oxford. (*DAB* 18: 256–257; *DNB* 19: 258–266; *DSB* 8: 216–222.)

The term at Leipzig had ended late in March, and James Cattell resumed his travels, visiting Switzerland and the Dardier family (see 1.19). He also met John Addington Symonds (*DNB* 19: 272–275), a distinguished English man of letters and a longtime resident of Switzerland (Journal, 7 May 1882). From Switzerland he went to southern France, and from Nice, early in April, he submitted his essay to Johns Hopkins. Then he visited Rome, Bologna, Florence, and Venice. He "looked up every church and picture in Venice" (Journal, 17 August 1882). From Italy, he returned to Switzerland and France. Meanwhile, in America, things were happening.

1.22 From Elizabeth Cattell, William Cattell, and Henry Cattell, Easton, 10 June 1882

My own dear Jim:
This house is in a great state of excitement. Dr Gilman sent to you this morning your appointment for the fellow of the Johns Hopkins University. Your Papa will write to Dr Gilman signifying your acceptance, but you had better write to President Gilman. I rejoice with you for I know how much you desired it & it is a big compliment. Your Papa worked hard for you. I enclose the list of Fellows (as you may possibly know some of the young gentlemen,) also your appointment.[1] Be sure now, and come as early as you can, I know you will want to spend a little time with us,

[1]Twenty fellows were appointed: four each in mathematics and languages; three each in chemistry, physics, and biology; and one each in logic, history, and philosophy (*JHU Circulars*, 1879–1882, p. 52). There is no evidence that JMC knew any of the other nineteen fellows.

and we will feel it is very hard to give you up the first of September & yet best for
you to be at Baltimore at the opening of the term. Come in the Nova Scotian the
first of August, and we will meet you at Halifax.[2] Buy your ticket through to Baltimore your Papa says it is just the same price as to Halifax, your books you can then
store until you go to Baltimore as it will be very nice for you to have them. I am so
excited I find it almost impossible to write or think, but these few lines will give you
more pleasure than anything. I have written you since you left home. We will have
you on this side of the Atlantic, and you will seem nearer to us, but we will hardly
know how to give you up so soon. God bless our dear boy, you little know how we
long to see you.

Your loving Mama

My dear Son
    I add my Congratulations! and all the more because, while I left no stone unturned
to secure the result, you have really won the appointment on yr own merit. And I
feel sure you will prove to the authorities of the University that you are worthy of
the honor.
    But more than all, do I crave for you the benediction & peace of our Lord—

Ever yr loving Father
W. C. Cattell

Carissimo Fratello
    Oggi una lettera di Johns Hopkins lesse. "James Cattell erci uno fellow."[3]
    Questa cosa mi fa il maggior pracere. Che tu sia felice. Suo devotessimo fratello

Enrico

[2]Steamers of the Allan Line left Liverpool
regularly for Halifax (Baedeker, *Great Britain*, p. xix). The Cattells had spent part of
the summer of 1881 at Digby, a resort on
the Bay of Fundy, across the peninsula of
Nova Scotia from Halifax, and hoped to
have JMC join them for a vacation.

[3]Note also the reaction of JMC's friend Henry
W. Rolfe (see 1.19): "Hurrah for the fellowship! What got it for you? Your general
knowledge (?) of philosophy, your two years
abroad, or your exposition of Lotze's view
on the atom?" (Rolfe to JMC, 11 July 1882).

1.23 From Elizabeth Cattell, Easton, 21 June 1882

(Several paragraphs of gossip have been omitted.)

Your Papa received yesterday a letter from Dr Gilman in which he says. "I knew you
would be pleased, but I cannot claim any part of the judgment which resulted in
your sons choice. The papers which he submitted with all akin to them, were submitted to a judge who knew nothing of any candidate, and he reported so favorably of
your son as "best of the bunch" that the choice fell upon him, and I was very glad.
When he comes to Baltimore, it will give me much pleasure to become his friend,
for his fathers sake, and for his own" I copy the letter knowing how much you would
be interested in it.

After a bit more travel in Europe, James Cattell sailed to meet his parents at Halifax. After some time at Digby, the family returned to Easton on 31 August, and James spent the next few weeks visiting friends, reading, and preparing for Johns Hopkins.

1.24 Journal, Easton, 20 September 1882

The people are sleeping in the town below me, and in a few hours they will awake and each will continue to follow his chosen bubble. Tired and weary I too will soon go to bed and in the morning will get up and start for Baltimore and a life that is all I could ask. But my life is only a desert that I cross until I come to the ocean that washes it. And I sit here restless and unhappy and see youth and health slipping from me, and alas I do not know what I want.

2

American Interlude:
Baltimore, September 1882–
October 1883

When James McKeen Cattell arrived in Baltimore as a Johns Hopkins fellow in philosophy, he found the school without a professor of philosophy but with three men teaching philosophy courses. President Gilman had found the chair in philosophy difficult to fill, particularly with a candidate of sufficient religious orthodoxy to satisfy the conservative citizens of Baltimore.[1] Through 1883, then, Johns Hopkins made do with the three lecturers, two of whom were important figures in the history of American philosophy.

The relatively minor figure Charles D'Urban Morris (1826–1886) was collegiate professor of Latin and Greek, and taught primarily in the small undergraduate college attached to the university, where he lectured on the classics. Cattell hardly got to know Morris, but later claimed to have found out that Morris had already played a major role in his life—apparently, he had been the judge of the theses submitted in competition for the fellowship.[2]

Charles Sanders Peirce (1839–1914), lecturer in logic, taught at Johns Hopkins from 1879 to 1884. Today recognized as one of the outstanding thinkers of nineteenth-century America, Peirce in the 1880s was appreciated only by a few of his colleagues and students in Baltimore. Many were repelled by his erratic behavior, and he finally lost his position at the

[1] Hugh Hawkins, *Pioneer: A History of the Johns Hopkins University, 1874–1889* (Ithaca, N.Y.: Cornell University Press, 1960), pp. 187–206; Ross, *Hall*, pp. 134–143; Albert L. Hammond, "Brief History of the Department of Philosophy, 1876–1938" (paper, 1938; Johns Hopkins University Archives). See also document 6.2.

[2] Hammond, "Brief History," p. 1; Cattell "Autobiography," p. 631; *Modern English Biography* 2: 979–981.

university as a result of some of the implications of his divorce.[3] Cattell did not study with Peirce—he had just completed a course in logic with Max Heinze, a man he respected—and mentioned him only once in his writings. But almost fifteen years later, when Peirce had retired to obscurity and poverty, Cattell helped support him. As William James, a mutual friend who served as a go-between, wrote to Cattell, "I shall be right glad of $10.00 for Peirce. He seems to have a few friends!"[4]

Much more important to Cattell was George Sylvester Morris (1840–1889), lecturer in history of philosophy. He had graduated from Dartmouth, and then studied at Halle and Berlin. From 1870 to his death he taught at the University of Michigan, first as professor of modern languages and then, from 1884, as professor of philosophy. From 1881 through 1884, he taught each fall at Johns Hopkins, where he hoped to be appointed professor of philosophy, and returned to Michigan each spring. Morris was a champion of idealism, and was probably America's leading idealist philosopher before Josiah Royce. These views, at least from Gilman's perspective, laid him open to challenges of his religious orthodoxy, making his appointment as professor of philosophy at Hopkins impossible.[5]

The competition for the Johns Hopkins chair in philosophy was not to be resolved until after James Cattell had left Baltimore, but the candidate who was eventually appointed, G. Stanley Hall, began teaching at the university as lecturer in psychology in January 1883, and Cattell worked closely with him. Hall (1846–1924) had graduated from Williams College and had studied for the ministry at the Union Theological Seminary in New York. He had also studied in Germany, taught at Antioch, lectured at Harvard, and even (in 1878) earned a Ph.D. in psychology from William James at Harvard. This background well suited him to Gilman's requirements, and his theological credentials ensured his appointment as professor at Baltimore. His later career was primarily identified with Clark University in Worcester, Massachusetts, and reflected, at least to some degree, several weaker points of his character.[6]

The one other member of the Johns Hopkins faculty who played an important role in Cattell's work there was Henry Newell Martin (1848–1896), the professor of biology. English-born and educated, Martin had concentrated his studies on physiology and was the first to earn a D.Sc. in the subject from Cambridge. In the early 1870s he had worked closely in

[3] Hammond, "Brief History," pp. 3–7; Max I. Fisch and Jackson I. Cope, "Peirce at the Johns Hopkins University," in Philip P. Weiner and Frederic H. Young, eds., Studies in the Philosophy of Charles Sanders Peirce, 1st series (Cambridge, Mass.: Harvard University Press, 1952), pp. 277–280, 285–286, 307–311; Christine Ladd-Franklin, "Charles S. Peirce at the Johns Hopkins," Journal of Philosophy, Psychology and Scientific Methods 13 (1916): 723–726.

[4] William James (1842–1910) to JMC, 13 December 1897 (Cattell papers).

[5] Hammond, "Brief History," pp. 1–4; Ross, Hall, pp. 135–138; Robert Mark Wenley, The Life and Work of George Sylvester Morris (New York: Macmillan, 1917), pp. 131–140.

[6] Ross, Hall, passim.

London with Thomas Henry Huxley, particularly in the development and instruction of laboratory techniques in physiology. He was a founding member of the Hopkins faculty, like Gildersleeve and Sylvester, and later played an important role in the development of the university's medical school.[7]

Others at Johns Hopkins were to be important to James Cattell, and he made many friends among his fellow students. He mentioned most of them in his journal, but one who was to be particularly important, both to Cattell and American life in general, appears in it only once. John Dewey had graduated from the University of Vermont (in 1879), taught high school in Oil City, Pennsylvania, for two years, and published several articles in the *Journal of Speculative Philosophy*. He had been Cattell's major competitor for the fellowship in philosophy, and Cattell later noted that if C. D. Morris had known more philosophy Dewey would have been awarded the fellowship. In any event, Dewey was able to borrow $500 from an aunt, and entered Johns Hopkins with Cattell in the fall of 1882.[8]

In September 1882, then, James McKeen Cattell's career took a new turn, and both he and his parents looked forward to his stay in Baltimore.

[7]*DAB* 12: 337–338; Donald Fleming, *William H. Welch and the Rise of Modern Medicine* (Boston: Little, Brown, 1954), pp. 82–84.
[8]*DAB* Supplement 4: 169–173; Cattell "Auto-

biography," p. 631; George Dykhuizen, *The Life and Mind of John Dewey* (Carbondale: Southern Illinois University Press, 1973), pp. 26–27.

2.1 From Elizabeth Cattell, Easton, 23 September 1882

(Several paragraphs of gossip have been omitted.)

We were delighted at receiving your letter on Saturday, it seemed like old times.[1] I do hope you will find your rooms comfortable, and I trust you have found a respectable family to rent your rooms from, but I suppose Mr Bright would know all about them.[2] Your Papa says that Biddle St. is a very pleasant part of Baltimore.[3]

[1]The letter mentioned has not been found.
[2]James Wilson Bright, Lafayette A.B. 1876 and Johns Hopkins Ph.D. 1882. At Lafayette he had worked with F. A. March, and his Ph.D. was in philology. In 1882 he was assistant in German at Johns Hopkins. He later studied in Germany, then taught at

Cornell, and finally, in 1891, returned to Johns Hopkins, where he taught until his death. (*Who Was Who in America*, 1897–1942, p. 139; *DAB* 3: 45.)
[3]JMC lived at 209 West Biddle Street, within easy walking distance of the university (*JHU Circulars*, 1882–1883, p. 19).

2.2 Journal, Baltimore, 3 October 1882

So here I am. There is no reason why I should not find it as tolerable here as any place else in the world. Indeed I can scarcely think of any improvement. The university and town give a good chance for both work and play. I like Profs Morris and Martin—they are both able men.[1] I'll have plenty of acquaintances—friends if you please. What a pleasant world it ought to be, and what a [three unclear symbols in Pitman shorthand] it is!

[1]JMC enrolled in three courses with George S. Morris: "Seminary: Science of Knowledge," "History of Philosophy in Great Britain," and "History of Philosophy (Hegel)" (*JHU Circulars*, 1882–1883, pp. 15–18; see also document 2.9). He also enrolled in Henry Newell Martin's "Animal Physiology," course, and participated in "Biology-Labora-tory Work" under the direction of Martin and William Thompson Sedgwick, Martin's assistant and later professor of biology at MIT (*JHU Circulars*, 1882–1883, pp. 15–18). Sedgwick, in particular, was known for his "neat and orderly systematic ways" (Hawkins, *Pioneer*, p. 145).

2.3 Journal, Baltimore, 5 October 1882

Since yesterday I have found a new world. I am as one who first drinks "the new strong wine of love."[1] After what I have felt, who could believe that there is any sense or realness in life? Yesterday at 2.30 I took twelve grains of hasheesh,[2] Stokes who was familiar with the drug took eight grains, and we went out to the park together.[3] After about an hour he began to act like a man slightly tipsy and kept this up all evening. During the afternoon I felt in excellent spirits—about the normal condition of many people—as a ride on horseback or a glass of wine may affect me. I would not of course have known that I had taken this drug. I went to supper and then to the library. Suddenly at 7.30 I felt the blood tingling through all my veins, as though I was taking a succession of electric shocks. What followed I cannot describe, cannot even think. Relations of time and space were lost. It seemed ages ago in another world that I had sat down on that chair. I got up and started for

[1]From Tennyson's *Maud: A Monodrama* (part 1, section 6, line 82).
[2]Purified alcoholic extract of *Cannabis sativa* (see *The Merck Index* [Rahway, N. J. : Merck, 1976], p. 602). Although "hashish" was the preferred transliteration from the Arabic, most nineteenth-century dictionaries accepted both spellings (see *OED*). The use of this psychedelic drug was known in western Europe from at least 1598 (see *OED*), and it was fairly common in America, being listed in the fifth revision of the *U. S. Pharmacopoeia* (Philadelphia: Lippincott, 1876), p. 23, and much "abused" (see Henry H. Kane, "A Hashish House in New York," *Harper's* 67 [1883]: 944–949; Fitz Hugh Ludlow, *The Hasheesh Eater: Being Passages from the Life of a Pythagorean* [New York: Harper, 1857]). In addition, it was discussed in medical circles and by such distinguished nineteenth-century philosophers as Herbert Spencer (see *EoP* 7: 523–527; *Principles of Psychology*, 3rd edition [New York: D. Appleton, 1895], vol. 1, p. 103—"It is a well known result of hashish to give an excessive vividness to the sensations"). Along with other drugs, it was the subject of much serious experimentation during the period (see John A. Popplestone and Marion White McPherson, "An Historical Note on Cannabis," *JSAS Catalog of Selected Documents in Psychology* 4 (1974): 68 [manuscript no. 660]).
[3]Harry Newlin Stokes, a fellow in biology at Johns Hopkins (see *AMS*, 1st edition [New York: Science Press, 1906], p. 310).

home. I walked on on on loosing all reckoning of space and time. On getting to my room I put on my slippers and gown and made myself comfortable on the sofa. I was perfectly happy—in complete harmony with the sum of things. I lay in a genuine trance.

I felt perhaps like a person in delerium. I seemed to be two persons one of which could observe and even experiment on the other.[4] I could feel the blood flow through my veins down to my feet and back to the heart and up to the brain. The physical sensations were delightful—the mental state none the less so. The whistling of a boy on the street sounded a divine symphony. Water was nectar—I was very thirsty. I felt myself making brilliant discoveries in science and philosophy, my only fear being that I could not remember them until morning. Time and space were all lost. When I got up to get a drink it seemed an unguessed time before I got back to the sofa. When I lifted the glass to my lips I carried it through untold distance. My blood flowed fire, all my body tingled with pleasant sensations. I was conscious of all my vital processes—breathing, the beating of the heart &c. When I slept my dreams were extremely vivid—mostly pleasant. I went to bed at 2.30 still strongly under the influence. In the morning I got up feeling delightfully—still feeling the effects of the drug. I was very hungry. The morning seemed extremely long, but every moment was pleasant. I did not feel like work, but in a delightful dreamy state. In the afternoon I worked—the day passed pleasantly. I fancied that for two or three days I could feel the stimulant. Every after-effect was delightful there has been no reaction!!!

[4]This reaction was similar to those reported by others experimenting with hashish in the late nineteenth century. See Charles Richet, "Poisons of the Intelligence—Hasheesh," *Popular Science Monthly* 13 (1881), 482–486, which despite its title argued that hashish "taken in moderate doses" could be "highly advantageous for a correct knowledge of intellectual phenomena," much as Cattell's reaction suggests. In the 1890s one psychologist, Edmund Burke Delabarre, even suggested that hashish be used as "a mental microscope and self-revealer to the psychologist." (Delabarre's work of the 1890s was not published until the 1970s, when John Popplestone published E. B. Delabarre and John A. Popplestone, "A Cross Cultural Contribution to the Cannabis Experience," *Psychological Record* 24 [1974]: 67–73.)

Cattell's experience with hashish did much to cheer him. Though he was still subject to fits of depression, they were much milder than those he had experienced in Europe.

2.4 Journal, Baltimore, 26 October 1882

I'm going to write something different from my usual melodramatic entries. Let it be granted: I know nothing. I believe nothing. Perhaps I bend towards agnosticism, fatalism, pessimism. Yet I am not miserable as I was one year ago, two years ago. Perhaps I am not miserable at all. Sometimes I feel badly enough, but I seldom suffer bitterly. It is merely ennue—world-weariness. But this goes away after a bit if I put myself at doing something. I often forget myself and feel pretty comfortable, *happy if you wish. This evening for example I am about what a person ought be. I slept ten hours last night, was hungry for breakfast. Read 200 lines of Plato and heard two lectures in the morning.[1] Talked with friends at dinner, and immediately after went to the biological laboratory. At 3.30 went out to the park and played football—was first choice in the toss up. Came home read a bit and went to supper, hungry and ready to talk. Read after supper some more (philosophy), rested a bit on the sofa, am now writing, will go to bed at ten. Thus I have been in good health, busy with change of occupation, not too much alone. Consequently I'm now in a good humor and have been pretty much all day. Life seems tolerably livable, and I don't trouble myself about death. Yes, I am certainly feeling better than I used to. All that suffering in Europe seems like a far off fearful nightmare. But how terribly real it was. At Göttingen; reaching perhaps the top of the climax at Vienna or Nuremberg. The folks probably helped me more than I knew. Still my suffering was intense, travelling, at Geneva, all through the winter at Leipzig.[2] Then I got better, have been on the whole improving ever since. Perhaps it was the travelling, bringing change and occupation for the mind and better health. Then I saw more of people— was less alone, at least less lost in myself. Aunt Nemie and Lizzie, afterward Zink-eisen did me good.[3] The charm of Italy helped me—still more the walk through Switzerland. Since then my life has been the best possible cure for such trouble. So it is then—explain it as I can. I have no faith, nothing to live for, still I am not all unhappy.*

[1] From September to December 1882, G. S. Morris's seminar on the science of knowledge concentrated on Plato's *Theaetetus* and Aristotle's *De Anima*, with special papers by students on Heraclitus, Parmenides, Empedocles, Democritus, and the Sophists and Anti-Sophists (Morris, "Memorandum respecting the work of the Philosophical Seminary, September–December, 1882," 12 March 1883, Gilman papers). JMC wrote a paper on Heraclitus for this seminar.

[2] See, for example, 1.15 and 1.16.

[3] "Aunt Nemie" is Helen McKeen Ferriday, half-sister of Elizabeth McKeen Cattell; "Lizzie" is her daughter Elizabeth. For more on Max Zinkeisen see 1.19.

2.5 Journal, Baltimore, 4 November 1882

I have taken hasheesh (11 grs) for the second time—at 6.30 this evening. It is now nearly ten. I am beginning to feel the effects of the drug. Nothing came suddenly. Reading has become uninteresting. I kept reading without paying much attention. Time seems long and every thing unnatural. I am not sure that I notice any physical sensations (except thirst).[1] After eleven. Nothing as marked as before—still very curious. It takes me a long time to write a word. I'm rather confused, have been reading.

[1] Other early American psychologists experimenting with hashish noted thirst (see Delabarre and Popplestone, "A Cross Cultural Contribution," pp. 67–73). See also 2.3.

2.6 Journal, Baltimore, 5 November 1882

The effects last night were neither so marked nor so pleasant as before. They were nearer to what alchohol gives, though peculiar enough—especially the time delusions. Stokes came to see me about 12. It was a long journey down stairs. Our conversation was curious—there seemed to be immense pauses between our sentences. The physical sensations were slight, waking dreams intensely vivid, it was certainly enjoyable. My feelings this morning were very pleasant.

2.7 Journal, Baltimore, 17 November 1882

What I write here would not interest strangers, scarcely acquaintances, and might pain those who love me—viz three—as it is altogether for myself that I keep these notes, and will perhaps destroy them some day, certainly will try to before I die. I write because it sometimes relieves my mind, and because the tracks of the footsteps by which I have come, keep in mind the way. I had a curious experience this evening which I record partly for scientific purposes, partly because the subject may have an influence on my life. I came to my room at 4 o'clock out of spirits & lay down on the sofa. I have only slept once in the afternoon since I have been here, so it is somewhat surprising that I went to sleep, still more so that I slept two hours & a half, but the surprising part of the affair was my dream—the most vivid dream I remember—it seemed to last through the whole time even when I was partly awake. I thought I was under the influence of hasheesh and all the physical & mental phenomena were extremely marked, even exaggerated. This is the third time I have experienced this, though before it has been less vivid. The interesting phenomenon was a sort of triple consciousness—was I awake, dreaming or intoxicated? I cannot describe this, it was very confused, and I am not in a clear state of mind yet—when I first awoke I was sadly mired, and have done nothing all evening. I can best give an example. I seemed to get up from the sofa and raise the window, apparently more under the influence of hasheesh as to time delusion &c. than I have really been. Now I seemed to keep considering and arguing and trying to test whether I was dreaming or really intoxicated. But this was not all, the question further was, supposing that I was intoxicated did I really get up and open the window, or was this a hasheesh dream, I lieing still. Now what a curious state of consciousness this was. In a dream I tryed to decide whether I was awake or dreaming. In a dream I knew that I was acting unnaturally and that things were unnaturally related and kept considering whether the incoherence of things was due to a natural dream, a hasheesh dream or the intoxication caused by the drug. All this seemed to last an eternity, I going through all sorts of experiences, forcing myself to act naturally in the presence of others &c. I seemed to know I was on the sofa, be this due to semi-waking consciousness or the course of the dream. I cannot now say whether I woke twenty times or not at all. I remember apparently awaking, and throwing off the influence of the dream or drug (I know not which)—this latter one really does when under its influence. Then I would seem to get up from the sofa—but this must have been dream. Part of the confusion is owing to the fact of the scene of the dream being laid where I really was i.e. on the sofa. It was certainly strange, perhaps it will look simpler in the morning.

2.8 Journal, Baltimore, 20 November 1882

I smoked a cigarette this evening—alone—"experimenting". What an idea! How completely this machine is run by habit—'tis the backbone of character. Will is not to be put along side of it.

2.9 Journal, Baltimore, 4 December 1882

Am just back from home where I spent Thanksgiving. Was at Phila a month ago at the Bicentennial.[1] Last Sat week was at Washington. We played (amusing) football with the Episcopal School at Alexandria. Then rest of the time I have been here. I have been doing pretty well as things go, better than might have been expected after the past few years. I hear ten lectures a week. Morris History of Eng. Phil (3) Seminary (2) Hegel's Phil. of History (1) Martin Physiology (4). I have read some Eng. & Ger. philosophy and several of Plato's dialogues, Apologia, Theaetetus, Phaedo. I work four or five afternoons in the week (say 15 hours) in the biological laboratory. I play football once or twice a week—am on the university eleven. I have learned to ride a bicycle & play lacrosse. Have been at two concerts (Nilsson & Thursby) and several operas, Bohemian Girl, Carmen Patience & Fra Diavola.[2] Always came away early (except from Carmen). In the evening am very generally in my room and go to bed early. Have read Bleak House.[3] Have plenty of pleasant acquaintances Bright, O'Connor, Stokes, Gould, Thomas, and others "too numerous

[1] A celebration of the anniversary of the settlement of Philadelphia.

[2] Christine Nilsson was a Swedish soprano (*Grove's* 6: 93–94). Emma Thursby was an Italian-trained American-born singer (*Grove's* 8: 455–456). *Bohemian Girl* was a popular romantic opera by Michael William Balfe (*Grove's* 1: 370–371). JMC had seen some of the first London performances of *Patience* and other Gilbert and Sullivan operettas the

previous year (see Journal, 22 May 1888). *Fra Diavola* was a light opera by Daniel Francois Esprit Auber (*Grove's* 1: 252–255).

[3] The novel by Charles Dickens best known in the nineteenth century for its scathing social criticism, particularly of the ways institutions degraded people (Edgar Johnson, *Charles Dickens, His Triumph and Tragedy* [New York: Simon and Schuster, 1952], pp. 762–782).

to mention"[4] *I have been invited to Pres Gilman's & Prof C. D. Morris's, and made one call. So much for what I have done—now what do I want to do? Alas! I do not know. About the same I suppose. Live mostly to myself and work a good deal. Keep on thinking about what it all means. As far as I know now it will be best to try gradually & without struggling to put away all appetites, desires & passions—to make my virtues quiet submission to the necessary course of things, and pity for my fellow-sufferers.*

[4]Bernard F. O'Connor was a French-educated fellow studying Romance languages (*JHU Circulars*, 1879–1882, p. 189). Elgin Ralston Lovel Gould, a graduate of the University of Toronto (1881) and a fellow in political economy at Johns Hopkins (Ph.D. 1886), later taught at Columbia University and the University of Chicago (*Who Was Who in America*, 1897–1942, p. 473). Henry M. Thomas was Baltimore-born member of a distinguished Quaker family, later professor of neurology at Johns Hopkins (*Who Was Who in America*, 1897–1942, p. 1229). His father James Carey Thomas, an important Baltimore physician, was a trustee of the university (Helen Thomas Flexner, *A Quaker Childhood* [New Haven, Conn.: Yale University Press, 1940]).

2.10 Journal, Baltimore, 8 December 1882

I wanted to take opium tonight but could not well get this and so took hasheesh again—17 grs at 6.30.[1] I wish I could describe all effects in full, as I may not take the drug again & the subject is one of real psychological importance. The subject of stimulants has always been of great interest to me—and no wonder for the practical bearings of the questions are of the utmost importance and its relations to philosophy evident. I have never used any stimulant—except perhaps a little wine & beer and these I have not taken regularly nor as a stimulant. I have never drunk tea or coffee—I scarcely eat meat. It has been my theory that a stimulant could not create, could only draw upon supplies laid up for the future. I am scarcely in a condition to speculate on the philosophic side of the subject. I now feel plainly the beginning of an abnormal condition—8 o clock.[2] Indeed I imagined that I noticed something wrong at supper an hour ago. How can I describe this feeling. Time is rather long, things seem somehow unnatural. My mind is confused instead of being made clear. I cannot write clever things—nor could I say them. Thus I differ from an intoxicated man, who thinks he is saying & doing clever things, be they ever so silly. Opium is said really to sharpen the wit.[3] Perhaps a small dose of hasheesh 3 to 5 grains

[1]JMC's difficulty in obtaining opium seems odd, as in the nineteenth century this drug had a long tradition of medical and popular use (*U.S. Pharmacopeia*, 5th revision, p. 41; *Merck Index*, p. 890; Eric T. Carlson and Meribeth M. Simpson, "Opium as a Tranquilizer," *American Journal of Psychiatry* 120 [1963]: 112–117; Althea Hayter, *Opium and the Romantic Imagination* [Berkeley: University of California Press, 1968]). One of JMC's aunts, Esther Gilmore Cattell Fithian (William Cattell's sister) was in fact being treated for pain at this time with opium (Elizabeth Cattell to JMC, 23 September 1882, and 2.19).

[2]The handwriting in this entry was sloppier than usual.

[3]See, for example, Hayter, *The Romantic Imagination*, pp. 132–150.

might have this effect.[4] I am physically tired as I skated this afternoon for the first time this year, and was still stiff from fencing & jumping yesterday. I feel a slight tingling sensation in my arms & legs.[5] I think I am somewhat nervous (starting at noises &c) and suspicious. Still I feel comfortable in my room. Senses seem abnormally acute. 9 o'clock Have been reading—Shakespere. It is hard to keep my attention fixed, but still I can appreciate what I read. Activity throws off the influence. I notice I make mistakes in writing, perhaps I do not pay close attention.[6] The effects are not especially pleasureable. My pulse is normal. (10) The effects are pleasanter. At first I was carried at discord with things, now I feel in harmony. Perhaps the most curious phase of all is my being able to watch and experiment on myself. I am two persons one intoxicated the other sober. (One)—I can write with ease. I have been dozing. Dreams of course very vivid. It is not very pleasant, nor very satisfactory. I do not seem to be nearly so much under the drug as the other times. It makes at least a decreasing geometrical ratio 4:2:1 I feel almost like writing 64:8:1. Now I am not sure that I feel anything. I am not, nor have I been especially thirsty.

[4] Others during this period made similar observations (Popplestone and McPherson, "An Historical Note on Cannabis," pp. 3, 5; Delabarre and Popplestone, "A Cross Cultural Contribution," pp. 68–69, 71).

[5] JMC had first written "feeling" rather than "sensation."

[6] For example, JMC had first written "throws" of the previous sentence as "through."

2.11 Journal, Baltimore, 17 December 1882

I took ½ gr of morphia[1] (=3 grs of opium, 75 drops of laudanum[2]) this evening at 6.50—now an hour after I feel slightly the drug, though not enough to know that I had taken it. My face is flushed, my legs feel weak. When I talk I am [three words unclear] (Stokes is here has likewise taken ½ gr). He feels the effects more than I. Pulse (mine) 90, fifteen minutes ago it was normal. Stokes was over a hundred half an hour ago. 9.30 nothing worth noting, temperature and pulse about normal. From its effects I would not now know, nor at any time in the evening have known that I had taken opium. I am in good spirits, but nothing extraordinary. Stokes has just gone—was slightly more animated than usually. 11.30 am going to bed, feel slightly depressed & nervous. Also somewhat sick at the stomach.

[1] Morphine, or morphia, as it was often called in the nineteenth century (see OED), is "the most important alkaloid of opium"; laudanum, first prepared, apparently, in the sixteenth-century, was "the simple alcoholic tincture of opium" (Merck Index, p. 890; OED). The fifth revision of the U.S. Pharmacopeia (pp. 229–321, 314) gave several preparations of morphine and laudanum. Despite the addictive properties of morphine and laudanum (as opposed to hashish and other cannabis derivatives, which are described clinically as "psychedelic" [Merck Index, p. 222]), these drugs were used extensively in the nineteenth century, and appear in such novels as The Moonstone and Uncle Tom's Cabin (Hayter, The Romantic Imagination, passim). For an example of other nineteenth-century experiments along these lines, see Herbert Spencer, "Appendix: On the Actions of Anaesthetics and Narcotics," in Principles of Psychology, vol. 1, pp. 631–635.

[2] JMC had first written "morphia."

2.12 Journal, Baltimore, 31 December 1882

I stand at the gateway and look backward and forward. Behind me lies the country through which I have come, rockly, dangerous. There are a few flowers it may be, but there is no single spot I care to visit again. What do I see before me? Alas, a country flat & sterile, with naught of beauty or grace; above is a dull leaden sky. And through this land I must walk until my strength fail me and I fall by the wayside. I have no hope or faith and my burden is heavier than I can bear. I do not suffer as bitterly as I used to, sometimes I think I am growing dull and sluggish. This time last year I wanted to kill myself,[1] now I want to go to bed. I cannot believe any thing, if I follow my reason I come always to absurdities, I cannot believe that anything is, not even myself—it is all a chaos of nonsense. But apart from this philosophic scepticism—made the more true by the fact that I don't act as though it were true—my life is sad and dreary. It is not only my own life that is a failure, all life is vain and unmeaning. My only virtue is pity—pity for myself—pity for the innumerable host that toils & suffers and dies. My only religion is a vague pantheism. The universe of things flows on resistlessly; for a little while we rise from out the ocean of being, and are as waves breaking on an unkown shore.

[1] See document 1.16.

While Cattell continued his studies and his experiments with drugs, his parents continued their concern with his future.

2.13 From Elizabeth Cattell, Easton, 8 January 1883

(Several paragraphs of gossip have been omitted.)

I suppose you noticed that D[r] Krauth died last week, he is the Professor your Papa was anxious to have you assist at the University, feeling some day you might fill his chair.[1]

[1] Charles Porterfield Krauth (*Who Was Who in America*, 1607–1896, p. 299) had been professor of moral and intellectual philosophy at the University of Pennsylvania from 1868. A Lutheran minister, Krauth had taught within the older tradition of American philosophy.

2.14 Journal, Baltimore, 10 January 1883

Of my various experiments with stimulants, that of day before yesterday was the most remarkable. I inhaled the vapor of ether for two hours.[1] It is utterly impossible to describe the experience. I was very drunk almost, perhaps quite, unconscious. It seemed absolutely impossible for me to put the bottle aside until every drop of the ether had evaporated (2 ozs)[2] Every sensation was delicious—something like the effects of hasheesh but more intense. There were delightful physical sensations—they can be feebly described as electric shocks. I seemed to live a great deal in a short time. My mind was active. I seemed to be making brilliant discoveries—as I remember them they were commonplace enough, but not nonsensical. For example at first the odor of the vapor was unpleasant, afterwards it became pleasant. I noticed this and rightly attributed it to the pleasant effects it produced and went on considering that usefulness was the ground of beauty &c&c Time was long, abnormally so. I seemed to be supremely happy. I imagine I would have kept on inhaling the vapor, if I had thought that it would have killed me in the end. When it was all gone I was very drunk staggered as I crossed the room. I slept well enough. Next morning I had a slight "Katzenjammer."[3]

[1] Ethyl ether had been used as an anesthetic since the middle of the nineteenth century, and had been inhaled for its exhilarating effects throughout the century (*Merck Index*, pp. 500–501; *U.S. Pharmacopoeia*, fifth revision, p. 76; L. R. C. Agnew, "On Blowing One's Mind [19th-century Style]," *Journal of the American Medical Association* 204 [1968]: 159–160). Like other psychedelics and hallucinogens, ether was studied during the period with regard to its effects on sensation and perception (see Spencer, *Principles of Psychology*, vol. 1, pp. 610–612).

[2] JMC had first written "perhaps 4 ozs."

[3] A hangover, or the symptoms of one (*OED* Supplement).

2.15 From William Cattell, New York, 12 January 1883

(This letter was dated "Friday 10:30 P.M." Several paragraphs of gossip have been omitted.)

I heard Prof. Morris again this evening & had a very pleasant talk with him about you[1]—he said he often wished he knew you better: and once called on you, but you were not at home—said he w^d certainly see you before he left for Ann Arbor. It seems to me there can be no doubt about yr getting the Fellowship next year (if we stay)[2] tho yr last letter seemed to put the contest for it as if it were de novo.[3] But w^d not Dr Morris mainly decide it? However we can talk about this when you come home next week.

[1] William Cattell was in New York "on some college business" on 10 January 1883, and stayed through 12 January expressly to meet George S. Morris, who was also in New York (William Cattell to JMC, 10 January 1883).

[2] The elder Cattell was apparently considering resigning the presidency of Lafayette College (see 2.40).

[3] This letter has not been found.

Prof. Morris gave us a really grand lecture tonight:—one of the few lectures that (of late) caught me at the start & carried me straight through.[4] *In some places however it was too deep for me.*

[4]In January 1883 Morris delivered eight lectures at the Union Theological Seminary in New York, the "object" of which was "to institute a comparison between what may be termed the philosophic presuppositions of Christianity, and the historic results of philosophic inquiry" (*JHU Circulars*, 1882–1883, pp. 36, 58).

2.16 To Parents, Baltimore, 17 January 1883

Mamma's letter tells me you expect to go to Phila. today, and as I am not sure you saw my last letter before leaving home I write a few lines anent Dr. Krauth's death.

I do not know how his place is to be filled, or indeed if it is to be filled at all, but if they get a young man I would like very much to be the one. Of course you know my reasons as well as I myself. If the place is once filled I cannot wait till it is again vacant. There are few or no positions equally desirable for me. I do not think it would be wise even if it were possible for me to stay on here. I would rather be at Phila than at New York, Ithaca &c and even if vacancies should occur at these universities, it is likely that you could not get me in. You understand the difficulties in the way of finding a place to teach philosophy (especially for a young man) and the disadvantages of studying this and teaching something else. I send you this as I thought it possible you might hear or do something in Phila. If you want to see me I can meet you or come home. I would enjoy ever so much a visit from you.

I won't write more as you may never see it any how.

2.17 From William Cattell, Philadelphia, 18 January 1883

(This letter was dated "Thursday 10:00 P.M." Several paragraphs of gossip have been omitted.)

Yr mother and I reached the city about 10 o cl this morning. Your letter anent the position in the University reached us on Tuesday: and I lost no time in calling upon Mr. Fraley,[1] *Pres^t of the Board who was out of town; upon my friend Mr. Welsh*[2]*— the first time he was not in but I found him later in the day (had a long talk with him; he did not know of the appointment)*[3] *then on Dr. Pepper*[4] *twice—the last*

[1]Frederick Fraley (*DAB* 6: 574–575) was a public-minded merchant and banker who had long served the University of Pennsylvania.
[2]John Welsh (*Who Was Who in America*, 1607–1896, p. 570) was a Philadelphia merchant, sometime U. S. diplomat, and trustee of the university.
[3]William Cattell added these parenthetical clauses as a footnote.

[4]William Pepper, provost (chief executive officer) of the University of Pennsylvania from 1880 to 1894 (*DAB* 14: 453–456). A member of a distinguished Philadelphia medical family, Pepper had graduated from the university in 1862, had taught medicine there from 1868, and later was to play an important role in JMC's life. See also 2.19; 6.8–6.10.

time at 8³⁰ finding him home & that the position has been filled There was one of
their own graduates[5]—trained by Dr K—who had given remarkable promise in this
Department—they appointed him at once—& he enters upon his duties Feb. 1.—

So my dear son that matter is settled; and as yr dear mother says "all for the
best". I try to think so. When I heard of the Dr's death my heart failed me. You
were not yet ready for such a place: and I knew they wᵈ want a man at once; and it
was wise in the Trustees to take one of the Dr's own students.

[5] George Stuart Fullerton (*DAB* 7: 66–67) was
a graduate of the University of Pennsylvania
(A.B. 1879, A.M. 1882) and of Yale (B.D.

1883). He too was to play an important role
in JMC's life. See 6.20.

2.18 Journal, Baltimore, 20 January 1883

I wonder if I will ever learn to let them alone, these "uralte, ewige Räthsel."[1] It
seems to me that I do nothing else, but brood on them. I see no escape except
death or madness. Could I perhaps carry through an experiment? I think I shall try.
Suppose that for a month I try a phase of quietism—though not really it itself. My
food is plain enough, I scarcely ever eat meat, but I will still further cut down my bill
of fare, and will eat as little as possible. I will go to bed at 10 and sleep nine hours,
perhaps 9½. I will read in my room until lecture time, then after lectures in the
library until dinner, then laboratory or exercise, lecture at five, tea at 6.30, then
write and read until 10. My meals consist of—Breakfast, oatmeal & milk perhaps
corn bread or some such in addition—dinner, vegatables, rice, potatoes, beans, mac-
caroni, hominy—perhaps plain desert as an orange—tea bread butter & milk—
perhaps stewed fruit. I will live altogether to myself—will be conventionally pleasant
to acquaintances, but must not try to say clever things. I will work at one special
thing—a paper on Clough which I will publish or read before the metaphysical
club.[2] Above all I must keep from mental (emotional)[3] excitement, as much as
possible. I do not know how far I can keep to the spirit of this, nor how satisfactory
the results would be any how. If I could get into a condition altogether neutral &
colorless, I would be well suited. Perhaps then I might give up the work—that would
be genuine quietism, or I might try the other pole for a while "action, passion, talk."
Well suppose I try something like the above until Feb 20ᵗʰ.

[1] Ancient eternal riddles.
[2] Arthur Hugh Clough was an English lyrical
poet. His poems reflect a concern with ethical issues and "bear the mark of spiritual ag-
itation caused by religious doubt" (*Oxford
Companion to English Literature*, p. 177). The
club, organized by Charles S. Peirce, met at
the university from 1879 to 1885. Students
and professors from all disciplines gave pa-

pers and held discussions in all areas of philosophy and in related subjects. Cattell had
been elected secretary on 14 November
1882. (Fisch and Cope, "Peirce at the Johns
Hopkins University," pp. 285–286, 371–374;
JHU Circulars, 1882–1883, p. 38.)
[3] JMC inserted this parenthetical word above
the rest of the line.

Meanwhile, illness in the family brought the Cattells into further contact
with the University of Pennsylvania. William Pepper was called in to treat

Esther Gilmore Cattell Fithian, William Cattell's sister, and Elizabeth Cattell met him while visiting her sister-in-law. Dr. Pepper is the subject of the following letter.

2.19 From Elizabeth Cattell, Easton, 22 January 1883

(Several paragraphs of gossip have been omitted.)

We met him one evening when he called to see Aunt Hetty he is very bright, and agreeable but I expected to see an old gray headed gentleman but he is very young looking and cannot be more than forty five years of age. Your Papa called to see him the first day we arrived in the City, he received him very cordially, and said he had had you in his mind if the young gentleman they elected had not been willing to accept the position of instructor. The young gentleman they elected was studying Theology at Yale, he was an old student of Dr Krauth's and of course he is obliged to step in at once, and carry on his classes. I think it would have been a very difficult position for you to fill, you so young, and I fear you would have been obliged to work, and study so hard you would have broken down, so do not feel too much disappointed. Dr Pepper intimated the evening he called upon Aunt Hetty if this young man did not please them, or if he did not fancy the position you should be elected.

By the end of January, Cattell, with the help of the drugs with which he had been experimenting, had left most of his depression behind and decided to devote most of his effort to laboratory work. Though not yet totally free of the mood swings that had troubled him so much in Germany, he was more satisfied with life and felt better. His work also improved.

2.20 Journal, Baltimore, 23 January 1883

I think I shall go to bed—9 o'clock. Unless I find it very desirable to publish something, I imagine I had better "save up" Clough and go to work on physiological psychology. I am thinking of devoting myself chiefly to the effects of stimulants—it seems to me four articles might well be written on this subject—from a physiological, psychological, ethical and metaphysical point of view. I was skating this afternoon. I'm feeling shall I say, happy or comfortable, i.e. well fed, lazy, sleepy & indifferent.

2.21 Journal, Baltimore, 26 January 1883

I drank a glass of extract of tea—equal to 10–20 cups of ordinary tea this evening. I have taken full notes. Last Sat. I smoked my first cigar. Sun. the 14 I took 100 drops of laudanum. I have in another place notes on these. It is twelve and I might as well go to bed though I will scarcely be able to sleep.

2.22 Journal, Baltimore, 28 January 1883

It seems to me I have changed greatly in the past year, especially in the last couple of weeks—almost like "conversion" or what I imagined might happen to me if I should fall in love. I don't understand it at all. I feel inclined to look up people instead of avoiding them. Feel inclined to be social & pleasant, instead of misanthropical & scornful. Feel inclined to believe rather than to doubt. I know nothing, but I feel that I can live & work & love & die; & have faith to believe that it is all well.

Cattell's experiences during this period were similar to those later described by Erik Erikson in his discussions of what he calls identity formation. Erikson writes of the process as the "selective repudiation and mutual assimilation" of previous personality traits, based in part upon "the establishment of a good relationship to the world of skills and to those who teach and share new skills." Most importantly, developing "self-esteem . . . grows to be a conviction that one is learning effective steps towards a tangible future . . . within social reality which one understands." This trial-and-error movement towards the emergence of a "healthy personality" may be contrasted, according to Erikson, with the process of "ego diffusion" that often precedes it. Ego diffusion is described as the "inability to concentrate on required or suggested tasks," often tied to a "self-destructive preoccupation with some one-sided activities" peripheral to one's major goals. Again, this portrait describes well many of Cattell's experiences during his first months at Johns Hopkins. See Erik H. Erikson, "Identity and the Life Cycle: Selected Papers," *Psychological Issues* 1: 1 (1959), especially "Growth and Crises of the Healthy Personality" (pp. 50–100) and "The Problem of Ego Identity" (pp. 101–164).

In early 1883, Cattell's life seems to have calmed down and become rather routine.

2.23 Journal, Baltimore, 5 February 1883

I resumed work in the laboratory today, cutting sections of dog's brain.[1] Dr. Hall began his lectures.[2] He impresses me favorably—not a man of much strength perhaps, but of considerable ability and enthusiasm. I am reading Carpenter's "Mental

---

[1] Much of Martin's physiological instruction was based upon dissection, and he explicitly recommended one year later that students learn how to section the brains of mammals. (Henry Newell Martin and William A. Moale, *Handbook of Vertebrate Dissection*, vol. 3, "How to Dissect a Rodent" [New York: Macmillan, 1884], pp. 226–228).

[2] During his second semester at Johns Hopkins, as G. S. Morris had returned to Ann Arbor, JMC enrolled solely in Martin's course in animal physiology. He attended Hall's lectures in "Psychology: Advanced Course" while participating in laboratory work in both subjects. (*JHU Circulars*, 1882–1883, pp. 90–93.)

Physiology." [3] *I intend to devote myself mainly to this subject during the rest of the year. I am feeling quite well & contented in a quiet way. I am not sure that it will last, but I am so used to watching & studying myself, that I ought to be able to direct my thoughts & actions in the way, which is most likely to make this state of mind lasting.*

[3] William Benjamin Carpenter (*DSB* 3: 87–89) was a distinguished Scottish physiologist. His best known book, *Principles of General and Comparative Physiology* (1839), went through at least five editions by the 1860s; in 1874 its "Outline of Psychology," which had grown tremendously, was split off and published separately as *Principles of Mental Physi-*

ology: With Their Application to the Training of the Mind and the Study of the Morbid Conditions. JMC owned the sixth edition of this book (London: Kegan Paul, 1881). On 26 September 1882, Carpenter had given the address officially opening the academic year at Johns Hopkins (*JHU Circulars*, 1882–1883, p. 12).

2.24 Journal, Baltimore, 6 February 1883

Looked up Lotze & J. Cook. [1] *Heard Dr. Martin lecture on the kidneys.* [2] *Read Anstie on stimulants in the Peabody.* [3] *After dinner worked in the laboratory: heard Dr Hall lecture on the sense of touch.* [4] *Exercised in the gymnasium. Ate too much supper. Worked up this evening Cook on Lotze. This afternoon in the laboratory I inhaled a little chloroform. (note elsewhere).* [5] *Now to bed, "So geht's im Welt."*

[1] Flavius Joseph Cook (see 1.17).

[2] See Martin, *The Human Body: An Account of Its Structure and Activities and the Conditions of Its Healthy Working*, 3rd edition, revised (New York: Holt, 1884), chap. 27, "The Kidneys and the Skin," pp. 404–422.

[3] Francis Edmund Anstie, *Stimulants and Narcotics: Their Mutual Relations With Special Researches on the Action of Alcohol, Aether, and Chloroform* (London: Macmillan, 1864). The Peabody Institute, founded by George Peabody (1795–1869), provided Baltimore with an art gallery and a "large and handsome" reading room for a library of "118,000 well-selected vols." and was starred by Baedeker (*United States*, p. 246). See also *JHU Circulars*, 1882–1883, p. 151.

[4] Hall's lectures in psychology at Hopkins stressed sensation and perception, and his laboratory work in the following two or three years concentrated on the sensitivity of the skin (Ross, *Hall*, pp. 153–158; G. S. Hall and H. H. Donaldson, "Motor Sensations on the Skin," *Mind* 10 [1885]: 557–572; G. S. Hall and Jujiro Motora, "Dermal Sensitiveness to Gradual Pressure Changes," *American Journal of Psychology* 1 [1887]: 72–98).

[5] See 2.14; *U. S. Pharmacopeia*, 5th revision, pp. 20, 78, 227; *Merck Index*, p. 272; Herbert Spencer, "Appendix: Consciousness Under Chloroform," *Principles of Psychology*, vol. 1, pp. 636–640.

2.25 Journal, Baltimore, 7 February 1883

I called on Mr. Trippe this evening. [1] *He is a man one likes to meet genuine; generous, fond of his family, his church, old memories, his wife: fairly well satisfied with himself & the world—a philistine if you please, but of the best type. Called on*

[1] Andrew Cross Trippe was a distinguished Baltimore attorney and a graduate of Lafay-

ette College (*Who Was Who in America*, 1897–1942, p. 1253).

Mrs. Thomas—she was out, but I talked for a while with Henry.[2] Examined a human brain in the laboratory & wrote out notices for the Metaphysical Club.[3] Exercised from 5–6.30 or rather from 5.15–6 in the gymnasium.

[2] Mary Whitall Thomas, well known in Baltimore for her philanthropic work, was the wife of James Carey Thomas and the mother of Henry M. Thomas (JMC's classmate at Johns Hopkins) and Martha Carey Thomas, later president of Bryn Mawr College. See 2.9; *NAW* 3: 446–450; Helen Thomas Flexner, *A Quaker Childhood*.

[3] The next meeting of the Metaphysical Club was on 13 February 1883. Papers were presented by Cattell, Hall, and Albert Harris Tolman. Tolman, a graduate student in philology, presented "A Review of M. Hopkins' 'Outline Study of Man.'" (*JHU Circulars*, 1882–1883, p. 94.)

2.26 Journal, Baltimore, 13 February 1883

I commenced work this morning in a new physiologico-psychological laboratory.[1] Spent the time with Drs Hall & Hartwell in getting apparatus ready.[2] This evening I read—or rather spoke on "Mr Cook of Boston" before the Metaphysical Club.[3] My remarks were not appreciative. I will keep my notes.

[1] Hall had established this laboratory, with the help of Cattell and other students, in a private home adjacent to the Johns Hopkins campus (Ross, *Hall*, 154; JMC, "Early Psychological Laboratories," *Science* 67 [1928], 543–548).
[2] Edward Mussey Hartwell was a graduate of Amherst College (A.B., 1873). He also held a Ph.D. in biology from Johns Hopkins (1881) and an M.D. (1882) from Miami Medical College. From 1883 to 1891 he was an instructor and director of the gymnasium at Johns Hopkins. (*Who Was Who in America*, 1897–1942, p. 530; *JHU Circulars*, 1882–1883, pp. 19, 37.)
[3] On Flavius Joseph Cook see 1.17 and 2.24.

2.27 Journal, Baltimore, 16 February 1883

I have no very startling entry to make. Experimented in psychology. Fenced in the gymnasium, ran a mile in 5.12. Wrote to Harry. Read Walt Whitman & Tennyson.

2.28 Journal, Baltimore, 20 February 1883

I took a long walk in the country & park this morning—the day was lovely in the "fullness of the Spring".[1] Dr. Hall began his public lectures on pedagogy.[2] I read the first 75 pages of Taine "De l'Intelligence"[3]—as also Keats & Tennyson.

[1] From Tennyson's *Locksley Hall* (line 36).
[2] Hall had given a series of public lectures in pedagogy in January 1882, which led to his appointment as lecturer the following year. He repeated the lectures in 1883 at President Gilman's urging. (Ross, *Hall*, pp. 134–138; *JHU Circulars*, 1882–1883, pp. 63, 89.)
[3] Hippolyte Adolphe Taine, a French philosopher, had been interested in psychology and was greatly influenced by physiological studies and positivistic ideas. His book *De l'Intelligence*, first published in 1870, developed a theory of the mind that was physiologically informed. However, the influence of this work in America, at least in comparison with that of Wundt's ideas, was negligible. (*EoP* 8: 76–77.)

2.29 Journal, Baltimore, 22 February 1883

Father came on to see me yesterday. He arrived at nine & we went to Pres Gilman's. Met Pres & Mrs. White.[1] Saw lawn-tennis tournament today. Attended the university exercises at 4 o'clock.[2] Fine speech by S. Teacle Wallis.[3] O'Connor took his Ph.D. We were invited to dinner at Mr. Garrett's.[4] They live elegantly. I really enjoyed myself. We went at 7 to university reception & talked to every body for an hour.

[1] Andrew Dickson White (*DAB* 20: 88–93) and Mary A. Outwater White. A. D. White was president of Cornell from 1867 through 1885 and a distinguished American diplomat during and after his academic career. He was one of the leaders in the movement that led to the emergence of the modern American university, and was a good friend of Gilman from the 1850s. William Cattell had met him and corresponded with him earlier. (Veysey, *The Emergence of the American University*, pp. 69, 304–306.)

[2] Each year, 22 February was celebrated at Johns Hopkins as the anniversary of the University, with formal speeches, distinguished guests, and receptions (*JHU Circulars*, 1882–1883, p. 88).

[3] Severn Teakle Wallis, (*DAB* 19: 385–386), a distinguished Baltimore attorney was deeply involved in civic affairs. Wallis's talk "The Johns Hopkins University in its Relation to Baltimore" was soon reprinted in *JHU Circulars*, 1882–1883, pp. 107–116.

[4] John W. Garrett (see 1.17).

2.30 Journal, Baltimore, 23 February 1883

Spent the day with father. Were at a company at Prof. Gildersleeve's this evening,[1] talked promiscuously. Father left for home at midnight.

[1] Basil L. Gildersleeve (see 1.17).

2.31 Journal, Baltimore, 26 February 1883

Lecture of physiology—the spinal cord. Took a long walk from 11–1.15 which I enjoyed. Looked up Helmholtz on sound & music.[1] Lecture—sense of sight. Gymnasium, Mr. & Mrs Florence in "The Might Dollar"—very fair but left at end of the second act. I received Clough's works this evening[2]—imagine I had better go to work on a paper to be printed.

[1] Hermann von Helmholtz (*DSB* 6: 241–253) had been Wundt's senior colleague at Heidelberg before 1871. His work of the 1850s on the sensory physiology of hearing had culminated in his resonance theory, presented most comprehensively in his *Lehre von den Tonempfindungen als physiologische Grundlage für die Theorie der Musik* (Brunswick: Vieweg, 1863).

[2] JMC had these books sent express from New York by a friend who worked for the bookseller Dodd, Mead & Company (J. M. Leiper to JMC, 24 February 1883).

2.32 Journal, 27 February 1883

One day is much like another with me—except subjectively. I heard my two regular lectures today & worked in the laboratory. This afternoon I took a long walk. This evening instead of supper I ate nearly half a pound of chocolate. It has not stimulated me, but has put me in rather a pleasant state of mind.

2.33 Journal, Baltimore, 28 February 1883

Was out at the Bay View Insane Asylum with Dr Hall & Thomas.[1] We got specimens of handwriting & composition. I am out of sorts today.

[1] The Bay View Asylum was officially the Psychopathic Wards of the Baltimore City Hospitals, located outside of the center of the city (*Directory of State and Private Institutions for Mental Diseases in the State of Mary-* *land* (1926), p. 20; see also Douglas Carroll, "The Bayview Asylum: Some Socio-economic Influences [1866–1890]," *Maryland State Medical Journal* 15 [1966]: 101–103).

2.34 Journal, Baltimore, 6 March 1883

I received a telegram Friday evening which took me to Phila. Sat. morning. Mother was being treated by the dentist & father had to leave her over Sunday. We were much afraid the trouble might be serious, but the dentist says it is over—he called it necrosis. I was with mother nearly all the time. I returned yesterday night arriving at 1.30. Today I worked as usual. This evening I called on Mrs Gildersleeve, Mrs & Miss Gilman & Mrs & Miss Garrett.[1] Wrote out account of J. Cook on Lotze which I shall send to the "Nation."

[1] Eliza Fisher Gildersleeve, Elizabeth Dwight Woolsey Gilman (D. C. Gilman's second wife) and Alice Gilman (the elder of two daughters of his late first wife), and Rachel Harrison Garrett and Mary Elizabeth Garrett. Mary Elizabeth Garrett later played an important role in women's education in Baltimore and elsewhere and in the establishment of the Johns Hopkins University Medical School (Margie H. Luckett, *Maryland Women* [Baltimore: by the author, 1931]; *NAW* 2: 21–22).

2.35 Journal, Baltimore, 7 March 1883

Sent communication to Nation. Worked up averages for "bilateral function."[1] Took a walk & exercised in gymnasium. Heard G. W. Cable lecture on "Literature—it's

[1] Hall and Hartwell, using JMC and other students as subjects, carried out a few experiments on bilateral asymmetry of such functions as arm movements and reactions to aural stimuli. They published the results as "Bilateral Asymmetry of Function," *Mind* 9 (1884): 93–109 (see Ross, *Hall*, p. 155). The paper itself does not mention the help of any student, and presents no data, citing merely "tables far too extensive to point here" (p. 103).

influence on Man's public life." [2] *Saw Strauss's "Der lustige Krieg" by Thalia Co.—*
very fair; I am living Bohemian—eat mostly fruit & oysters. I think I am feeling
better than I have these three years or more—am almost in high spirits. I am not
contented though even to be happy.

[2] George Washington Cable, popular Louisianan essayist and novelist (*Oxford Companion to American Literature*, p. 126). The lecture was part of a series of talks by Cable on the general subject of "The Relation of Literature of Modern Society." (*JHU Circulars*, 1882–1883, p. 63).

When James Cattell visited his mother in Philadelphia, William Cattell—
hoping to learn something that would enable him to get his son a suitable
position—asked for his opinions as to how philosophy should be taught in
America. James's letter in response to this request is among his most significant.

2.36 To William Cattell, Baltimore, 9 March 1883

(Two paragraphs of gossip have been omitted.)

You ask me how I think the philosophical courses of a university ought to be ar-
ranged. If a university can afford to devote only three thousand dollars a year to its
philosophical department it seems to me it would be far better to have three young
men as instructors than one older man as professor. If, as the university grows it
becomes able to devote more money to its philosophical department, one of these
instructors will probably have shown himself worthy to be made professor, and with
him will be associated younger men as instructors. A change has taken place in
philosophy during the past twenty years, somewhat similar to the change that took
place in the natural sciences during the preceeding twenty years. We now only hope
to arrive at a true philosophy by careful observations and inferences from ascer-
tained facts; we no longer believe that a Hegel can, neglecting facts and experience,
evolve certainty from his consciousness and intuitions. Philosophy has become a sci-
ence, and must be studied and advanced like the other sciences. A young man is
probably better able to teach philosophy and contribute to its advance than a man
trained in the older methods. Of course we all know that this view of philosophy has
been carried to an extreme in these days, but it is only a man trained in scientific
methods, who is able to take a worthy stand against materialism and sensationalism.
Our college students read the papers and magazines and are apt to drift with the
current toward agnosticism and materialism, as their professors have little knowledge
of the doctrines of certain popular scientists and less sympathy for the state of mind
these are apt to bring about in college boys.

It is impossible for one man to be a master of the whole field of philosophy, and
it is not right to expect him to teach it all. I have spoken of the change which has
come over philosophy as over the natural sciences. Some years ago the same profes-
sor taught physics, chemistry, botany, zoology, geology and what not, but if we asked

a man to do this now-a-days he would want to know whether we considered him a professor or a school-teacher. But philosophy is quite as broad a field as the natural sciences, and its different departments are fully as independent of each other. There ought to be one man to teach metaphysics—say about what is contained in Pres. Porter's "Human Intellect".[1] Then there ought to be another to devote himself to the history of philosophy—a most important department. In Germany there are four courses of lectures given on the history of philosophy to one on metaphysics. Empirical psychology is altogether unlike either metaphysics or the history of philosophy— it is a pure science. The man who teaches it and tries to advance it must be acquainted with physiology, anatomy, histology and physics. Logic and ethics are both distinct departments, but might if necessary be united with metaphysics.

I imagine that if a university would thus collect about it young men, all devoted to philosophy, but each to a separate department, not only would philosophy be well taught, but the instructors would be teaching themselves, and training themselves to become professors and men to whom we may confidently look for advancing philosophy which is the science of knowledge and life.

[1] Noah Porter, president of Yale from 1871 to 1880 was a representative of the older, non-empirical tradition of American philosophy. His *Human Intellect: With an Introduction Upon Psychology and the Soul* (New York: Scribner, 1865) went through many editions. (See also Herbert W. Schneider, *A History of American Philosophy*, 2nd edition [New York: Columbia University Press, 1963], pp. 211, 215–216.)

Cattell continued pursuing his interests in lectures, laboratory work, literature, music, and stimulants. Meanwhile, his father continued looking for a good position for him.

2.37 Journal, Baltimore, 10 March 1883

Editor Nation writes he will print my communication "with pleasure."[1] The papers all contain notices of father's report, mentioning his possible resignation. I read about "Faust": Taylor, Hartmann, Lewes &c.[2] Ran a mile in 4.57. Saw Dr. Carpenter's mesmeric experiments—truly wonderful.[3]

[1] Edwin Lawrence Godkin (*DAB* 7: 347–350) was an Irish-born lawyer and social commentator who had edited *The Nation* since its establishment in 1865.
[2] See William Taylor, *Historic Survey of German Poetry* (London: Troutlel and Wortz, 1830); Edouard von Hartmann, *Das Unbewusste vom Standpunkte der Physiologie und Descendenztheorie* (Berlin: Duncker, 1872); George Henry Lewes, *The Life and Works of Goethe: With Sketches of His Age and Contem-* *poraries* (London: D. Nott, 1855).
[3] Possibly William Carpenter, an English-born adherent of a flat-earth theory who settled in Baltimore, espoused spiritualism and other similar causes, and regularly challenged the university to disprove his views (Thomas William Herringshaw, *Encyclopedia of American Biography in the Nineteenth Century* [Chicago: American Publishers Association, 1898], p. 195; *Pre-1956 Imprints* 96: 339).

2.38 Journal, Baltimore, 12 March 1883

Am very drunk. Lectures—laboratory practice gymnasium. Drank fifteen glasses of beer or there abouts—then three whiskeys—Heard Planquette's Chimes of Normandy then German Kräuden.

2.39 Journal, Baltimore, 14 March 1883

I drank that alcohol partly for experiment, partly for the emotional affects, not intending to carry either as far when I began. Yesterday I suffered slightly from nausea. In the evening Metaphysical Club of which I am secretary & "manager." Papers by Prof Martin, Dr. Sedgwick & Mr Pierce.[1] Heard Dr. Brooks lecture on heredity today.[2] Took 7½ grs of caffein this evening—notes in second book.[3]

[1] Martin spoke "On the Development of Sight in the Lower Organisms," and Sedgwick "Concerning Perception and Reflex Action in the Frog." Peirce is not listed as having spoken. (*JHU Circulars*, 1882–1883, p. 94; Fisch and Cope, "Peirce at the Johns Hopkins," p. 373.)

[2] William Keith Brooks (*DAB* 3: 90–91) was an associate in biology at Johns Hopkins. Later, as professor of biology, he was an important figure in the American biological community (*Who Was Who in America*, 1897–

1942, p. 144; Dennis M. McCullough, "W. K. Brooks' Role in the History of American Biology," *Journal of the History of Biology* 2 [1969]: 411–438). This talk on heredity was one of six that Brooks gave, with other natural scientists, as part of a series of twenty-seven "Advanced Lectures in Biology" (*JHU Circulars*, 1882–1883, p. 127).

[3] JMC kept another journal for a short time, which he labeled "Notes Stimulants." See also 2.21, 2.24.

2.40 From William Cattell, Philadelphia, 16 March 1883

(This letter was dated "Friday." Four paragraphs of gossip have been omitted.)

I have had long conversations with two Trustees of the University on the subject: and if they do enlarge with a force of young men—two or three—I feel sure they will offer you a place.

2.41 Journal, Baltimore, 17 March 1883

Called on Dr Craig, Dr Wilson (out) & Mrs Thomas Thursday evening.[1] Gounod's Faust[2]—as they call it—last night—rather poor. Played base ball this afternoon the

[1] Thomas Craig (*DAB* 4: 496) was a graduate of Lafayette with a Johns Hopkins Ph.D. in mathematics. In 1883 Craig was an associate in mathematics at Johns Hopkins; he eventually became professor. "Dr Wilson" is prob-

ably Henry Parke Curtis Wilson (*DAB* 20: 326), a graduate of Princeton with an M.D. from the University of Maryland who specialized in gynecology in Baltimore.

[2] See 1.6 and 5.5

first time these three years. At the club this evening. I have discovered some work in the psychological laboratory that may be of interest.[3] My communication was printed in this week's Nation.[4]

[3]JMC's experiments involved measuring the time required for an individual to recognize and name letters of the Latin alphabet. He pasted standard-sized letters on the vertically mounted drum of a kymograph, and placed the drum behind a screen with a vertical slit. He varied the width of the slit, and then determined the maximum speed of rotation of the drum, for each slit width, at which his subjects were able to recognize and name the letters correctly. In this way, he thought he was able to measure the duration of mental processes. Among the subjects of this experiment were JMC, Hall, Hartwell, Bright, Dewey, Thomas, Joseph Jastrow (see 6.10),

and Henry Herbert Donaldson (see 6.83). (JMC, "Über die Zeit der Erkennung und Benennung von Schriftzeichen, Bildern und Farben," *Philosophische Studien* 2 [1885]: 635–650, translated by Robert S. Woodworth as "On the Time Required for Recognizing and Naming Letters and Words, Pictures and Colors" [*Man of Science*, vol. 1, pp. 13–25].)
[4]JMC, "Hermann Lotze," *The Nation* 36 (no. 924, 15 March 1883): 232. As soon as it was published, JMC sent copies of this issue to several of his friends (see J. M. Leiper to JMC, 20 March 1883; M. S. Bailey to JMC, 23 March 1883).

2.42 From Elizabeth Cattell, Easton, 31 March 1883

(Several paragraphs of gossip have been omitted.)

Your Papa had a letter from Prof Morris, but he was very guarded in expressing himself, and your Papa thinks he can make no use of it.[1]

[1]This letter has not been found. However, in a "Memorandum respecting the work of the Philosophical Seminary, September–December, 1882," dated 12 March 1883, Morris mentioned the work of only one student—John Dewey—by name, and praised his study of Empedocles. Later, in a letter to Gilman (23 May 1883), Morris noted that

because JMC's primary work at Johns Hopkins was in experimental psychology, with no "pretense of exhibiting specifically philosophical knowledge or ability," he could not comment on JMC's achievements and had to defer to somebody "thoroughly at home in the psycho-physic laboratory."

James Cattell visited his family in Easton late in March, and spoke with his parents about his employment prospects as a philosopher. He returned to Baltimore with his brother Harry, who was in his senior year at Lafayette.

2.43 Journal, Baltimore, 1 April 1883

Harry and I were at Washington yesterday looked through the Capitol, Corcoran Art Gallery, White House, Smithsonian Institute &c. We shook hands with the President. He looks as if he ate and drank too much and had to suffer for it[1]—but

[1]Chester Alan Arthur (1830–1886), who was then suffering from Bright's disease, a serious kidney ailment, and lived through his term of office (1881–1885) knowing that he

was a dying man (Thomas C. Reeves, *Gentleman Boss: The Life of Chester Alan Arthur* [New York: Knopf, 1975], pp. 317–318, 483).

perhaps it's work and worry. Attended symphony concert last night. Heard Dr Jones preach this morning.[2] Harry is not well tonight.

[2]John Sparhawk Jones, a graduate of the University of Pennsylvania and of Princeton Theological Seminary, pastor of the Brown Memorial Presbyterian Church in Baltimore from 1871 through 1889 (*Who Was Who In* *America,* 1897–1942, p. 649; J. F. P. Boulden, *The Presbyterians of Baltimore: Their Churches and Historic Grave-Yards* [Baltimore: W. K. Boyle, 1870], pp. 120–123).

Cattell resumed his interests in Baltimore, and seems to have entered a period of enthusiasm. In particular, he became more active socially.

2.44 Journal, Baltimore, 3 April 1883

Heard Modjeska last night in "As You Like It" a wonderful piece of acting—not Shakespere's (by that I mean my) Rosalind perhaps, but truly magnificent.[1] She is the finest actress I have seen after Bernhardt. This evening I took a large amount of hasheesh—I did not weigh it but it must have been over 75 grs.[2] I have scarcely been affected, there was slight exhileration (=2 glasses beer) and now I am slightly confused—not enough however to attract my own attention, if I did not know that I had taken the drug. I imagine I had better let it alone hereafter.

[1]Helena Modjeska was a Polish actress who learned English specifically to play Shakespeare (*Oxford Companion to the Theatre,* p. 646).

[2]JMC probably misestimated or used a weak hashish mixture, as a dose of 75 grains of the drug is usually fatal (see Popplestone and McPherson, "An Historical Note on Cannabis," p. 7; Cattell "Autobiography," p. 632).

2.45 Journal, Baltimore, 8 April 1883

I have accomplished nothing worth mentioning—have been carrying on my experiments with the revolving drum[1]—looked up Mach on Bewegungsempfindungen.[2] Played ball yesterday & was at concert in the evening. This evening I have taken caffein[3]—the old idea—get myself excited, decided on some course of conduct and quietly carry it out. I am nervously excited now certainly—my hand trembles as I can scarcely write. My mind seems clear, and I feel well and strong—but I fear no caffein can solve the riddle of the world & what I am to do in it. The only thing I can say is, knowledge may be impossible, but I have certain instincts as to what is good and true. Granted, but these instincts at different times would lead me in exactly opposite directions. Still have I not a guide? When I am at my best can I not

[1]See 2.41.

[2]Ernst Mach (*DSB* 8: 595–607) was an Austrian scientist best known for his development of energetics and for his work on the physiology of sensation. The book JMC refers to here is *Grundlinien der Lehre von den Bewe-* *gungsempfindugen* (Leipzig: 1875), a study of the sensation of motion.

[3]Caffeine, used today as a stimulant (*Merck Index,* p. 1625), was listed in the fifth revision of the *U. S. Pharmacopoeia* (p. 21) as part of the standard *materia medica.*

will well what must be done, and then carry it out through all weakness and weariness? Now I seem to see clearly enough what ought to be done. I ought not to kill myself, even though I should feel that it would so be best. That would be final and could not be changed. Neither should I give up the struggle of living for quietism, however much I may be inclined to it. This would be final, I could not return to the strife. No, my course is surely simple enough. I must use all my will and self to force me into the thickest of the fighting. After a while I can fall out into quietism or death. Now I am full of nervous strength and energy, and what I decide to do now must—will be done. I look forward to final quietism, perhaps suicide—but first I will drink deep of the missed joy & pain of living. I will be what most I can. I will force my infinitesimal self as far from zero as the sum of my strength and will makes possible. So let it be. In a few years I will go back to Easton to be mystic and pantheist, to muse and brood on the mystery of things, to sink into God, which is the sum and mystery of things. But now for a little while I will love and hate, will suffer and toil and win.

2.46 Journal, Baltimore, 9 April 1883

I am going to make a remarkable statement. The 24 hours, from 7 last evening to 7 this evening were the happiest I remember. Last night I was wonderfully excited. I felt to the full the above as I wrote it. I read Shelley's Adonais with extreme delight. See notes on "Stimulants". All day I have been feeling wonderfully well. It was a joy to walk to breakfast, to eat it, to write a letter home, to listen to lectures, to work in the laboratory. I rejoiced in running a mile, in taking a cold bath. I have worked to unusually good effect all day. It is now after one, I am working up Sensations of motion, equilibrium & rotation, which I must speak on tomorrow.

2.47 Journal, Baltimore, 10 April 1883

That caffein experience was truly wonderful. Today I still feel above my normal line. I cannot understand it at all. I spoke for half an hour before the class on Sensations of motion, rotation & equilibrium with special reference to Mach.[1] I exhibited a pigeon with the semicirculars canals cut.[2] Tonight Metaphysical Club.[3] A young lady Miss Savage at the place where I am boarding drank 500+ grs of chloral—purposely killing herself, I suppose.[4]

[1] Probably Martin's class in animal physiology (see 2.24).
[2] "Experiment shows that cutting a semicircular canal is followed by violent movements of the head in the plane of the canal divided· the animal staggers, also, if made to walk; and, if a pigeon is thrown into the air, cannot fly" (Martin, The Human Body, p. 588).
[3] The club that night heard John Dewey on "Hegel and the Theory of Categories" and

Joseph Jastrow on mechanical logic (JHU Circulars, 1882–1883, p. 94).
[4] Chloral hydrate is a white crystalline powder used as a hypnotic and sedative (U. S. Pharmacopoeia, 5th revision, p. 25; Merck Index, p. 260. JMC added the name of this otherwise unidentified young woman after he had written the rest of this entry. Instead of "purposely," he had first written "intending to."

2.48 Journal, Baltimore, 11 April 1883

I have been feeling unusually well all day. If a reaction were to follow equal to the apparent action of that caffein, I doubt if I could live through it. I have not been in such good condition and spirits for three successive days these five years as far as I can judge. I made several important advances in my work in psychology this morning. This afternoon I played ball.

2.49 From William Cattell, Easton, 12 April 1883

(Several paragraphs of gossip have been omitted.)

As to the legacy:[1]—that is more than a month old. I have talked with Dr. Pepper about it. Of course it is to endow a chair and the income will go to one man. It is a relief to the University;—but the Dr. says they are still too poor to enlarge in this Department of Philosophy. We have had several talks about the matter; he says the pressure for enlargement in other directions is very great & he sees no immediate prospect of doing any thing more for Psychology than the old plan; and this not only from the more urgent pressure of other interests, but from a real indifference to any enlargement of Dr. Krauth's Chair. The "business" men of the Board "don't see the use of it"—I have written at some length to Dr. Boardman & to Mr. Welsh on the subject:[2]—showed the letters to Dr Pepper and he approved of all the suggestions but admitted that even he had other "enlargements" of the course which he w[d] push, in preference, if they had the money. So I dont know whether there will be an opening in Philad[a] soon. The thing to do is to make whatever interest I can with my personal friends in the Board:—and wait for results. Meanwhile you sh[d] make a good impression at Johns Hopkins. The Trustees of the University—sh[d] there be an opening—w[d] be sure to apply at Balt. about you;—and this independent of written Testimonials which are, nevertheless, valuable. I am sorry Dr. Morris was so "cautious". I suppose it is his habit. Dr. Gilman will be the main authority and I trust you will make a good impression upon him. It is marvellous (sometimes humiliating) how personal relations will influence our judgment as to the learning & ability of men. I shall write to Prof. Martin this week. Had I better write to Prof. Hall? If so send me his full name.

[1] Henry Seybert, the son of early-nineteenth-century scientist Adam Seybert, had left the University of Pennsylvania $60,000 "to endow a chair of philosophy on the condition that an additional sum be used to support the activities of a commission of the University appointed to investigate modern spiritualism" (George Daniels, *American Science in the Age of Jackson* [New York: Columbia University Press, 1968], p. 222; Frank McAdams Albrecht, Jr., "The New Psychology in America, 1880–1896" [Ph.D. diss., Department of History, Johns Hopkins University, 1960], p. 173; Edward Potts Cheney, *History of the University of Pennsylvania, 1740–1940* [Philadelphia: University of Pennsylvania Press, 1940], p. 319).

[2] George Dana Boardman was pastor of the First Baptist Church of Philadelphia from 1864 through 1894 and sometime chaplain at the University of Pennsylvania (*Who Was Who in America*, 1897–1942, p. 111). The Mr. Welsh referred to is John Welsh (see 2.17).

2.50 Journal, Baltimore, 13 April 1883

I am getting worried—I feel so well. Last night and again tonight, I summoned up my resolution and good clothes and made calls. Last night Mrs Morris (out), Mrs & Miss Gilman, Miss Garrett. Tonight Mrs Sedgwick (out), Dr Wilson, Mrs Gildersleeve (sick) and Mrs Martin.[1] As to these people Mrs Morris is rather odd, the Professor I like very much. I'm rather doubtful about the Gilmans. Pres Gilman has not taken as much interest in me, as he might have done. Sometimes he shows marvellous tact in conversation, but not always. He is without sympathy or magnetic power. I sometimes wonder where his strength lies—however he certainly has great executive ability. I think he is troubled with nervousness, like father. Mrs Gilman's manners & conversation are faultless, always easy and pleasant. Sometimes I imagine I might like her, sometimes I think no one could know or like her very well. I confess I was more pleased with Miss Gilman that first night than I have ever been since. Still I have undoubtedly made her my Baltimore "favorite". I suppose I will always understand what that means. Mr & Mrs Garrett are in the South. Mr. G is a man of strong will & much self assertion. As he has accomplished much, one admires him & is ready to listen to him. Mrs. G. is very quiet. I like Miss G. She is perhaps 26, is fond of study, art &c[2]—what might not appear very attractive in others, is really pleasing in her. She & Miss Rogers go with me to the theatre next week.[3] I guess I've written enough for tonight.

[1] "Mrs Morris" is the wife of Professor C. D. Morris (see introduction to section 2); "Mrs Sedgwick" is Mary Katrine Rice Sedgwick; "Mrs Martin" is Hetty Cary Pegram Martin, wife of Professor Martin.

[2] Mary Elizabeth Garrett and three of her friends, including Julia Rogers, spent much time in the early 1880s reading, studying art, and in general trying to provide themselves with the equivalent of a college education, which was unavailable formally in Baltimore for women (*NAW* 2: 21–22).

[3] On 21 April 1883 JMC accompanied Mary Elizabeth Garrett and Julia Rogers to a performance of *As You Like It*, with Lily Langtry (Journal, 21 April 1883). Emilie Charlotte Langtry (*Oxford Companion to the Theatre*, p. 547) was an English actress who, though never a great performer, was extremely popular. JMC noted that "she looked well and her acting was not at all bad."

2.51 Journal, Baltimore, 23 April 1883

I heard Marie Geistinger tonight in Trompette (Lecocq).[1] Bailey has been elected Master Orator.[2] I keep working on my psychological experiments. I need not describe them here, as I give an account of them at the next meeting of the Scientific Association and dare say they will be published.[3]

[1] Marie Charlotte Caecelia Geistinger was an Austrian soprano best known for her roles in Strauss operettas (*International Cyclopedia of Music and Musicians*, p. 806). Alexandre Charles Lecocq (*Grove's* 5: 103–104) was the composer of more than fifty comic operas, including many (such as, apparently, *Trompette*) that were never published.

[2] Morton Shelley Bailey had been JMC's best friend at Lafayette. He graduated with JMC in 1880, and then went to Colorado to read law. He had political ambitions (he was later elected to the Colorado State Senate) and had asked JMC to help him win this position as Master Orator in the hope that it would help his political career. He eventually served as a justice of the Colorado Supreme Court. (Wilbur F. Stone, ed., *History of Colorado* [Chicago: S. J. Clarke, 1918], vol. 2, pp. 776–778; *Denver Post*, 16 May 1922, pp. 1, 13.) The Master Orator for the class of 1880 was to speak at the 1883 commencement on the occasion of his class's receipt of the master's degree.

[3] The Scientific Association, like the Metaphysical Club, met monthly, but was less specialized and was open to students and faculty from all areas. There is, however, no record that JMC presented anything at the next meeting of the association, which was held on 2 May 1883. (Hawkins, *Pioneer*, pp. 113–114; *JHU Circulars*, 1882–1883, p. 156.)

2.52 Journal, Baltimore, 27 April 1883

With the exception of a couple of hours yesterday afternoon I have been in good spirits for the past two days. Last evening I heard the rehearsal and this evening the final rendering of Gounod's Redemption: it was extremely well given with a fine chorus and Thomas's orchestra—by far the best music I have heard since last Winter. This evening I was with Miss Garrett and Miss Rogers—I met Miss Garrett yesterday at an exhibition of decorative art and she invited me to go with her.

2.53 From Elizabeth Cattell, Easton, 25 April 1883

(Several paragraphs of gossip have been omitted.)

Your Papa, and I both think it would be splendid if you could be elected Assistant in Psychology, and a great honor for one so young. But dear Jim do not get spoiled, and allow me here to deliver a little lecture you know it comes from one who loves you dearly. I fear you do sometimes try D[r] Hall, for you have the power of making a person feel very angry, it is not so much in what you say, as the way in which you say it, and I would try, and be careful. You speak of D[r] Halls being changeable, there is nothing so trying to a McKeen (and you have the McKeen nature) but you must learn to be patient with them.

2.54 Journal, Baltimore, 4 May 1883

For the past three days I have been in splendid condition, working pretty hard & thoroughly enjoying things. Played cricket & tennis Wed., lacrosse yesterday and tennis today. Was at Scientific Association Wed evening,[1] called on Mr & Mrs Trippe, Dr & Mrs Wilson & was at university reception for ten minutes last night.[2] Tonight instead of supper I ate half a pound of sweet chocolate which has put me in splendid condition for work. I have written letter to Pres Gilman giving account of my years work and asking for renewal of the fellowship and also thesis to be presented with this[3]—the thesis gives an account of my work in experimental psychology.

[1] At the Scientific Association on 2 May 1883 JMC heard papers on physical and chemical topics, as well as a talk by Hartwell, "On the Relation of Bilateral Symmetry to Function," based on the work he had been doing with Hall (see 2.35). (*JHU Circulars*, 1882–1883, p. 156.)

[2] On Andrew Cross Trippe and Caroline Augusta McConky Trippe see 2.25; on Henry P. C. Wilson and Alice Brewer Giffin Wilson see 2.41.

[3] The Johns Hopkins regulations relative to fellowships called for "the preparation of a thesis, the completion of a research, the delivery of a lecture . . . to give evidence of progress," and allowed for the reappointment of fellows (*JHU Circulars*, 1879–1882, p. 52).

2.55 Journal, Baltimore, 8 May 1883

I continue to feel quite well and comfortable—my state of mind being careless and sceptical. Saturday it rained when I wanted to play athletic games. In the evening I played whist and drank beer. Yesterday I was at Barnum's & saw the immortal Jumbo.[1] Tonight I was to have read a paper on Lotze at the Metaphysical Club but the other speakers Profs Gildersleeve & Remsen took up all the time.[2] I work fairly well—it is now 2.30 I have been writing since eleven.

[1] Phineas Taylor Barnum brought Jumbo, a huge African elephant, to the United States in 1882, and exhibited him on tour until the elephant died in 1885.

[2] Gildersleeve spoke on "Rhythm in Classical Languages," and Ira Remsen, professor of chemistry, delivered a paper that would have interested Cattell: "Wundt's Logic of Chemistry." This talk was an abstract of Wundt's "Die Logik der Chemie: Eine methodologische Betrachung" (*Philosophische Studien* 1 [1883]: 473–494), a discussion of the ideas of John Herschel, John Stuart Mill, and others regarding chemical processes. (*JHU Circulars*, 1882–1883, p. 156.)

2.56 Journal, Baltimore, 16 May 1883

Saturday I took a tremendous dose of hasheesh—more perhaps than has ever been taken 200 grs. of Squib's extract.[1] I took 75 grs at 7.30, 85 grs at 10 and 40 grs at 12.30. I was only exhilerated until I lay down on Stokes' bed at 2, but when I awoke at 5 was quite intoxicated. I walked home having all the hasheesh delusion, but after eating some oranges, went to sleep, and slept apparently normally until two. Then I got dinner & went out to the park. I was rather dazed during the evening & not in especially good condition for work Monday. Tuesday I saw Herrmann the magician & tonight saw Flotow's Martha given by the Germania Mennaschor. This afternoon I played ball, the University nine beating the Stars horribly 25 to 2. I have been feeling first-rate today.

[1] Squibb's extract of *Cannabis indica*, a commercial preparation of the drug. See Cattell "Autobiography," p. 632. See also 2.44.

This rather pleasant period was about to be brought to an end by an unexpected academic setback.

2.57 From William Cattell, Easton, 19 May 1883

Dr Hall has written me a very pleasant letter about you—says he has greatly enjoyed the association & work with you.[1] He says "He has worked upon an original theme with unusual independence & originality which, when his results are completed & written out will be a contribution to the sum of the worlds knowledge upon a very complicated psychological theme"—

But he is very anxious that you shall not seek any Professional work now—or any teaching that will withdraw you from these studies—with three or four more years of student life he says of you "I sh^d feel extremely confident of eminent academic life and work" &c—

And of course this is just what I want you to do—either here or in Leipzic. You might secure a position as Instructor in some College where the teaching w^d not be laborious or burdensome—but still it w^d be better to have yr whole time & strength for yr studies & investigations.

I will show you Dr Hall's letter when you come home.

[1] This letter has not been found. However, two days earlier, Hall wrote to Gilman: "I think the order of merit in the candidates for fellowship in philosophy is as follows: 1. Dewey, 2. Veblin, 3. Jastrow, 4. Cattell. The others are out of the question." (Gilman papers.) JMC later wrote that "Dr. Hall has not acted honorably towards me," and detailed the way in which he was misled (see 5.18). Dorothy Ross summarized Hall's actions as follows: "The best that can be said for Hall in this affair is that still insecure in his position and driven to appease all around him, he praised Cattell when with Cattell and agreed to Dewey's superiority when with Morris and Gilman. . . . The worst . . . is that . . . he felt safer without strong intellectual challengers around him and set about to eliminate them." (Ross, *Hall*, p. 140.) The "Veblin" mentioned was Thorstein Veblen (1856–1929), later an important economist and social philosopher (*DAB* 19: 241–244; *IESS* 16: 303–308).

2.58 Journal, Baltimore, 28 May 1883

Things have been going on about as usual. Dr. Hall stopped lecturing Friday the 18th & Prof. Martin Tuesday the 22nd. I've been taking things rather easily— worked a bit in the biological laboratory & read a bit. Played tennis several times— was at Annapolis Sat the 19th with the nine. The caffein was not so satisfactory as before—for an hour or so I felt full of will & strength, but soon cared more for idleness & dreams. Next morning I was rather used up. Pres. Gilman indicated today that I would not be reappointed as fellow—not pleasant news, even though I am not in love with the university.[1]

[1]Like Hall, Morris (the other judge) ranked Dewey first and Veblen second, with three other candidates not mentioned by Hall in the next three spots. He then commented that the theses by JMC and Jastrow were not strictly philosophical (Jastrow's was on logic). For his comment on JMC's see the note to 2.42. See also Morris to Gilman, 23 May 1883 (Gilman papers).

2.59 From William Cattell, Easton, 29 May 1883

(A postscript relating to personal business has been omitted.)

Yr letter just recd—the Ante scriptum has of course given us no little excitement—I can hardly believe it possible that you will not get the Fellowship & yet, from what you say, it looks as if Dr. Gilman & the Facy had decided against you.[1]

In my present nervous state many things assume an importance altogther beyond what is really due to them;—this may be one of these. But certainly I shall feel very indignant;—and I must be careful not to do or say anything to make matters worse. And so must you. In fact this is why I write. Be very careful to make no remark about it. Preserve a dignified silence. It will be easy to make bad worse.—If you are not reappointed it will be very hard for me not to think there has been something unfair in the decision;—very different from what I shd have felt had you not recd the appointment last year: but to express this will not mend matters;—it will as I have said, make them worse. So be very careful.

Write to us at once when it is decided. Even the suspense makes me very nervous.—

[1]This letter has not been found, but its contents are clear from this response.

2.60 From William Cattell, Easton, 29 May 1883

(A postscript relating to personal business has been omitted.)

Yr despatch just arrived[1]—unfortunately I left Dr. Halls letter with Hetty to show Dr. Pepper. She said she w^d mail it to me;—this was last Monday (a week fr. yesterday) We have not rec^d it. I have telegraphed Aunt Hetty to mail it to you;— and am haunted by the fear it may be lost. I hope for the best.

I copied a sentence from it for another purpose:—which I enclose;—tho' I suppose it will hardly serve yr purpose. It was the best thing in it—the only "quotable" sentence.

I wrote to you this morning after rec^g yr letter. We were both excited by what you wrote—the ante-scriptum. *Yr dispatch has still further excited me.*

I infer that the question has been decided & that you wish to have this letter to show that Dr. H. was friendly to you; and not in hopes of its re-opening the questions. How I wish I c^d see you & I at first determined to go at once to Balt;—but sent the telegram instead.[2]

[1] Not found.
[2] The following day (30 May), William Cattell wrote to Gilman, quoting from the latter's correspondence with him about his son's original appointment (see 1.23) as well as from Hall's letter (see 2.57). In this letter William Cattell wrote "of course I do not write to reopen the question. . .but I write to you in view of our personal friendship of 30 years to beg you to let me know frankly if there was anything in my son's conduct . . . that lost him the reappointment." (William Cattell papers.)

2.61 From Elizabeth Cattell, Easton, 30 May 1883

(One paragraph of gossip has been omitted.)

Your letter of the 28th just received. Both your Papa, and I feel that there has been great injustice done to you. Your Papa is very much excited, and I fear the effect upon him, but I try to keep calm, and feel in the end it may all turn out for the best. I think if I was in your place I would go to Leipsic and take my PhD. I fear with you, that Dr Gilman has imagined you much more intimate at Mr Garrett's than you really were, but he has no right to let a thing of that kind influence him.[1] I cannot understand why he treated you as he seems to from the first. Dr Gilman wrote your Papa a real nice letter upon hearing about his resignation. Your Papa says if Mr Garrett was only at home he would write to him.

[1] In January 1883 Garrett split with Gilman and the rest of the trustees and publicly questioned Johns Hopkins University's investment policies. From that time until his death, Garrett, though officially still a member of the board, did not actively take part in university affairs. JMC had been friendly with Garrett, but it is doubtful that this had any bearing on the loss of his fellowship. (Hawkins, *Pioneer*, pp. 4–5, 105, 239; Gilman, *The Launching of a University and Other Papers* [New York: Dodd, Mead, 1906], pp. 35–36.)

We cannot tell you how much we sympathize with you, but try, and keep calm, and cool, this is your first disappointment, but I feel so thankful we are able to let you go on with your studies, without the aid you have received from the University. If in any way you had been beholding to D^r G for the $500, I would want to send him the check for the amount at once. We will give you a hearty welcome at home, baggage, and all. Papa joins me in love.

James McKeen Cattell did not "keep calm." He blamed his failure on Gilman, and told various members of the Johns Hopkins faculty that he was not reappointed because Gilman disliked him. He also sought to be appointed "Fellow by Courtesy"—a title usually reserved for senior scholars studying at the university and for Johns Hopkins graduates working elsewhere (*JHU Circulars*, 1879–1881, p. 87; ibid., 1882–1883, pp. 19–20; 1883–1884, p. 18)—to "save me some mortification, and be less unpleasant for father" (JMC to Hall, 30 May 1883 [Gilman papers]). In doing so, he claimed to have the support of Professors Martin, Gildersleeve, G. S. Morris, and Sylvester. Morris recommended this appointment "without hesitation and with pleasure" (Morris to Gilman, 31 May 1883 [Gilman papers]), but Hall begged off and sent Cattell's request to Gilman.

All this was too much for Gilman, who complained to William Cattell that "your son attributes his failure to receive a reappointment as a Fellow to a motive on my part which I utterly repudiate" and demanded an apology (Gilman to William Cattell, 4 June 1883 [Gilman papers]). It is not clear whether one was forthcoming.

James Cattell spent the summer in Easton, reading and studying, helping his father give up the presidency of Lafayette College after twenty years, and making plans to return to Leipzig in the autumn to study again with Wundt.

2.62 Journal, Easton, 12 June 1883

I have not written any thing since I have been home as I am never in my room in the evening. I had quite a time before leaving Baltimore over the fellowship. Dewey has been appointed for next year—he is a good man, taught in the university this year—but I was scarcely treated fairly. Letters I have received & written tell the story fully, so I need not repeat it here.[1] I had plenty of books to pack up & bring home. Sat I took Hasheesh—very satisfactory & strange. I came home Monday (the 4th). I have been living very quietly & regularly. I work in the morning & play tennis in the afternoon and evening. Go to bed at ten and get up at seven. I see almost no one except our own family & the fellows who come to play tennis.

[1] These letters have not been found.

2.63 Journal, Easton, 26 August 1883

I felt badly—physically—for about ten days, but the last three days have been all right again. My pulse was below 60 constantly & I felt very weak & generally used up. I could not work and scarcely take exercise. My life is uneventful and pleasant enough. I am a good deal with mother and Harry. Mother has read Morris's Sigurd the Volsung to me.[1] I breakfast at eight and work until dinner at one. Then loaf for an hour, and play tennis the rest of the day, going swimming or riding Harry's bicycle once in a while. In the evening I stay down stairs. I have nearly finished an analysis of Lotze's Grundz. der Psychologie, which I intend to translate this winter.[2] As to my state of mind, I am simply drifting. I am never thoroughly miserable, as I used to be, and am seldom troubled even with malaise. Under these circumstances I can read, talk & exercise with some pleasure and am comfortable—perhaps I should say happy. But sometimes I think myself only happy when stirred and excited, so that happiness and misery come very close together. I am not sure but that I would think it better if I struggled & suffered more.

[1] *Sigurd the Volsung, and the Fall of the Niblungs*, an epic poem by William Morris based upon Norse mythology (*Oxford Companion to English Literature*, pp. 561, 755).
[2] Lotze's *Grundzüge der Psychologie* (Leipzig: S. Hirzel, 1881) was based upon notes taken of his last lectures at Göttingen by his son.

Two English translations of this slim volume were published as *Outlines of Psychology* before JMC finished his: by Clarence Luther Herrick (Minneapolis: S. M. Williams, 1885) and by George Trumbull Ladd (Boston: Ginn, 1886). See also Cattell "Autobiography," p. 631.

2.64 Journal, Easton, 14 October 1883

I spent Tuesday night in N.Y. going to Boston Wednesday night & returning home Sat. The dentist was the attraction in N.Y. and Dr. Hall in Boston. Saw Blake, Armstrong & Royce.[1] The travelling rather bored me, & the prospect of two years

[1] Lucien Ira Blake (*Who Was Who in America*, 1897–1942, p. 104) was an electrical engineer who had studied with JMC at Göttingen (see pp. 30, 39). Andrew Campbell Armstrong (ibid., p. 30) was a graduate of Princeton College (A.B., 1881) who in 1883 was studying at the Princeton Theological Seminary. He had earlier written to JMC requesting information about Wundt (Journal, 15 April 1883), and later studied at Berlin. From 1888 until 1930 he taught philosophy and psychology at Wesleyan University, where he directed many men toward ca-

reers in philosophy. (See Geraldine Joncich, *The Sane Positivist: A Biography of Edward L. Thorndike* [Middletown, Conn.: Wesleyan University Press, 1968], pp. 53, 70–75.) Josiah Royce (*DAB* 16: 205–211) was a graduate of the University of California (A.B., 1875) who had earned a Ph.D. in philosophy from Johns Hopkins (1878) for work with G. S. Morris. After teaching for a few years at California he went to Harvard in 1882, where he eventually became the leading American idealist philosopher.

given over to walking up & down on the face of the earth is not over-pleasant. I have not done much during the past month. I worked in the morning & played tennis in the afternoon. I think I have been as happy as I can hope to be, but the mere weariness of living sometimes seems more than I can bear. I do not know what the end will be; but after all it does not much matter. My life is only a little thing & for a little while.

3

Experimental Detail:
Leipzig; November 1883–May 1884

In the fall of 1883 James McKeen Cattell again left America, accompanied by his parents and his brother. William Cattell, who had retired the previous summer from the Lafayette presidency, was feeling the strain of the preceding twenty years as well as the trauma of the actual retirement. He and his wife planned to rest and recuperate at Davos, the Swiss resort, where they would be joined by Elizabeth Cattell's half-sister Helen McKeen Ferriday ("Nemie") and her two daughters Helen and Elizabeth ("Lizzie"). Meanwhile, the sons were to continue to Leipzig, where Jim, now fully committed to experimental psychology, would resume his studies with Wundt. Harry, who had graduated the previous June from Lafayette, planned to study chemistry with Professor Adolf Wilhelm Kolbe at Leipzig, where Kolbe had set up what was reputed to be one of the largest and best laboratories of the period.[1]

Though James Cattell had been at Leipzig, he had never gotten to know it as well as he had Göttingen. And Leipzig was very different from Göttingen. It was a large and important city, with almost 200,000 residents, and was expanding rapidly during this period.[2] Cattell was looking forward to his studies at the university; apart from the university, the most attractive aspect of the city to him was the excellent music and theater available. Leipzig was the home of the Mendelsohn Conservatory, which attracted students from all over the continent, Britain, and the United States.[3] Cattell

[1]*DSB* 7: 450–451; Max LeBlanc, "Das Physikalisch-Chemische Institut," in *Festschrift zur Feier des 500 Jährigen bestehens der Universität Leipzig* (Leipzig: S. Hirzel, 1909), vol. 4, part 2, pp. 85–106; Ernst Beckmann, "Das Laboratorium für angewandte Chemie," ibid., pp. 107–122.

[2]Baedeker, *Northern Germany* (6th edition 1877), p. 244: population 177,000. Ibid. (10th edition, 1890), p. 343: population 225,000.
[3]*Grove's* 5: 119–121; *Königliche Conservatorium*.

grew friendly with many of these young men and women. Leipzig was also well known for the concerts held at the *Gewandhaus*, the market hall of the linen merchants of Saxony.[4] However, in opera, the form of music Cattell enjoyed most, Leipzig was less than excellent. The Leipzig opera company of the 1880s specialized in popular, undemanding works, whereas Cattell was most taken with the operas of Richard Wagner (which were then the subject of much debate and controversy). Cattell was able to hear several Wagnerian operas in Leipzig, but heard more on trips to Dresden and Berlin. Leipzig also had the New Theatre, where the works of the great German dramatists, such as Goethe and Schiller, were performed.

Leipzig was to be the setting for James Cattell's first mature successes, and for much of his life he would look back fondly on his years there.

[4]Baedeker, *Northern Germany* (6th edition, 1877), p. 243; *Grove's* 5: 120 Burgess, *Remi-*   *niscences of an American Scholar*, pp. 107–109.

3.1 Journal, Leipzig, 23 November 1883

πάντα ῥεῖ οὐδὲν μένει,[1] *To think of it; here I am in Leipzig! We sailed from N.Y. on the Servia Wed. Oct 31st arriving in Liverpool early Friday Morning (the 9th).[2] The ship was tossed about somewhat but it was a pleasant passage for this season of the year. I was only sick two days. We knew some people & saw most of [Here Cattell listed eighteen people, much as he had noted his shipboard companions in 1.1]. We went direct to Bailey's London,[3] the Keith's going with us and the Ormistons joining us later. We spent ten days shopping & sightseeing and left Tuesday, father & mother going to Paris and thence to Davos, and we coming direct here via Rotterdam. We spent Wed. night on the boat & Thurs. night on the cars. We were fortunate in finding what seems to be a very nice boarding place here. It is the family in which McLean & McCoy lived two years ago[4]—there are father, mother, two daughters (circa 23 & 12) and a son in the university. One American a*

[1]"All things change, nothing stays still," a fragment from Heraclitus, recorded by Plato in *Cratylus* (see G. S. Kirk and J. E. Raven, *The Presocratic Philosophers: A Critical History With a Selection of Texts* [Cambridge University Press, 1969], p. 197). JMC had studied Heraclitus the previous year at Johns Hopkins (see 2.4n.1).

[2]The *Servia*, Cunard's first all-steel ship, displaced over 7,000 tons and was over 500 feet long. Launched in 1881, it was one of the first liners to carry mainly passengers and little or no cargo. (Tute, *Atlantic Conquest*, pp. 32–33.)

[3]A good West End hotel (Baedeker, *London*, pp. 7–8).

[4]Jim and Harry Cattell lived at Arndstrasse 53, away from the center of the city, with the Gerhard family mentioned in 3.7 (*Personal-Verzeichniss der Universität Leipzig*, no. 104, winter semester 1883–84, p. 47). George Edward MacLean (*DAB* Supplement 2: 419–420), a graduate of Williams College and Yale and a Congregational minister, had studied philology at Leipzig from 1881 to 1883. He later taught at the Universities of Minnesota and Nebraska and was president of the University of Iowa from 1899 to 1911. JMC had mentioned the otherwise unidentified McCoy during his previous stay in Leipzig (see 1.19).

Prof Bissell lives here, but is about to leave for home.[5] *We have beautiful rooms and the table seems good. I had thought to take rooms, & live to myself, but likely enough its better for both Harry & me that we are together. I know as yet nothing about the university, but we shall probably matriculate & I will hear Wundt & Heinze.*[6] *I must work harder than I have since freshman year in college. My program is to study mostly empirical psychology, reading a little on the history of philosophy. I want to work on Lotze, translating his Psychologie*[7] *& writing some review, and to work and write on stimulants & my Baltimore experiments*[8]—*then of course other things may turn up. As to my mental state I think it is better & stronger than for many years.*[9] *I am usually fairly comfortable, sometimes miserable, once in a while happy—this is as it must be and should be. I grieve that my will and feelings are not stronger, but who am I that I should dwell in the seventh heaven once a week, is it not enough? I need no better guide than to try to*[10] *live as I would like to see my brother live. Whatever I may be in theory, today at least I am not practically a sceptic.*

[5] Allen Page Bissell had been a professor at Blackburn College, Carlinville, Illinois, before coming to Leipzig in 1882 to earn his Ph.D. His dissertation, *The Law of Asylum in Israel, Historically and Critically Examined* (Leipzig: Theodor Shuster, 1884), includes his personal data.

[6] On Wundt see introduction; on Heinze see 1.15.
[7] See 2.63.
[8] See 2.41.
[9] JMC had originally written "for some time."
[10] "try to" was inserted.

3.2 To Parents, Leipzig, 2 December 1883

(Two paragraphs of gossip have been omitted.)

Here I am writing to you from Leipzig as of old—fortunately though, I do not have to send the letter to America. There is a good deal I might tell you, but I hope you will be here before so very long to see and hear all for yourself. Everything has gone with us as well as we could have wished, and much better than I expected. We are very well pleased indeed with our boarding-place. I have never lived in such nice rooms away from home, and the table is excellent. What is more important than these—we like the family and are made to feel at home. Though you would not be perfectly satisfied perhaps, I am sure you would like it here better than in Hotel Hauffe. There are several other very good rooms Harry and I can take, and you can have those we now occupy. The manager of the hotel says he will not take you for less than $5 per day and here it will cost only $1.60—both rooms and table being better. However I do not think you had better come as long as you enjoy yourselves at Davos, as you will have a good chance to get tired of Leipzig—we will very likely stay here eighteen months. It is on the whole better for Harry (if he studies chemistry) to stay in the same laboratory and I will want to be here next winter, so we had better settle down here for the rest of our stay in Europe. I do not think you would suffer so much from the cold at Davos or even here as at San Remo, notwithstanding the difference in the height of the thermometer.

We were matriculated without any trouble, and are busy with our university duties. Harry hears a lecture every morning at eight, and works most of the day in the laboratory. I have twelve (not very important) lectures a week—all by Wundt and Heinze, and work as much as I please in Wundt's laboratory. I was fortunate in getting into it. Oddly enough Berger[1] was making some experiments for Wundt, and we will work together.

We have found a lawn-tennis court within a stone's throw of our house. Though we knew no one in the club we got into it without any trouble, as I was a member when here before.[2] We have played several times, but it is getting too late for out-of-door play. I have been at the theatre several times, they gave all Goethe's plays (XII). They are playing Tannhäuser tonight, Mama.

[1]Gustav Oscar Berger was the son of a minor civil servant from Sangerhausen, a small city located between Leipzig and Göttingen. He had been educated at the *Gymnasium* in Sangerhausen, and had enrolled at the University of Leipzig in 1880. JMC apparently knew him from his previous stay at Leipzig, and the two grew to be close friends. Berger's Leipzig dissertation (see 3.12n.13), as published separately, includes his personal data. [2]The club referred to was probably the Misnia, the student *Corps* to which Cattell belonged in 1881 (see 1.15).

3.3 To Parents, Leipzig, 16 December 1883

(Three paragraphs of gossip have been omitted.)

We are delighted to learn that you arrived safely at Davos and are comfortably settled with Aunt Nemie. As far as I know, Papa, you could find no better resting-place. It does not matter as much if you do find it a little tedious, you know some doctors put nervous people in bed or shut them up in a dark room. My prescription would be; make your body as strong and healthy as possible (ergo: proper diet, plenty of sleep and constant out of door exercise) and let your mind rest (i.e. freedom from work or excitement).[1] £3 please. I think too you will enjoy it at Davos, when you get used to the life and acquainted with the people. You can read novels, Dickens for example. You might read some Latin, so as to feel at home when you have to work on Lactantius;[2] I should think you would find an exquisite pleasure in rereading Homer or the Dramatists.

Harry and I are getting along very nicely, nothing very important having happened or being likely to happen. Harry spent most of the day in the chemical laboratory, and I most of my time working on psychology. Berger is going to make his examination in psychology and physics, and Wundt will give him some work as

[1]JMC's advice is similar to that prescribed by Silas Weir Mitchell, a Philadelphia neurologist well-known for his "rest cure" for overwork (see *DAB* 13: 62–65, *DSB* 9: 422–423, and especially Mitchell, *Wear and Tear, or Hints for the Overworked* [Philadelphia: Lippincott, 1871]). [2]Lactantius was an early Christian apologist and "the most classical of all the early Christian Latin writers" (*Oxford Classical Dictionary*, pp. 575–756).

an *"Arbeit"* and this we propose doing together.[3] Indeed we are now doing original work, finding under certain conditions the reaction & apperception times—if you know what these are.[4] I spend pretty much all day Wednesday and Saturday in the laboratory. Fortunately I do not have to pay anything for this, not even for what I use up and break, where as Harry has already had to pay $20. I am reading Wundt's *Physiologische Psychologie,*[5] noting the sections corresponding with Lotze's *Psychology.*

[3]Wundt would regularly assign *Arbeiten* (works) to his students as part of the formal opening ceremonies of the laboratory each semester (see "In Memory of Wilhelm Wundt," *Psychological Review* 28 [1921]: 166–169, 185–186; George M. Stratton to George H. Howison, 4 November 1894, 8 May 1895, and 24 November 1885 [Howison papers]).
[4]In the middle of December JMC and Berger were carrying out measurements of what Wundt called the *Unterscheidungszeit,* which usually was translated as "discrimination time" and which JMC later translated as "perception time." In this letter JMC exaggerates his success, as in his own notes he called his results "fragmentary" and wrote of "faulty method" and "sources of error" (see 3.12).
[5]*Grundzüge der physiologischen Psychologie,* 2nd edition (Leipzig: Engelmann, 1880), 2 vols.

3.4 To Parents, Leipzig, 30 December 1883

(Several pargraphs of gossip have been omitted.)

Christmas week has passed quickly and as pleasantly as we could have expected. Christmas day we all went to Pastor Puchek's for supper[1]—they had another tree and innumerable toys for the children. We have as you know fine music here—I have been attending the novelettes and tomorrow hear the famous violinist Joachim.[2] Tomorrow afternoon Prof. Wundt is going to explain to Berger and me an *"Arbeit"* with which Berger expects to pass his examination. If we get valuable results they will be published under our two names. I am getting a piece of apparatus made, which I have invented in order to carry on the work I began in Baltimore.[3] I have written to Dr. Hall, hinting that I would like to have my notes and papers.[4] I am not

[1]This man has not been identified, at least in part because the transcription of his name from Cattell's handwriting is uncertain.
[2]Novelettes were short concert pieces, "the musical equivalent of a romantic story," introduced in the middle of the nineteenth century by Robert Schumann, a composer with strong ties to Leipzig (*Oxford Companion to Music,* pp. 693, 932–933). Joseph Joachim was a world-famous Austrian-born violinist "surpassed by no one" (*Grove's* 4: 642–643).
[3]This instrument was JMC's "gravity chronometer," which would become a standard instrument in most late-nineteenth- and early-twentieth-century psychology laboratories (see Michael M. Sokal, Audrey B. Davis, and Uta C. Merzbach, "Laboratory Instruments in the History of Psychology," *JHBS* 12 [1976]: 59–64). A screen was supported by two vertical columns on which it could slide. In the screen was an adjustable slit. When an object was placed behind the screen and the screen was allowed to fall, the operator could control the amount of time the subject saw the object by varying the width of the slit (see figure 8). See also JMC "Über die Tragheit der Netzhaut und des Sehcentrums," *Philosophische Studien* 3 (1885): 94–127; JMC, "The Inertia of the Eye and Brain," *Brain* 8 (1885); 295–312; and *Man of Science* 1: 26–40.
[4]Hall had borrowed the notes on the experiments JMC had done the previous spring in Baltimore. This loan was later to cause much tension between the two men (see 5.18).

anxious that he should publish a paper, which would give him credit, which he does not deserve. If magazine editors do not see fit to print my work under my name it need not be published at all. It looks as though I had found several serious mistakes in results published by Wundt, but of this I cannot be sure until I make further experiments.[5] I will have to work hard during the rest of the semester as I am moving on a number of different lines having little in common. Experimental psychology lies about as far apart from philosophy as chemistry from Greek.

[5] For JMC's criticism of Wundt's work see 3.12.

3.5 To Parents, Leipzig, 6 January 1884

(Two paragraphs of gossip have been omitted.)

I am working hard and I think successfully. There were four of us, working at a subject in experimental psychology, I being the last to join, and one, who was to publish the results under his name, having been working on the subject for about a year. I told you Wundt was going to give Berger and me a subject; we were pleasantly surprized when he told us that we could keep on with the same work, that the others had not been very successful, and would have to take up a new and less important subject.[1] This was of course quite a compliment to us. We hope to have results worth publishing in two or three months. It would be quite nice for me to be the joint author of a German paper. We work every day and very hard. Day before yesterday we worked, with only twenty minutes intermission, from eight in the morning until after seven in the evening. We have a room to ourselves, and can come when we please and do what we please. Besides the work given us by Wundt, we are making some experiments I suggested, and which we hope will give interesting results.[2] Of course I have lots of other work before me. I can now continue my Baltimore experiments, and must do a good deal of preparatory study before I will be ready to translate Lotze's psychology.

Harry and I played tennis New Year's Day and we had skating for two days, but the ice has again melted. I heard Joachim twice last week, and this week expect to see Rienzi, Wagner's first opera. Harry and I get along nicely together, and are in every way comfortably and pleasantly fixed. Lectures will begin in a day or two, and Harry will go to work in his laboratory.

[1] The two men dropped were the otherwise unidentified A. Müller and D. Krauss (see 3.12; *Personal-Verzeichniss der Universität Leipzig*, no. 104, winter semester 1883–84, pp. 77, 90).

[2] These experiments were related to the influence of the intensity of the stimulus and the attention of the observer on the reaction time (see 3.12).

3.6 To Parents, Leipzig, 20 January 1884

(Three paragraphs of gossip have been omitted.)

I spend four mornings and two afternoon's working in Wundt's laboratory. I like Berger very much, he works hard, has good ideas and—what is equally important— is ready to follow my good ideas. He does most of the averaging & table-making, which is no small or easy work. On the whole I like Wundt, though he is inclined first to disparage our ideas and then adopt them, which is rather agravating. It is of course possible that we may not get along well together, as differences of fact and opinion will occur, and I dont know how to give up when I think I'm right. German Professors are not used to "indocile" students, as it usual here in the Professor's presence to "hold

Awe-stricken breaths at a work divine."[1]

But as I said I respect and like Wundt, and as he mostly lets us alone every thing may may move along smoothly. Our work is interesting. If I should explain it to you you might not find it of vast importance, but we discover new facts and must our- selves invent the methods we use. We work in a new field, where others will follow us, who must use or correct our results. We are trying to measure the time it takes to perform the simplest mental acts—as for example to distinguish whether a color is blue or red. As this time seems to be not more than one hundredth of a second, you can imagine this is no easy task. In my room I am continuing my Baltimore experiments, and here too I measure times smaller than 1/1000 of a second. With all this and my lectures I have as you may suppose but little time for reading or writing.

Though in good health, I do not feel strong as I am used to taking more exercise, than one can easily[2] find here. Still we must walk three to six miles a day, and I have begun to take fencing lessons, and will try to exercise twice a week in the gymnasium with the "Turnverein". It seems colder this evening, but we have had no frost for more than two weeks. I wish we could have skating, as cold weather would be healthier. My second dancing lesson happens this evening so I am not there.

[1]From Tennyson's *Maud* (part 1, section 10, lines 16–17). See also 2.3.
[2]JMC had originally written "usually" instead of "easily." The *Turnhalle* was located on the other side of the city from JMC's residence (see 5.1 and figure 6).

3.7 To Parents, Leipzig, 27 January 1884

(One paragraph of gossip has been omitted.)

It seems to me it must be doing you both good at Davos, and yet Mama, if the elevation still troubles you you ought not to stay any longer. Perhaps when you come away you will find for all that your health has been improved. Though we want to see you so much, I can scarcely advise you to hurry in coming here. We will proba- bly have at least a year together in Leipzig, and I do not want you to get tired of it.

But I really think you will enjoy the life here. As I have often said no one could want better board and rooms. The family too is very nice. Herr Gerhard was for many years consul at Alexandria, he collects and sells butterflies and has a collection worth over ten thousand dollars. He is very queer, but I like him. I think you will like Frau Gerhard and the daughters. Frau G. talks English, French and Italian. You, Mama, will of course study German, and I imagine you will not be at a loss to know what to do. Papa can study the German educational system here to perfection, and can use the library. There is also a fine reading room, with a long list of English periodicals, including the London Times and News, the Weekly Tribune and Nation, the Athenaeum, Academy, Saturday Review, Punch, Atlantic Monthly &c. I think you could spend an hour or two a day very pleasantly there, Papa. Then I imagine you can see as much as you wish of German society. We can hear beautiful music, and perhaps can once in a while see something like "Lohengrin" or "Krieg im Frieden".[1]

With us things move along quietly and pleasantly. I am very busy and so of course time goes quickly. I work every day in the laboratory and make psychophysic experiments at home. Prof. Wundt in his Psychology gives an account of experiments he made proving that it takes over $1/10$ second to see what color a thing is; we have made over two thousand experiments and find this time to be O.[2] I am thinking of writing a book on Psychometry.[3] Did you think you would ever have a son, who would write a book on Psychometry?

I am sorry that Mr. Scott's feelings are hurt, but I hope Papa can explain matters. Please do not let them take the painting until it is very different from what we saw.[4]

[1] *Der Krieg und der Friede, ein Lust spiel über die jetzigen Zeitlaufe*, a popular comedy (*British Museum Catalog, Compact Edition*, vol. 14, p. 415).

[2] As JMC later noted, "one cannot react so quickly that he does not know the color" (5.15), and "[I] have come to doubt the possibility of determining U (as distinct from W)" (see 3.12, where these terms and abbreviations are defined).

[3] According to the *OED*, this word, whose etymology well suited it to describe Cattell's work, was first used in 1879 by Francis Galton (Galton, "Psychometric Experiments," *Brain* 2 [1879]: 149–162; Galton, "Psychometric Facts," *Nineteenth Century* 5 [1879]: 425–533, reprinted in *Popular Science Monthly* 14 [1879]: 771–780). JMC's use of this word

here, in a letter that reflects his acquaintance with English periodicals, strongly suggests that he was reading Galton's work—which later was to be extremely important to him—as early as January 1884. See introduction to section 6.

[4] Soon after William Cattell retired as president of Lafayette, several alumni living in New York—including Charles Payson Gurley Scott, who was then teaching philology at Columbia College—arranged to have a portrait of him donated to Lafayette (Skillman, *Biography of a College*, vol. 2, p. 45). The first version of the portrait was totally unsatisfactory in the eyes of the Cattell family for some unspecified reason. The task of clearing up these "complications " was to fall to JMC (see 5.8).

3.8 From William Cattell, Davos, 27 January 1884

I address you this separate letter,[1] because your mother & I have had our anxiety as to your relations with Prof. Wundt much increased by yr reference to him in yr last letter.[2] If in this matter I trusted alone to my own judgment, so often clouded of late with groundless apprehensions, it might seem as if I was giving myself needless anxiety. But dear Mama, from yr very first reference to Prof. W. & without a word from me, has felt intensely anxious & has often expressed her great apprehension lest, instead of securing him as a friend you w^d say or do something that w^d first alienate him & then (alas! Such is human nature in the best of men) lead him to place something in yr way—Of course on other grounds than any personal reason, such as want of "docility" &c—

I fear you cannot understand how deeply we feel this danger. Both of us are distressed at the very thought of your having—even if you are in the right—a repetition of the ill feeling wh. it is evident was brought about between some of the authorities[3] & yrself at Balt.—We both fear you are not as cautious in this matter as you should be. You speak in yr letter of not being able to accept the views of the German Professors as to their students being "docile"—but (apart from the letters of Gilman & Gildersleeve in which I could read much between the lines) in the very kind letter you rec^d fr. Dr. Hall at Easton he used this very word with ref. to you:— he was favorably impressed with yr abilities & had the highest expectations of what you c^d have accomplished had you only been more "docile"—Parental partiality is proverbially blind, especially where a son is so loving & thoughtful & dutiful as you are: yet dear Mama & I cannot but fear that there is some real ground for apprehension here; that what you may regard as "independence," or even "self-respect" in yr relations to yr Professors, we shd call by another word;—and I should be wanting in parental love & faithfulness if I did not earnestly warn you against it— for your own sake.

But do not forget that your parents are necessarily involved in all that you do—I am sure that so loving & dutiful a son will not overlook this. And the circumstances, especially with ref. to myself, are peculiar. I shall never entirely recover from the shock which the Balt. trouble gave me. I never have a nervous depression that it does not again come to the front as on the first;—not only the same feelings of keen disappointment, and of indignation at what still seems to me injustice on the part of Gilman involving a betrayal of our long friendship—but, I am sorry to confess, there is the same anger & bitterness that I know to be unworthy[4] of a Christian and against which (tho' it makes me so wretched) I struggle at such times in vain. In my better moments, the great disappointment of my hopes at Balt. and for the plans based upon yr success there, takes the form of a prayer that your career at Leipzig will show these men that they made a mistake in dropping you from the Fellowship for a second year—this much of a "revenge" cannot be wrong

---

[1] JMC's mother also wrote him on this date (letter not included).
[2] See 3.6
[3] William Cattell first wrote "your Pro" (the beginning of "your professors") instead of "the authorities."
[4] William Cattell first wrote "unchristian" here.

for you or yr mother or myself to cherish:—and a conspicous succuss at Leipzig too w^d have a great effect in winning a position for you in Philad^a—though any word from Balt. would naturally be against you. I had therefore hoped you w^d be able to take yr Ph. D. this year: or that you would make such a favorable impression upon yr Professors that, through Gregory or others it would become fama clamosa at Johns H. & elsewhere at home and effectually dispose of what Dr. Gildersleeve said, that yr career at Johns H. was a great disappointment to the Fac^{ys} But, my dear son, if, instead of this there should be a repetition at L.^{6} of the Balt. experience— however much yr Professors might be at fault—it w^d be an irreparable blow to us. You may, indeed, with yr youth & health be able to keep from worrying over it: and you may finally surmount the obstacles it will place in your path, but for your dear mother & myself who are getting old it will be a lasting sorrow.

So my dear son let me beg you to act with great circumspection in all your relations to the Professors for our sakes as well as your own. Ah how happy & grateful we both are that I have no need to write & urge you as to habits that have blighted the fond hopes of many a parent. Thank God our dear boys are so pure & good;—as well as kind & loving to us;—and that our only fear is lest you may from a mistaken view of independence offend & alienate your Professors!—Yet I w^d emphasize this. You write of having found errors in some of Prof. W's published investigations. If he is really wrong & you have discovered it, to publish this with becoming modesty when you are no longer under his instructions will of course do you honor. Now it is dangerous ground. No matter if Prof. W. is convinced that he is wrong; it will require great tact for you to let him & others know it without wounding his amour propre: and engendering unkind feelings towards you. And no honor that you could get from any such correction of his errors could possibly com pensate for the loss of his kindly personal interest in you. It w^d be far better for you, while a student, to be as "docile" as the most exacting Professor could demand.

I have written you a long letter—& yesterday & today have been poor days with me:—but dear Mama was anxious that I sh^d write. She has read what I have already written and says it expresses just what she feels. I know you will appreciate our devoted love for you that leads me to call yr attention to this matter, & to urge you to act with great caution—for yr own sake as well as ours.

May God ever bless you my dear son.

[5]On Caspar Rene Gregory and Basil Lanneau Gildersleeve see 1.17. *Fama clamosa:* noisy rumor, or scandal.

[6]"at L." was inserted above the rest of the line.

3.9 To Parents, Leipzig, 6 February 1884

(Two paragraphs of gossip have been omitted.)

Harry and I will doubtless have a good deal to tell you when you come, but our life is so even and quiet that nothing happens, deserving the dignity of being told about in a letter. After you have been here two days, you will know about what we have been doing for two months. Prof. Bissel, who, as Harry perhaps told you, has just *taken his Ph.D, sails for America next Wednesday, leaving Leipzig Monday, so his room will be given up at exactly the right time. I was invited to dinner by Prof. Heinze, the Rector, last Sunday. I like him very much, and he is a good lecturer. I hear his course on the History of Modern Philosophy, as you know. We work regularly in the laboratory, but are annoyed because we cannot have a single hour with a clear sky in order to determine photometrically the intensity of the light we use.[1] I am beginning to make other experiments by lamp light, because I cannot get a good and regular light in the day time. You see I want to prepare you for Leipzig weather.*

I thank you, Papa, for your letter anent my relations with Wundt and other people. No one, not even I myself, has my welfare so much at heart as you and Mama, and by following your advice I would probably always be more successful and happy than in following my own impulses. The only question is whether a life of uniform success and happiness would not be as undesirable as it is impossible. I admire men who struggle, and suffer, and fail, rather than those who are always comfortable and in a good humor. It is perhaps better to fight one's way through life, than to slip through. As to this special case my relations with Wundt could not be pleasanter, nor do I see any reason why they should change. Both Berger and I intend to take our Dr.'s degree here, and would not want to get into trouble with Wundt.

[1]JMC apparently determined this intensity (see 3.12).

The elder Cattells soon joined their sons at Leipzig, and the family enjoyed the next few months together. Jim and Harry continued their studies, and the parents rested.

Jim resumed his use of stimulants—at first only tobacco, caffeine, and chocolate, but after a while morphine and alcohol (Journal, 25 February 1884 and 29 February 1884). This behavior bothered him, and though he was "not ready to say that I should never use drugs as stimulants," he did "look to asceticism as a remedy." He "did not eat anything but bread and milk for four days, and would have kept it up for a week if [he] had not been invited by Prof. Wundt to supper" (Journal, 15 March 1884).

The use of drugs was not the only thing that distracted James Cattell from his work. One evening he noted that "the sexual appetite does not seem to trouble me now," but that night was apparently unusual. For example, a young English woman flirted with him for a while, and the affair disturbed him. She was Margaret Bright, the seventeen-year-old daughter of James Franck Bright, master of University College, Oxford. Cattell had met her in January at a dance given by one of the richer families of Leipzig, and was taken with her intelligence and the wide scope of her reading. He met her at other dances and at the opera, but she would not go with him to dinner. The longest entry in Cattell's Journal (15 March 1884) was devoted to an account of their relationship. Cattell was clearly infatuated with her—

he noted "I dreamed about her all night"—and once left a box of flowers for her. But she soon grew tired of his attention, and a bitter exchange of letters soon ended their relationship (see Journal, 16 March 1884 and 18 March 1884).

Meanwhile, Cattell continued his work, and, at the end of March, resolved to "try not to eat anything during April but bread, milk and soup" (Journal, 31 March 1884).

3.10 Journal, Leipzig, 4 April 1884

That experiment lasted 2 ⅔ days. On a specious pretext I drank a cup of chocolate yesterday evening. I am sorry to say I smoked a cigar this evening. Today however I am in good spirits, as I thought of a new piece of apparatus this morning, which will be of great value in my psychometric work. Probably the best thing I have done in psychology so far, was inventing the gravity apparatus which enables me to see letters, words &c for a short & measurable time.[1] The new apparatus will be a modification of this, & will simply close a current when a word &c is exposed to view[2]—this will do away with the cumbersome electrical machinery we have been using for the past four months, and which all my predecessors have used. It will also enable me to carry on my experiments alone.

[1] See 3.4.

[2] JMC attached to the vertical columns of the gravity chronometer a small key, or switch, arranged so that the falling screen closed it the moment the slit revealed the object beneath the screen (see figure 8). (JMC, "Psy-chometrische Untersuchungen," *Philosophische Studien* 3 [1886]: 305–335, 452–492; "The Time Taken Up By Cerebral Operations," *Mind* 11 [1886]: 220–242, 377–392, 524–538.)

Later that April, James Cattell visited Berlin but did not enjoy his stay. He took what he called "an excessive dose of hashish," twenty-four grains over two and a half hours, and was discombobulated for a while (Journal, 18 April 1884). He returned quickly to Leipzig to resume his experiments and found that his mother was not well. William Cattell had gone off on a short trip of his own; James wrote to him about his mother's condition, which was not serious. He closed by noting that he was "glad to be home again with Mama, Harry, and my work," to which his brother added, "especially the work" (JMC to William Cattell, 17 April 1884).

Two days after returning to Leipzig from his short trip, James Cattell took time to outline a schedule for the next few months.

3.11 Journal, Leipzig, 20 April 1884

(Half of this entry, devoted to JMC's ramblings about his emotional state, has been omitted. This entry was dated "Sun., Apr. 18th '84," but as 18 April 1884 was a Friday, it was probably written on 20 April.)

The summer semester begins this week. I suppose it were impossible to live without some object, some hope, yet it seems we can live on objects and hopes we know to be delusive and worthless. I make my plans just as though I thought there to be some reason why I should do one thing more than another. Well, perhaps there is, mental unbelief, cannot do away with instinctive faith. I do not know whether it were better to seek quietism or action-passion, & if the latter to seek the maximum in the present, or spread over my lifetime. Of these three ways of living my reason recommends the first, my higher instincts the last, yet I seem to follow the second. However now I will make out a program guided by my reason & instincts & we will see how far I can follow it during the semester. I will get up at 6.30 & be ready for work at seven. Then I will work hard for an hour and a half, writing or reading what requires clear thinking. At 9.15 I will hear Heinze lecture on the History of Ancient Philosophy four days in the week. I will walk to the university by way of the woods and Johanna Park.[1] After the lecture I will make psychological experiments either in the laboratory or in my room until 1.30 dinner time. Of course a twenty minutes walk comes in here. Wed. & Sat the lecture falls out, but I will take exercise for from one half to one hour. After dinner with the exception of the 20 minutes walk I will read perhaps in the Lesehalle until Heinze (Psychology) lectures at 3.15.[2] Then I have an hour free & will exercise, read or work. At 5.15 I will hear Wundt on History of Modern Philosophy & stay out of doors until supper time (7 or 7.30) walking home perhaps by way of the park. After supper from 8 or 8.30 I will read, study or write until 10.30 bed-time. Wed. & Sat. afternoons I will spend out of doors. Now for the very numerous exceptions. Sunday is an exception. I want to "keep" Sunday better than I have for some years. I propose sleeping late, as long as I sleep soundly. I will spend the morning writing letters and in my journal. After dinner I will take a long walk—always the same I think, to the Grosse Eiche & through the Rosenthal.[3] Then a concert or church.[4] After supper I will read the Bible or Shakespeare. The evenings will be spent oftener as exceptions than according to rule. I propose going to the theatre twice on the average, once to a play and once to an opera. I may attend Prof. Wundt's seminar on Logik Tuesday evening. Perhaps

---

[1] As usual, JMC enjoyed the exercise of a longer than necessary walk; a route to the university via the *Scheibenholz* (woods) and Johanna Park was twice as long as a direct route (Baedeker, *Northern Germany*, pp. 242–243).

[2] *Lesehalle:* reading room.

[3] The *Grosse Eiche* (great oak) was a monumental tree in the large, heavily wooded preserve known as the *Rosenthal*. This tree was la-

beled *Friedetis Eiche*, or oak of peace, on at least one map of the city. See 5.1 and *Leipziger Adress-Buch für 1901* (Leipzig: Albergander Edelmann, 1901).

[4] English Episcopal services were held twice each Sunday, and American evangelical services at least once a week (Baedeker, *Northern Germany* [10th edition, 1890], p. 343).

one evening a week I will be invited out, attend a concert, Kneipe, celebration or something of the sort. I would like to exercise in the gymnasium with the Turnverein. In summer when it gets dark so late I will sometimes stay out of doors—in a boat for example.[5] While mother & father are here I will spend part of the evening with them. All this looks as though I would not read or work much in the evening. I will sometimes miss lectures, sacrificing them to work or exercise. For example I may spend the whole morning or afternoon on a walk or boat-ride; I may make my experiments from 8.30 or 9 to 12 and then play tennis; from two to five or from four to seven I may row, play tennis, ride or something of this sort. Again I may miss a lecture if I am in a good mood for writing, or want to hurry with my work. When & how often this will happen will depend on the nature & importance of the lecture, the weather, my health, my work &c. I am troubled by my eating and the use of stimulant drugs. There are three courses: abstaining, moderation & excess—while I am hesitating between the first two, I follow the last. My ideal for myself is asceticism, perhaps only milk & cereals. But I am not strong enough for this. I suppose I had better aim at moderation & no stimulants, except perhaps a cup of chocolate & a cigarette Sunday afternoon. Sundays I want to write here a tolerably full account of the week gone by.

[5] As Leipzig is located on three rivers (the Elster, the Pleisse, and the Parthe), boating was a common recreation in the city throughout the spring and summer (Baedeker, *Northern Germany*, p. 244).

Despite his detailed plans for spending the summer of 1884 at Leipzig, James Cattell soon discontinued his experiments to devote his time to other matters (see 4.1). Later (see 5.6) he would prepare a document summarizing much of what he had done in 1883–1884. It is among the most significant of his records, and is presented here as 3.12 (see also figure 9).

The experiments Cattell describes were related to what he and Galton called "psychometry" (see 3.7) and others called "the time-relations of mental phenomena."[1] Cattell tried in these experiments to measure the time taken by various thought processes. The basic methodology followed was first outlined in 1868 by Frans Cornelius Donders, a Dutch physician, and by other physiologists (see 5.57), and expanded and modified by Wundt in the various editions of his *Grundzüge der physiologischen Psychologie* and elsewhere.[2]

[1] Joseph Jastrow, *The Time-Relations of Mental Phenomena* (New York: N.D.C. Hodges, 1890). JMC later reviewed this book (*Educational Review* 1 [1891]: 189–190). See also Theodule Ribot, *German Psychology of Today: The Empirical School*, chap. 7, "The Duration of Psychic Acts," tr. James Mark Baldwin (New York: Scribner, 1886), pp. 250–286.

[2] Frans Cornelius Donders, "Die Schnelligkeit Psychische Processe," *Archiv für Anatomie und Physiologie und wissenschaftliche Medizin* 6 (1868): 657–681 (English translation: "On the Speed of Mental Processes," *Acta Psychologica* 30 [1969]: 412–431). See also Thom Verhave, "Origins of Psychometry: A Review Article," *JHBS* 8 (1972): 352–356; Robert S. Woodworth, *Experimental Psychology* (New York: Holt, 1938), pp. 298–310; Wilhelm Wundt, *Grundzüge der physiologischen Psychologie*, 2nd edition (Leipzig: Englemann, 1880), vol. 2, pp. 219–247.

Cattell, while following the lead of these men, changed their techniques significantly to develop a methodology of his own that in many ways represented a great advance in experimental design.[3] His development of these modifications is evident from the documents.

Donders and Wundt had used as the basis of their procedures the reaction-time experiment. The simplest form of this technique involved presenting the subject with a stimulus (often a flashing light) and calling upon him to react in a specific way (usually by lifting a finger from a telegraph key) as quickly as possible. The first few reactions varied greatly in time, but the subject soon became accustomed to the procedure, and his "simple reaction time"—held to exclude all mental, or cerebral, activity[4]—could thus be measured easily. Both Donders and Wundt had realized that conditions had to be carefully controlled for meaningful results to come from this procedure. Cattell began to investigate the effects of several of the conditions related to the design of this experiment, including attention and the intensity of the stimulus. Here he went beyond Wundt, but not as far as he was to go the following year (see 5.15).

To examine the more complex mental processes, both Donders and Wundt had extended the simple reaction experiment, but in different ways.[5] The differences between their approaches is not important here—Cattell's friend Gustav Berger discussed them in depth in his doctoral dissertation[6]—but both had introduced what they held to be the processes of "discrimination" (*Unterscheidung*) and "choice" (*Wahl*) into their experimental procedures. Wundt measured the time taken up by these mental activities in this way: For discrimination, after the subject's simple reaction time was determined he was presented with one of two possible stimuli (perhaps a red light and a green light) and instructed to react as soon as he knew which stimulus was presented. As in all of Wundt's experiments, introspection played an important role—that is, the subject had to determine just when he discriminated which of the two stimuli was present, and hence just when he should react. In Wundtian terms, he had to determine just when he apperceived the stimulus in his *Blickpunkt*. This procedure led to the measurement of what was known as the discrimination-reaction time. The discrimination time (the time needed for a mind to discriminate between two different stimuli) was determined by subtracting the simple reaction time from this

[3] JMC, "Psychometrische Untersuchungen," *Philosophische Studien* 3 (1886): 305–335, 452–492; JMC, "The Time Taken Up By Cerebral Operations," *Mind* 11 (1886): 220–242, 377–392, 524–538; *Man of Science*, vol. 1, pp. 41–94.

[4] Wundt always interpreted the results of these experiments from a mentalist perspective. JMC, however, was gradually to develop a physicalist point of view, as seen in the

English title of his dissertation. See 5.123 and 5.142.

[5] Woodworth, *Experimental Psychology*, pp. 302–305.

[6] Gustav Oscar Berger, "Über den Einfluss der Reizstärke auf die Dauer einfacher psychischer Vorgänge mit Besonderer Rücksicht auf Lichtreisse," *Philosophische Studien* 3 (1886): 38–98.

time. Choice was introduced into the procedure by calling for one reaction to one stimulus and a second reaction to another stimulus (for example, lifting one finger for one colored light and another for the second color). Again, introspection was called for to determine when the choice was made, and the time needed for making the choice was determined by subtracting the simple reaction time from the choice reaction time.[7]

In November 1883 Cattell was assigned by Wundt to perform several experiments relating to discrimination times and choice times. He began his work with pleasure, keeping a laboratory notebook on the unbound sheaves of paper (*Hefte*) favored by German students.[8] He soon found reason to criticize Wundt's definitions and procedures (primarily those regarding introspection and the concept of apperception), and to introduce modifications in laboratory technique. Also, when he did not use the German terms, he wrote of the "perception time" rather than the "discrimination time," and the "will time" rather than the "choice time." This usage illustrated an effort to reflect his own views on what his experiments showed and to simplify what he felt to be overly elaborate German terminology (especially that used with regard to apperception). In the fall of 1884, for his own use, Cattell prepared an abstract of his laboratory notebook, which he entitled "Journal of Experimental Work in Psychology" (see 5.6). This document cited the various *Hefte* of his original notebook, but primarily it outlined what he did, what he found, and why he thought Wundt's procedures worthy of criticism. Also at this time Cattell apparently sketched a statement of some of these criticisms, which he may or may not have shown to Wundt. This statement was similar to one he did present to Wundt that fall, which set forth these criticisms in detail and presented Cattell's plans for the future.[9]

Throughout his experimental work, James Cattell used several instruments common to late-nineteenth-century physiological and psychological laboratories. One that should be mentioned here was known in Germany as the Hipp chronoscope and in England as the Wheatstone-Hipp chronoscope. A timing device capable of measuring milliseconds, it was used throughout Europe during this period.[10] It originally was built by Mathais Hipp in 1842, and was a modified version of an instrument designed by Charles Wheatstone around 1840. From that time the chronoscope had been continually modified and improved upon. In 1883 it was still not perfected; for example, Cattell found that it was quite delicate and was greatly affected by minor changes in temperature and the strength of the current flowing

[7] Wundt, *Grundzüge der physiologischen Psychologie*, vol. 2, pp. 247–256.
[8] Hart, *German Universities*, pp. 61–62.
[9] See 5.15.
[10] Beatrice Edgell and W. Legge Symes, "The Wheatstone-Hipp Chronoscope: Its Adjustments, Accuracy and Control," *British Journal of Psychology* 2 (1906): 58–88. See also Edward Bradford Titchener, *Experimental Psychology: A Manual of Laboratory Practice* (New York: Macmillan), vol. 2 (1905), part 2, pp. 326–335.

through its electromagnets.[11] Work with this apparatus required much time and attention, and this document makes clear just how demanding the laboratory practice of the period was.

[11]See Michael M. Sokal, Audrey B. Davis, and Uta C. Merzbach, "Laboratory Instru-   ments in the History of Psychology," *JHBS* 12 (1976): 59–64.

3.12 Journal of Experimental Work in Psychology

Nov. 21st 1883.

Worked for the first time in Wundt's laboratory. Again Dec. 1st & 5th with Berger, Müller & Krauss.[1] They had been working for sometime but had accomplished nothing. They were trying to determine the "Unterscheidungszeit" by Wundt's method.[2] Objections to this (Unterscheidungszeit by Wundt's method written Apr '84.)[3] The "scema" was R(3) + Ru₂(6) + R(3). Ru − R should give U.[4] Results fragmentary (Heft 2 & 3 numbered in red ink)

Dec. 8, 12 & 15

My results put together in table dated Dec. 15th. Worthless as far as Unterscheidungszeit goes. (Heft 3, 4 & 5)

Jan. 2nd '84—3, 4, 5 & 7th

Müller & Krauss being dropped Berger & I continued the above method. Besides faulty method, there were sources of error in the apparatus. The two currents were not always simultaneously closed, and the stronger the current the shorter the time recorded by the clock.[5] This latter we only discovered later—it made all the absolute times uncertain, but did not affect U.[6] The Unterscheidungszeit we measured

[1]Gustav Oscar Berger was JMC's close friend (see 3.2 and 3.5); Müller and Krauss have not been identified.

[2]JMC later described "Wundt's method" as "let[ting] the subject react as quickly as possible in one series of experiments, and in a second not to react until he has distinguished the impression, the difference of the time in the two series giving the perception-time [i.e., discrimination-time] for the series" (*Man of Science*, vol. 1, p. 65).

[3]See 5.15, which was a later statement of these objections. Still later, JMC noted "I have not been able myself to get results by this method; I apparently either distinguished the impression and made the motion simultaneously, or if I tried to avoid this by waiting until I had formed a distinct impression before I make the motion [i.e., in Wundtian terms, apperceived the sensation], I added to the simple reaction, not only a perception [i.e., a discrimination], but also a volition [i.e., a choice]" (*Man of Science*, vol. 1, p. 65). (See also postscript.)

[4]K was a measurement of simple reaction time, Ru₂ a measurement of discrimination reaction time, and U the discrimination time (*Unterscheidungszeit*). This "scema" involved three simple-reaction-time experiments, followed by six discrimination-reaction-time experiments, followed by three simple-reaction-time experiments.

[5]The Hipp chronoscope, or "clock," had two electromagnets, one to start the timing process and the other to stop it. Each magnet was supplied by its own current source. This arrangement allowed for greater flexibility, but also led to the problem JMC noted. JMC later solved this problem in part by installing adjustable springs on the armature controlled by the magnets (*Man of Science*, vol. 1, p. 43; Edgell and Symes, "The Wheatstone-Hipp Chronoscope," p. 60). See also 5.47.

[6]The error was constant in all measurements, so that it appeared in both R and Ru₂. Hence, the subtraction of R from Ru₂ would remove the error and leave U unaffected.

about as accurately as could be expected with the method used. (Heft 6, 7 & 8) I have Berger's results tabulated. 360 reactions, R 194.93 Ru$_2$ 198.59 U$_2$ +3.66[7] Also for colors.[8] Mine must be someplace.

There were also experiments made at my suggestion on the influence of attention & fatigue (Heft (unnumbered)) also a slip dated 31/12 '83.

Jan. 15$\underline{th}$

Changed the method in two respects suggested by me. The series were made long (15)R & (15)Ru.[9] In the other series there may have been <u>constant</u> Errors. The series were afterward corrected—the five measurements varying most from the corrected average being rejected. The corrected averages I think are the most correct—but if sufficient experiments are made there is little difference between the corrected & uncorrected averages, especially as to U

Jan. 17, 18, 19, 21, 23, 25, 26, 28 (Heft 10, 11, 12, 13)

Same method using six colors. Fully tabulated. Curves for colors also drawn. Each of us made 720 reactions. Averages[10]

	R	mV	Ru	mV	U$_6$	R	mV	Ru	mV	U$_6$
B	193.8	20.4	198.	25.2	+4.8	194.1	11.4	198.2	13.9	+4.1
C	160.4	25.1	165.2	23.7	+4.8	163.1	13.7	163.9	12.7	+0.8

All reactions received 15 in a series.   *5 reactions varying most from the corrected average rejected 10 in a series.*

The remarkable coincidence of our both haveing +4.8 is shown to be accident by the corrected tables.[11] The small + value is probably not U$_6$ but is due to the fact that one reacts more quickly to an awaited and exactly perceived stimulus.

[7]The unit omitted here is the millisecond. JMC later introduced the symbol "σ" for millisecond, in a false analogy with "μ" for micron, which he took to be defined as a thousandth of a millimeter and not as a millionth of a meter (JMC, "The Inertia of the Eye and Brain," *Brain* 8 [1885]: 295–312; *Man of Science*, vol. 1, pp. 26–40). See also Edwin G. Boring, "σ," *Science* 71 (1930): 362–363, which indicates the possible confusion between this use of σ and its use as the symbol for standard deviation.

[8]The *Arbeit* assigned by Wundt apparently included the measurement of the discrimination time for each of several colors. JMC's technique for presenting such colors is discussed below in note 29.

[9]The "scema" was now fifteen measurements of simple-reaction time, followed by fifteen of discrimination-reaction time.

[10]In this table mV is mean variation, defined as the average of all of the variations of each measurement from the mean of all the measurements. B stands for Berger, C for Cattell.

[11]This is not true. Berger's uncorrected U$_6$ was actually +4.2 milliseconds, as 198.0 − 193.8 = 4.2. Arithmetic was not JMC's forte. More important, the fact that Berger's U for colored light (4.1 milliseconds) differed only slightly from his U for white light (3.66 milliseconds) gave JMC the results that led him to claim Wundt was incorrect in claiming that it took one-tenth of a second (or 100 milliseconds) to perceive a color. See 3.7.

In Heft 10 are our first "Wahlversuche" in 11R & Ru by Wundt & Lorenz.[12]

Jan 31, Feb 1 & 2,   (Heft 13 + 14)

Reaction time on four intensities of light. Not, I think tabulated.

Feb. 8, 9, 11, 12, 13, 14, 15, 18, 20, 22.

Heft 15, 16, 17, 18, 19

Reaction time on eight intensities of light. Fully tabulated and valuable.[13] *Curves drawn with ordinates for both percent of light passing through the glass (accurately determined by mean of photometer*[14]*) and varying according to the psychophysic law,*[15] *glass absorbing 99.6% 97.6 91.5 60 27 0 +2 Dan el. + lens.*[16]

[12]*Wahlversuche:* choice experiments. JMC means that these are the first he has done by the method suggested by Wundt and Gustav Hermann Lorenz, a fellow student and the *Famulus* (attendant) of Wundt's Institute for Experimental Psychology. The *Famulus* apparently held a position midway between that of the *Aufwärter* or *Diener* (a servantship held in 1883–84 by a man named Hermann Hartmann) and that of the *Asistent* (a laboratory post, first occupied by JMC in 1885–86). See 5.109 and *Personal-Verzeichniss der Universität Leipzig,* no. 104, winter semester, 1883–84, p. 26.

[13]JMC later published this work separately ("The Influence of the Intensity of the Stimulus on the Length of the Reaction Time," *Brain* 8 [1886]: 512–515; see *Man of Science,* vol. 1, pp. 103–106), and it laid the groundwork for the experiments on which Berger's dissertation was based (see Berger, "Über den Einfluss der Reizstärke auf die Dauer einfacher psychischer Vorgänge mit Besonderer Rücksicht auf Lichtreize," *Philosophische Studien* 3 [1886]: 38–98). In fact, these tables, with the figures rounded off, were published in Berger's dissertation (p. 83). However, JMC presented different figures, probably because he based his published data on fewer experimental trials.

[14]The type of photometer JMC used has not been identified. See 3.9.

[15]According to Gustav Theodor Fechner's "psychophysic law," sensation (a mental quality) was related logarithmically to stimulation (a physical quality). JMC's uneven scale of stimuli was an attempt to get at a regular

scale of sensation. For Berger's use of this "law" in his experiments with sound stimuli, see "Über den Einfluss der Reizstärke," p. 85.

[16]JMC's light source was the "light produced in a Geissler's tube by an induction current from six Daniell cells" (*Man of Science,* vol. 1, p. 103). In the first five intensities noted here a smoked glass was placed in front of the tube; in the sixth no glass was interposed between the observer and the tube; in the seventh and eighth two Daniell cells were added to current source for the Geissler tube, and lenses were substituted for the smoked glass (ibid., pp. 103–104.) The Geissler tube and the Daniell cell were two standard pieces of apparatus in late-nineteenth-century scientific laboratories. The latter, invented by John Frederic Daniell (*DSB* 3: 556–558), professor of chemistry at King's College, London, around 1836, was a modification of the standard voltaic cell that provided a regular and continuous current (W. R. Cooper, *Primary Batteries: Their Theory, Construction and Use* [London: "The Electrician" Printing and Publishing Company, 1901 (?)], pp. 201–205). The former, invented by Heinrich Geissler (*DSB* 5: 340–341), a glassblower and mechanic connected with the University of Bonn, around 1858, was a glass tube, filled with an inert or inactive gas, and two electrodes, which permitted current to pass through the gas. The color of the light emitted by the tube depended on the gas within it. JMC began preparing the following table with different values, which he crossed out.

	99.6	97.6	91.5	60
B	337.9—26.5	265.1—17.9	237.8—16.3	229.8—14.9
C	280.7—29.9	204.7—17.4	190.1—16.3	177.6—14.2

	27	0	+2 Dan. el.	+ lens
B	222.3—14.9	225.1—16.8	207.4—17.7	197.6—16.0
C	177.1—15.3	173.2—12.8	165.2—15.8	158.0—18.8

corrected as above

	99.6	97.6	91.5	60
B	337.8—14.1	263.8—9.6	236.6—8.8	231.8—8.1
C	279.4—15.8	203.2—8.7	191.7—8.7	179.9—7.6

	27	0	+2 Dan. el.	+ lens
B	222.0—7.9	223.4—8.6	208.0—9.4	198.0—8.9
C	178.9—8.4	174.0—7.0	164.6—9.1	159.4—9.9

On each intensity there were 10 series of 15 reactions made in all 1200 each. One series on every intensity was made each day the order the intensities being varied. The total[17] average for each day is also calculated and curves drawn has no special value as it depended in the strengt of the current passing through the clock.[18]

Feb. 12, 14, 15, 18, 22, 25   (Hefte 17, 18, 19)

Same days as above, we experimented on the effects of attention.[19] The grades being take (1) greatest possible attention (2) normal attention, 3 attention distracted by adding as rapidly as possible 17 to 17. Tables fully made up and curves drawn.

	strained at.	nor. at.	adding
B	219.1—15.5	231.3—17.1	274.6—28.1
C	188.0—17.3	162.3—15.6	182.9—19.0

	strained	normal	adding
	217.4—8.2	227.5—9.0	271.7—12.9
	186.5—10.0	162.9—8.7	181.3—10.2

corrected

[17]The word "total" was added above the rest of this line.
[18]See Man of Science, vol. 1, p. 104.
[19]JMC published these experiments as part of his dissertation "Cerebral Operations" (Man of Science, vol. 1, pp. 60–64). The reaction-time values reported in that paper differ from those presented here, probably because the published data were from a smaller number of experimental trials.

Ten series on each grade—in the uncorrected values 15 reactions in a series, as above. Light six Dan. elements. 60% absorbed by grey glass. The corresponding figures in the above table being B 229.8—14.9 (cor) 231.8—8.1 C 177.6—14.2 (cor) 179.9—7.6 attention there being "normal." For me normal attention is a state of expectancy, without any strain or concentration. My thoughts drift in all directions.[20]

Feb. 27, 28, 29, March 3, 6

Same: stimulus being a momentary electric shock on the forearm.[21] (II) in table below. Fully tabulated

	strained	normal	adding
B	190.3—13.2	195.2—12.4	220.4—16.2
C	176.8—13.9	179.6—15.0	214.5—21.2

Corrected

strained	normal	adding
191.1—7.3	194.4—7.1	219.1—9.2
176.6—7.7	180.2—9.0	213.9—11.4

notice—B's times much shorter than for light. mine about the same. Both for light & electricity B's curve ⟋ C for electricity also so for light ⌄ This perhaps due to C reaction being more thoroughly reflex than B— therefore requiring little attention. C's reaction on light was more reflex than on electricity as several thousand[22] light reactions had been made.

Corresponding figures for table below

B 193.5—13.7   Cor. 193.1—7.2 ⎫
C 185.7—18.4   184.6—11. ⎭ attention normal

Feb. 28, 29, Mar. 3, 6, 7.   Hefte 19 last leaf 20, 21, 22

Influence of the strength of an electric stimulus on the length[23] of the reaction time. Fully tabulated[24]—curves drawn. Minumum—near the "Reizschwelle"[25] is possible. Maximum (IV) strong enough to be painful. No constant owing to changing sensitiveness of the skin. Intensities II & III were graded as nearly as possible psychophysically equidistant from I & IV & from each other.

[20]JMC later described normal attention as when "the subject expected the stimulus and reacted at once, but did not strain his attention or make special haste" (*Man of Science*, vol. 1, p. 60).

[21]"I.e., an induction shock of moderate intensity" (*Man of Science*, vol. 1, p. 60).

[22]JMC first wrote "many thousand" here.

[23]JMC first wrote "strength" here.

[24]Berger presented this table, with the figures rounded off, in his dissertation ("Über den Einfluss der Reizstärke," pp. 84–85). JMC also mentioned these experiments later (see *Man of Science*, vol. 1, pp. 104–105).

[25]*Reizschwelle*: stimulus threshold.

10 series of 15 reactions on each grade.
Order always varied.

	I	II	III	IV
B	211.9—16.8	193.5—13.7	188.1—11.9	190.4—10.3
C	194.0—18.8	185.7—18.4	162.1—13.6	161.0—13.7

Corrected

	I	II	III	IV
B	213.0—8.3	193.1—7.2	187.1—6.5	190.6—6.0
C	192.9—9.8	184.6—11.0	162.7—8.	160.8—7.8

Notice B's times much shorter than for light. C's about the same. For both mV smaller. This may be due to physical causes i.e. the apparatus. It is possible the Geisler tube was not immediately, nor regularly illuminated.

March 12, 13.   Heft. 23

Went back to "Unterscheidungszeit" which was the subject Wundt had given us to investigate. The colors[26] were this time changed both for R & Ru[27] and were in both cases unknown In one series, we reacted as quickly as possible, in the other tried first to distinguish the color. This was an improvement in method. However it is impossible to determine U in this way—owing to the fact that one cannot react so quickly that he does not know the color. See my paper written for Wundt.[28] From eight series (15 reactions in a series) each on R & Ru

	R	Ru	Ru − R
B	281.8—20.8	290.7—19.6	+ 8.9
C	238.2—25.7	241.4—21.2	+ 3.2

I do not know why the times were so long. Probably the stream running through the clock was very weak. We discovered afterwards that the clock was magnetised on one side. The direction of the current was changed after each reaction, therefore the electro-magnet alternately[29] let the [30] [one word unclear] go more quickly or slowly, and mV was made large. After the Christmas vacation, when the electro-magnet was probably not magnetized mV is unusually small.

March 15th   Heft 23, 24

The intensity of the light was reduced so that the colors could just be distinguished. We thus get a value for U but it was more properly due to the hesitation caused,

[26]JMC produced colored light for stimuli by placing a colored glass between the Geissler tube and the observer before the tube was lit. To prevent possible clues as to the color of the stimulus, the room in which the observer sat was dark (*Man of Science*, vol. 1, pp. 68–69, 80, 104–105).
[27]JMC first wrote "U &" here.
[28]See 5.15.
[29]JMC first wrote "therefore was alternately stronger & weaker" here.
[30]JMC first wrote "drew the" here.

less one could not tell one color from the other—even though he had distinctly seen it. Thus I could scarcely tell blue from violet, and green & yellow were too much alike.

	R	Ru	Ru − R	
B	314.4	325.0	+10.6	4 series (60 reactions)
C	255.3	274.3	+19.5	in each average.

March 24, 25, 26, 27

The same—continued after the vacation. 16 series on each R & Ru for both B & C.

	U
B	+9.0
C	+14.1

March 29, 31, Apr. 1, 2.   Hefte 25, 26.

Unterscheidungsversuche with four colors (red, yellow, green, blue) white & violet being rejected). Scema 5R −5Ru −5R −5 Ru & inverted. The one reacting did not know the color—as above. These results can be directly compared with Wundt's.[31]

24 series, 20 reactions to the series.

	U.
B.	+7.1
C.	+1.3

—2 sq.   Heft 26, end

Regulation of clock. First "Wahlversuche"

May 6th 8, 9.

Wahlversuche. Red light was reacted on with the right hand, blue with the left.[32]
Scema[33] (5)Rred (5)Rw (5)Rblue (5) Rw & Rw, Rb, Rw, Rr   Rr, Rw, Rb, Rw
Rw, Rr, Rw, Rb

25 series 500 reactions each

	R.	Rw.	W.
B	151.2	275.5—28.5	124.3
C	115.3	278.3—42.8	163.0

End of Winter's work I going to America.

[31] Wundt, *Grundzüge der physiologischen Psychologie*, vol. 2, p. 251.
[32] In this way JMC explicitly introduced the element of choice into the reaction process.
[33] Each "(5)" in this line was added above the rest of the line.

4

Cattell as Advocate:
America, May 1884–September
1884

James McKeen Cattell had not
planned to return to America in May 1884. His parents and his brother
were still with him in Leipzig, and the family had hoped to travel together
for a while. But circumstances took Cattell back to America to act as an
advocate for his father, and thereby to discover new abilities within himself.

4.1 Journal, Leipzig, 10 May 1884

*It is certainly a most curious thing—this world. In my way I have unfaltering faith in
a providence—I do sincerely believe that if things could be moved a hair's breadth
from where they are the world would be ended. I have decided this evening to go to
America. The secretaryship of the church board of disabled ministers is vacant, and
has informally been offered to father.[1] I think it would have been better if it had
not been offered to him, but as it stands now it seems to me better that he should
accept it. The affair is however complicated, as father wrote withdrawing in favor of
Dr. Nixon, who wants the place.[2] I have all of a sudden decided that I had better go
home and see how things stand. Then I may be able to smooth over a mess about
father's portrait and look after selling our house &c.[3] From the common sense point*

[1] The Presbyterian Board of Relief for Disabled Ministers, and the Widows and Orphans of Deceased Ministers, was established in 1849, and chartered in 1876, when its endowment was about $135,000. (Alfred Nevin, ed., *Encyclopedia of the Presbyterian Church in the United States of America* [Philadelphia: Presbyterian Encyclopedia Publishing Co., 1884], pp. 83–84; Edgar Sutton Robinson, *The Ministerial Directory* [Oxford, Ohio: Ministerial Directory Co., 1898], pp. 134–135). A fundraiser with William

Cattell's skill and experience was just what the board needed.
[2] Jerimiah Howard Nixon (*Presbyterian Encyclopedia*, pp. 577–578; *Pre-1956 Imprints* 420: 213).
[3] JMC had to get changes made in his father's portrait without alienating its donors (see 3.7). Also, as the Board of Relief met in Philadelphia, if William Cattell were to become secretary the Easton house would have to be sold for capital to buy a home in Philadelphia.

of view it is the best thing that could happen to me. I am in rather a queer way here & cannot tell what the end would be. If life is to be taken seriously action is the best antidote for fantastic speculations & brooding world-weariness.

4.2 To Parents, aboard the S.S. *Aurania* off Queenstown, 18 May 1884

(Three paragraphs of gossip, including Cattell's list of his fellow passengers, have been omitted.)

I was glad to receive your letters. I hope, Mama, that by this time you are quite well, and will really enjoy your "cure" at Wildungen.[1] Then you must live out of doors all Summer. You too Papa must be especially careful of yourself, since there is a chance of your going home in the Fall. I wish you had to stay in Wildung until October—quite without "outings". I will be extremely anxious to hear from you. The mails leave London Tues. Thurs. and Sat. at eight o'clock. P.M. You will write I suppose care of Uncle Tom.[2] The Aurania sails from New York July 9th—I might however sail on the Gallia July 2nd—I scarcely know when to engage my passage—however it does not make so very much matter as in either case I would meet you in England, and we could stay together as long as you thought best. If you want to leave England before the middle of July, you had better let me know, as I can then sail on the Gallia—the trouble is I must engage my passage right away, before I know what I will have to do in America or what your plans for the summer are.

I will probably get to Easton Tuesday—possibly Monday, if we get into N.Y. Sunday or I decide to go to Easton immediately and return to N.Y. to see Dr. Shearer, Mr. Wood &c.[3] I do not much favor the telegram but will send it Tues. or Wed. to Brown, Shipley & Co. London,[4] you can instruct them as to what they shall do with it.

Let Harry use his discretion about engaging the rooms. He had better take them however by the month not for the winter.

[1] A town near Kassell, with "mineral spring, which contain iron and carbonic acid gas, and are beneficial in cases of bowel-complaints, diseases of the bladder, etc." (Baedeker, *Northern Germany*, pp. 310–311).
[2] Thomas Ware Cattell, William Cattell's brother.
[3] George Lewis Shearer (*Who Was Who in America*, 1897–1942, p. 1111) was a graduate of Lafayette College and the Princeton Theological Seminary who from 1872 through 1902 was corresponding secretary of the American Tract Society in New York. He was one of the Lafayette alumni who had donated the portrait of William Cattell to the college (see 3.7). "Mr. Wood" is probably George C. Wood, a New York investment banker with whom the Cattells regularly did business (see William Cattell papers).
[4] A London bank with close commercial and personal ties with Brown Brothers and Company, a Philadelphia bank (John Crosby Brown, *A Hundred Years of Merchant Banking: A History of Brown Brothers and Company, Brown Shipley and Company, and Allied Firms* (New York: privately printed, 1909).

4.3 Journal, aboard the S.S. *Aurania* at sea, 22 May 1884

I left Leipzig Monday the 12<u>th</u> at nine in the evening and arrived at Rotterdam at one next afternoon having stopped a couple of hours at Utrecht. The steamer sailed at six & I arrived next morning at seven in London. I illustrated my wisdom & self-restraint by taking 12 grs. hasheesh, which completely intoxicated me and the effects (not aftereffects but pure intoxication, characteristic & pleasant) of which I could feel for five perhaps seven days. All day Wednesday I was quite drunk, but seemed to act soberly enough, drew money, engaged my passage &c. Thursday evening I met Aunt Nemie & Lizzie who crossed from Paris. We saw G & S's Princess Ida & went to Streatham to see Helen. Friday at midnight I took the train for Liverpool & we sailed at noon. The Aurania is a new Cunard S.S. (this is its third voyage) well fitted up & with powerful engines. The sea has been rather rough, but I have suffered little from sea-sickness, not having missed a meal.

4.4 Cable to William Cattell, New York, 28 May 1884

Cattell Brownship

Ldn

Come

4.5 To Parents, Baltimore, 31 May 1884

This is not the first letter I have written you from Baltimore. I came here from Wilmington last evening, having gone from Philadelphia to Wilmington in the morning. I had a long and as you may suppose pleasant talk with Dr. Nixon. He was doubtless somewhat disappointed, but he would have been still more so, if he had been defeated in the election. Dr. Dripps, he says, was going to support him.[1] He will write to Dr. Dripps (at my suggestion) withdrawing in your favor. This will surely be pleasanter for him, than to be an unsuccessful candidate. I enclose a letter he wrote you. He sails July 14th direct to Belfast.[2]

I am sorry to say Prof. Hall is not here. He is lecturing at Williams College. I have just written him, and must see him if possible. He has been made full professor here, Prof. Morris was very anxious to secure this appointment. I have seen Thomas and several other personal friends this morning. I met Pres. Gilman; he tried

[1] Joseph Frederick Dripps was a Presbyterian minister, a graduate of the Princeton Theological Seminary, and the author of several inspirational books (see *Presbyterian Encyclopedia*, p. 198; *Ministerial Directory*, pp. 243–244; *Pre-1956 Imprints* 149: 165).

[2] Later in the summer of 1884, both William Cattell and J. Howard Nixon attended the meeting of the Presbyterian Alliance in Belfast as delegates of the Presbyterian Church in America (Biographical Sketch of William C. Cattell, p. 13).

to be very cordial.[3] He is not popular, or at all events has many enemies.[4] I have a good many friends and acquaintances here—more I think than any place else in the world. I expect to go out to the cricket and tennis grounds this afternoon. I must call on some of the professors. Mr. Garrett is at Long Branch.[5]

I expect to return to Easton Tuesday, and hope to find letters there from you

[3]Gilman himself "at once noted down" his conversation with Cattell. In doing so, he quoted Cattell asking about the "effect of returning here last year and if I had it would have not been very pleasant to me or to you either—We don't act from high motives.— Interrupted him, saying *what do you mean.* He said I will only speak for myself. I don't act from high motives. There the conversation broke off. I had previously told him that if he wished to say anything, say it in writing, and I would give him an official answer.—he said he did not want an official answer—he wanted my personal opinion as to what he had better do.—He said he should probably apply again for a Fellowship in a future year." (Gilman papers.)

[4]Apart from Garrett, whose break with Gilman (see 2.61) did not necessarily make them enemies, it is not known if anyone connected with Johns Hopkins was especially dissatisfied with Gilman.

[5]"One of the most popular watering-places in the United States," on the Atlantic Coast of New Jersey (Baedeker, *United States,* pp. 223–224).

4.6 To Parents, Easton, 4 June 1884

(Two paragraphs of gossip have been omitted.)

It seems I must always write to you after bed-time, having gotten up at five in the morning, with the same prospect for the next day. I have really been extremely busy ever since I landed, and have seen so many people and talked so much, that perhaps I won't know how to go back to a philosopher's life. I had a pleasant enough time at Baltimore. I have plenty of right good friends there and had satisfactory talks with Profs. Martin, Gildersleeve and Morris (latin). With Pres. Gilman however I had a less satisfactory interview. I suggested that if I did good work in Germany they should make me fellow by courtesy[1]—this however will scarcely be done while Gilman is there.

I spent last night at Lincoln.[2] The university held its commencement yesterday and there was a special free train run from Baltimore. The exercises were interesting— Uncle Alec made a speech—but of course I went to see Uncle Tom.[3]

I saw Dr. Poor in Philadelphia this morning.[4] He was Dr. Nixon's chief friend in the Relief matter, but says he would rather have you than Dr. N. or any one else.

[1]See 4.5.

[2]Lincoln University, in Oxford, Pennsylvania, was a school for Freedmen established in 1857 (as the Ashmun Institute) by the Presbyterian Church (*Presbyterian Encyclopedia,* pp. 427–429).

[3]On Alexander Gilmore Cattell see 1.14; on Thomas Ware Cattell see 4.2.

[4]Daniel Warren Poor had since 1876 been corresponding secretary of the Presbyterian Board of Education (*Presbyterian Encyclopedia,* p. 626; *Pre-1956 Imprints* 465: 252).

The election takes place next Tuesday. I send you Dr. Hale's last report, with the names of the members of the board.[5] I saw Dr. Knox this evening[6]—there is I think no doubt as to the election.

I go to New York tomorrow to complete matters as to the portrait. I want also to see Mr. Wood off, so that he can bring you the latest news from me.[7] Tomorrow evening or Friday morning I will go to Phila. and will try to see Dr. McFetridge.[8] Friday night I intend to start for Columbus. I received a really nice letter from Mr. Scott, and have reason to hope that the end of that affair will be better than the Beginning.[9] From Worthington I will probably go direct to Williamstown, where Dr. Hall writes he will be until the 16th. I am the more anxious to see him since he has been made professor at J.H.U.

[5]George Hale had since 1869 served as the first secretary of the Presbyterian Board of Relief (*Presbyterian Encyclopedia*, pp. 288–289; *Pre-1956 Imprints* 226: 620).
[6]Charles Eugene Knox had since 1873 served as president of the German Theological School in Newark, New Jersey (*Presbyterian Encyclopedia*, p. 404; *Ministerial Directory*, p. 348; *Pre-1956 Imprints* 300: 647).

[7]"Mr. Wood" is George C. Wood (see 4.2).
[8]Nathaniel S. McFetridge, pastor of the Presbyterian Church in Germantown, Pennsylvania, and a trustee of Lafayette College (Stonecipher, *Biographical Catalog of Lafayette College*, p. 140).
[9]"That affair" refers to the matter of the William Cattell portrait (see 3.7).

4.7 To Parents, Amherst, Massachusetts, 14 June 1884

(Two paragraphs of gossip have been omitted.)

You are perhaps surprised at the above reading. Indeed it seems to me I have been everywhere—during the past two weeks I have not spent two nights in the same place—my wanderings have been—Baltimore, Lincoln, Easton, Woodbury, cars, Worthington, cars, Niagara, Geneva, Cornell, cars, Williamstown, Amherst. As much for the nights, but I have also been in N.Y. Phila. Columbus, Buffalo, Albany &c. I left Worthington at two o'clock Monday morning and got to Niagara late in the afternoon. I was delighted far more than I had hoped. I wanted to stay a week, but left Tuesday afternoon and slept at Geneva. Next morning I went to Cornell. I wanted to see Rolfe (who is one of my best friends)[1]—as also the university. The position is as you know, Papa, wonderfully beautiful. If I take a position in a college and cannot be with you I think I would prefer Cornell to any other place. There is an opening too, as they will soon call an assistant to Prof. Wilson, who is old and not a very good man.[2] They say he reads his lectures on the History of Philosophy, backward the next year for the Philosophy of History. Thursday night, I spent on the cars

[1]Henry Winchester Rolfe (see 1.19).
[2]William Dexter Wilson (*DAB* 20: 349–350) was an Episcopal minister who taught Scottish realist philosophy at Cornell from 1868 to 1886 as the only member of the philosophy department. In 1882, a faculty colleague of Wilson's described him as a "bad, cunning, envious, malignant, plotting old man," and even an alumni historian of Cornell had to note that "weakest, perhaps, of Cornell's departments was Philosophy." (Morris Bishop, *Early Cornell* [Ithaca, N.Y.: Cornell University Press, 1962].)

arriving at Williamstown in the morning. I had several very pleasant talks with Dr. Hall. He was very complimentary and I imagine would make me assistant at Johns Hopkins if it were not for Pres. Gilman. Dewey by the way has been dropped, as Dr. Hall has now the selection of the fellow. When we came to talk of that paper, we had some trouble. Dr. Hall wanted to print it in a way not satisfactory to me, and so I asked him to give me back my papers & notes.[3] I do not know what the end of the affair will be. At all events I have done more work worth printing during the past fifteen months than Dr. Hall and the whole psycho-physic laboratory.

I could not get home (?) for Sun so came here to Amherst (it is on the way to N.Y.) yesterday.[4] I intended to see the university, but my chief reason for coming was that the Sturdivants are here. They have changed but little since we left them at Digby, and I have had a pleasant day.[5] Cornell, Williamstown and Amherst all have beautiful situations—coming second to Lafayette only.

[3] Hall's account of his meeting with JMC in a letter to Gilman (14 June 1884, Gilman papers) was more graphic: "Cattell just left me. He came all smiles and amiability and tempted me to introduce him here and give him introductory cards elsewhere. . . . Suddenly begun to talk with most insulting way and almost charged me with lying when on the spot without even a <u>shadow</u> of either basis or occasion. I do not know that I have ever in my life been so angry at a human being. He has been visiting many colleges, and I imagine from what he said—perhaps wrongly—that he has repeated his general vituperation of the University at B. and its officers which he managed to get in bits to me under my protest. I am convinced that he and his father are one in all this. He could hardly have had the wit or the boldness himself."

[4] The question mark seems to convey JMC's doubt that Easton was still his home.

[5] On Digby see 1.22. The Sturdivants were an otherwise unidentified Boston family whom the Cattells had met in 1882. Since September of that year JMC had been exchanging letters with Jessie Sturdivant, the youngest daughter of the family.

4.8 Journal, Easton, 22 June 1884

(Cattell's itinerary and list of his fellow passengers have been omitted.)

I have done fairly well every thing I have tried. Father has been unanimously elected to the Board of Relief. The portrait complication has been well settled. I have been of great service to father anent college affairs, expecially the Drown-King matter.[1] I have looked after house, my papers with Dr. Hall &c &c. I have seen most of the family & at least half the people in America I know well. I am utterly weary of hurrying & worrying, talking & persuading, but am, I dare say, at a more

[1] Thomas Messinger Drown (*DAB* 5: 460–461), a chemist with European training, had been professor of chemistry at Lafayette (1874–1881) before going on to a similar position at Massachusetts Institute of Technology. He eventually returned to northeastern Pennsylvania as president of Lehigh University. Despite (or because of) his reputation, his relations with his colleagues and William Cattell were not at all cordial, though he had several friends among Lafayette's trustees and faculty who regularly tried to arrange for his reappointment there. One of these was David Bennett King (*Who Was Who in America*, 1961–1968, p. 529), a professor of Latin at Lafayette until 1886, when he resigned as a result of his conflict with other colleagues (see Skillman, *Biography of a College*, vol. 2, pp. 45–57).

normal level than when I left Leipzig. But my mind is not really sane and healthy. I am tossed hither and thither by every wave of feeling & impulse. Things are somehow unreal & queer. I am as a man dreaming, a tree walking.

4.9 Journal, aboard the S.S. *Gallia* at sea, 5 July 1884

I attended commencement, making myself agreeable albeit very tired of the business. Every thing was delightful for father, and some of this must be charged on the credit side of my life ledger. I sold our house to Mr. Blair for the college ($16000).[1] I really kissed the wall when I left it Thursday evening.[2] I felt myself without a home for the first time in my life. Friday I went to Phila. and was entertained by Dr. Smith I have had my upper wisdom teeth pulled out. I stayed at Woodbury until Tuesday, sleeping Sat. night however with Brown in Phila. Tues. I went to N.Y. visiting Mackenzie at Lawrenceville.[3] I sailed Wed at one, and here I am. My five weeks in America were certainly a success—quite a surprising and startling success. I seem to be able to make acquaintances and friends and to do what I try. I suppose I ought to be proud of what I have done, but on the whole am rather disgusted with myself. My only joy is in the fact that I scarcely faltered in doing what I had resolved must be done. It drives me half mad to feel that after while I shall come to the conclusion that on the whole I had better settle down and make myself comfortable.

[1] John Insley Blair (*DAB* 2: 338–339) was a railroad investor who became a benefactor of Princeton University, Lafayette and Grinnell Colleges, and the Presbyterian Church. Despite Blair's gift, in 1926 a "Map Showing Real Estate Accessions to Lafayette College" listed the President's House as a grant from William Cattell. (Skillman, *Biography of a College*, vol. 1, insert.)
[2] JMC had lived in this house from ca. 1865 until he left for Europe in July 1880 (see Skillman, *Biography of a College*, vol. 1, pp. 298–299, which includes a photograph of the house).
[3] James Cameron MacKenzie (*DAB* 12: 93) was a Scottish-born Presbyterian minister who had graduated from Lafayette (1878) and who had, in 1882, established a preparatory school in Lawrenceville, N.J.

4.10 Journal, aboard the S.S. *Gallia* at sea, 11 July 1884

(Most of Cattell's list of his fellow passengers has been omitted.)

We have had an ugly passage for July, but at last are nearing Liverpool. I have not been seasick, as far as its most notorious symptom goes, nor have I missed a meal, but was quite miserable for two days, and have not been quite comfortable most of the time. The accomodations of the steamer are poor compared with the Aurania and Servia—the table has been especially bad. In spite of all this I have rather enjoyed the voyage. I am relieved at putting off all responsibility, whereas this very responsibility and activity has driven away to some extent introspective fantasies. Then there are very nice people on board. I got quite well acquainted with

Chauncey M. Depew, his wife, and her mother Mrs. Wm. Hegeman.[1] *Yesterday I had a long and pleasant conversation with Christine Nilsson.*[2] *I have talked a good deals with John S. Billings, who has offered to do several things which would be of great service to me.*[3] *I have also talked with Geo. Jones owner of the N.Y. Times and Prof Wm. Darling.*[4] *Among people not as distinguised I have made more acquaintances than usual.* [Here Cattell listed nine other individuals and families.] *I have shown more talent than might have been looked for in getting acquainted with the right people. Well, I shall be at Liverpool in a few hours, but am not yet nearly done with the longer voyage I have on hand. I think I have better control than ever before of internal and external conditions, but it seems to me scarcely worth the while to sail anywhere. It makes but little matter whether the ship sinks at the equator or the pole. Still I might as well try to do two things in the next six weeks, be of service to father and mother, and collect a maximum of health and strength.*

[1] Chauncey Mitchell Depew (*DAB* 5: 244–247) had served in the senate and as first U.S. minister to Japan. In 1884 he was officially connected with the New York Central Railroad.

[2] Christine Nilsson, the operatic soprano referred to in 2.9. Instead of "conversation," JMC had first written "acquaintance."

[3] John Shaw Billings, in 1884 Deputy Surgeon General of the United States and curator of the Army Medical Museum, was an important figure in the establishment of the Johns Hopkins Hospital and in the reform of the University of Pennsylvania Hospital (*DAB* 2: 266–269; Fielding H. Garrison, *John Shaw Billings: A Memoir* [New York: Putnam, 1915]). See also 5.46.

[4] For further information on George Jones see *DAB* 10: 171. "Prof Wm. Darling" may be William Lafayette Darling (*Who Was Who in America, 1897–1942*, p. 195).

James Cattell was pleased to return to his parents in England. He traveled with them through the lake district and discussed with them his plans for the coming year. In London he visited the International Health Exhibition at the South Kensington Museum (Journal, 10 October 1884). There he may have met Francis Galton, who was sponsoring an anthropometric laboratory as part of the exhibition (see Galton, "The Anthropometric Laboratory," *Fortnightly Review*, new series, 31 [1882]: 332–338; "On the Anthropometric Laboratory at the Late International Health Exhibition," *Journal of the Anthropological Institute* 14 [1885]: 205–221). This possibility is suggested by the fact that Cattell had just met John Shaw Billings, who was well acquainted with Galton (see Garrison, *John Shaw Billings: A Memoir*, pp. 248–250). By the middle of August Cattell had reached the Continent. He had planned to return almost immediately to Leipzig, but almost as soon as he left England his mother began writing to him with advice to the contrary.

4.11 From Elizabeth Cattell, London, 8 August 1884

(Several paragraphs of gossip have been omitted.)

We are delighted that you are so much better, but do not run any risk by returning to Leipzig too soon, and when you do return not overwork, but work at the very farthest not more than six hours a day, have Berger help you in every way, is the earnest wish of your Papa, and myself, and give him in money whatever you think will be right, and I would pay him well, as he will be more willing, and ready to help you, and you will be helping a deserving young man along in the world, but we do beg of you to take good care of yourself.

4.12 From Elizabeth Cattell, Ullswater, 12 August 1884

(Several paragraphs of gossip have been omitted.)

Dear Jim I remember you once asked me about signing your name. As we are going to Philadelphia where there are a number of Cattells, I should always write my name J McKeen Cattell, or James McKeen Cattell, which ever you fancy, but I should take one, or the other.

4.13 To Parents, Wiesbaden, 13 August 1884

(Two paragraphs of gossip have been omitted.)

I am very sorry to write what I am about to, and would not if I could find a pretext for what I am going to do. I have been quite sick for the past three or four days and this has led me to think I ought not to go back to Leipzig immediately. A couple of days sickness is of course in itself nothing, but it convinces me that I am not as well as I ought to be. I must write frankly if at all, and so must say that I have not been in good health all year. Of course I do not know what has been the matter with me, still less what caused it. It may possibly have been Arndtstr. drainage; but the fact most likely is that my constitution is not very strong and I at times overstrain it. I undoubtedly have a talent for worrying about things, that other people have just as much reason to worry about as I, but somehow they don't. The voyage to America seemed to do me good, but I was not at all well while there, and found it hard to keep some of the engagements I had made. The return voyage did not seem to help me much, but I felt better after the weeks at the Lakes. At one time I had decided to stay with you until September, but I was very anxious to get to work; and as my health seemed to be improving and as it would be ten days before I should begin work. I hoped it would not hurt me to do so, and of course intended to take good care of myself. But I have not been well since I left you, and Saturday at Cologne was taken quite sick. I suffered a good deal Sun. and Mon. but came on here by the boat. Mon. night I was in rather a bad way—was out of my mind for a while I think—but I slept well last night and feel well enough today. My present indisposi-

tion is a small matter—a cold and headache—and is I imagine about over, but I cannot hide from myself that it is scarcely safe for me to go directly to Leipzig and begin work. I am sorry now that I did not go with you to Scotland—however the best thing I can do probably is to walk for a couple of weeks. When I left Leipzig somewhat over two years ago, I felt about as I do now[1] and the three weeks I spent walking in Switzerland seemed to do me a great deal of good. I remember how I astonished Zinkeisen. We spent two weeks together at Florence and I ate scarcely anything; then we met again after I had been walking several weeks, and I had quite as hearty an appetite as he.[2]

I thought something of going to Switzerland now (I should like to see Harry,[3] but shall go to the Tyrol, as it is not so far away and is also nearer Leipzig. I do not know where I shall walk. I shall not of course climb high or dangerous mountains, nor overtire myself.

I trust you will not worry about what I have written—there is indeed no reason to. I am as well now as I was six months ago, or two and a half years ago, or four and a half years ago, or ten years ago. I can scarcely hope ever to be in perfect health, but after all this is no great matter. Those who are in perfect health do not get the slightest pleasure from it, and a little pain does no great harm. I want health because it will enable me to do things I try to do—but everything is relative and if I succeeded in one thing, I should try something I could not do.

I cannot help being sadly worried about my work. I do not like to give up or postpone things, and unless Dr. Hall and Prof. Wundt exaggerate my work is of importance. Dr. Hall even when provoked at me, told me my work is the best that has been done in England or America, and Prof. Wundt treated me with more respect than any one else in his laboratory—gave me the keys of his private apparatus, and told me he would admit any one to the laboratory I would recommend, though he had rejected a number of applicants. I am in no special hurry about my Ph.D. but should have credit for the original work I have done. You see some one may independently reach the same results as I, or some one may use my work. Dr Hall has not shown himself especially delicate in this matter.

Still it is better to look on the brightest side. A month now is not worth as much as a month in the Fall or Winter. It is too hot to work to the best success even were I well. There are no lectures. The library will I suppose be closed, and Prof. Wundt and Berger away. Then the time will not be lost. The Tyrol is well worth seeing. I shall be talking more German than at Leipzig. I must rest about four hours in the middle of the day, and will try to learn short-hand thoroughly. Yet I confess it is very hard for me to turn my face away from Leipzig, and I feel now like burning this letter, and following my original plans.

[1] The time of JMC's tour through Switzerland in 1882 was a period of great depression for him (see 1.19).
[2] See 1.19.

[3] Henry Ware Cattell was visiting several friends in Switzerland, including the Dardier family (see 1.12).

James Cattell did take his walking tour of the Tyrol, and it improved his health greatly (JMC to parents, 16, 20, and 24 August 1884). But before his parents received news of his recovery, they got his distressing letter (4.13). His father wrote immediately of their concern, and included several notes about inquiries he had been making on James's behalf.

4.14 From William Cattell, Blair Athole, Scotland, 20 August 1884

(This letter was dated "Wednesday 20th." Four paragraphs of gossip have been omitted.)

Yr letter reached us Sunday morning & your mother replied the same day.[1] How much we have both regretted you did not remain with us! Had we known as much as we do now about your general health, we should have urged you to stay & be with us in the highlands. We were only reconciled to parting from you because we knew you wanted to get to work at Leipzig: it is evident you are not in a condition to resume yr studies & you did wisely in going to the Tyrol and we beg you, dear son, do not go to Leipzic till you are thorghly well—and not before the hot weather is over. We have wondered whether it was not possible for you still to join us somewhere & be with us till we sail—this wd give us joy but we have thought you could best bring up yr strength by remaining in the Tyrol. If however you do not regain yr health satisfactorily by the trial—be frank with us—dont run any risks in order to hurry on with yr work—let us meet somewhere & talk over the situation— We are of course very anxious. As your dear mother read the first part of yr letter she exclaimed "I will go to him!"—And many times have we talked—& still often-er thought—about you & what was best to be done. Tho' in these days sickness is not much of a matter of concern—but what you say about your general health is what alarms us: —And we are unspeakably sad at the thought of leaving you on this side of the Atlantic—knowing that you are not well. We have even thought it might strike you as wise to take a long holiday—to go home with us, and spend the winter quietly doing such work as you may feel able to do—bringing with you yr apparatus. Savitz (or some other good man) can do the clerkly work for you[2]—& so if you can go on with yr researches, you will make good progress & can return to L. for your Ph.D.—yr presence will be a great comfort to us—& we shall often need you, in our many new "departures". But if this does not strike you favorably, I must again & again insist upon yr sparing no expense where your health is concerned—You are very precious to us, dear son;—once more let me beg you not to run the risk of a break-down in a land of strangers. Be very careful of yrself—if in God's Providence

[1] Elizabeth Cattell's letter (not included here) was a general expression of concern, without the detail of this letter.

[2] Charles Jennings Savitz had been William Cattell's private secretary from 1879 through 1883 (*Biographical Catalogue of Lafayette College*, p. 236).

sickness overtakes you, secure at once the very best medical attendance & the very best nursing—no matter what it costs. Don't overwork, dont be in a hurry to take yr degree.

We came to this little place yesterday with Mr. Wood & family—yr mother stood the ride splendidly in fact is beginning to be her old self—and I continue very well. If you were with us, it seems to me we shd enjoy every hour. At Edinboro there were no less than 13 American friends of mine at one time in our Hotel! It seemed almost like being at home. One of these was Dr. MacCracken—ex-Pres. of the University of Western Penn^a & recently elected Professor of Metaphysics &c in the University of the City of N.Y.[3] He hopes to have a <u>Department</u> of Philos^y there as Dr. McCosh has at Princeton:[4] he is very much interested in yr special work—& so is Dr. Hall[5]—the Chancellor and he hoped there w^d be an opening in the University for you. Of course I talked the matter up with him to the very best of my ability. I do not cease to hope you may live in Philad^a; but it is well to cast other anchors to the windward. Dr. McC. knows Dr Hall very well. He regrets that he himself is not 10 years younger that he might take up these studies in physiological psychology!

[3] Henry Mitchell MacCracken (*DAB* 11: 619–620).
[4] James McCosh was a Scottish-born Presbyterian minister, educated at both Glasgow and Edinburgh in the Scottish realist philosophy, who served as president of Princeton College from 1868 through 1888. Despite his traditional background (he was professor of logic and metaphysics at Queen's College, Belfast, for 16 years before coming to America), McCosh was a supporter of a natural theological, if not Darwinian, view of evolution. By 1882 he had helped to organize a "Wundt Club" at Princeton. (*DAB* 11: 615–

617; William Milligan Stone, ed., *The Life of James McCosh: A Record Chiefly Autobiographical* [New York: Scribner, 1896]; James Mark Baldwin, *Between Two Wars, 1861–1921; Being Memories, Opinions and Letters Received* [Boston: Stratford, 1926], vol. 2, pp. 199–200; T. W. Moody and J. C. Beckett, *Queen's, Belfast, 1845–1949: The History of a University* [London: Faber & Faber, 1959], vol. 1, p. 171; see also 5.56 and 6.171.)
[5] John Hall (*DAB* 8: 137–138) was chancellor of what is now New York University from 1881 through 1891.

4.15 To Parents, Unsere liebe Frau, Austria, 30 August 1884

(Four paragraphs of gossip have been omitted.)

I am sorry I gave you some worry about me, but I had to explain my change of plans. My health should not give you any uneasiness. I am probably in better health than most people who go about their dailey work, and dare say I could have gone back to Leipzig as I had intended, and worked all winter without hurting myself. Still it was wiser to come here. I was, especially when I left Leipzig, somewhat run-down, and it is safer to run no risk. In modern life we seem at times to use up the socalled "vital energy" faster than we create it. If the damage is not too great it can be repaired, as in your case, Papa: but on the whole people would do better work if they never overdrew their account. It is surprising how much work a man can do in three hours a day. If people should confine themselves to that many hours of mental work, and should spend five or six hours in physical work, the human race would be in a much better way five hundred years hence, than now seems likely, and each individual would now be happier. Darwin only worked three hours a day, and his work is not only of high quality, but also of great quantity. But I am theorizing at too great length.[1]

To come back to myself, my plans have not really been changed since I left you, nor has anything new occurred. I have only decided to take a longer vacation, and hope to make up the time by faster work next winter. I have never been in an especial hurry to take my Ph.D. Germans take the degree at an average age of perhaps twenty-six, and Americans are usually over thirty. American students graduate from college at an average of twenty-three or four, they must then spend three to four years studying a profession: and medical and law students can scarcely hope to make a thousand dollars a year until they have been practicing five years. I do not see why a teacher does not need as thorough a preparation as a lawyer or physician, and those who want to do original work must cover a very large ground in their studies. It is true the lawyer's or physician's profession is the better financial speculation—they may have to wait until they are thirty before they can earn anything, but at fifty can hope to make $10000 a year; still this does not make it any the less desirable for the teacher to be as thoroughly prepared as the lawyer or physician. Of course I know that a part of a teachers training is teaching, and that in teaching the teacher learns, but a professor who year after year teaches the same elements to successive classes of college students, settles down into a man like Prof. Youngman or Moore.[2] A very admirable man perhaps, but not an original thinker. To come back once more to myself. I am perfectly willing to try any position you, Papa, may find for me, and indeed would be very glad to make sure of a desirable one, such as at the Univ. of Penna. but if such a position does not turn up, I shall not worry about it for several years to come. The Ph.D. has always appeared to me a minor matter—but its most fitting place is at the conclusion of my studies. If you secure me a position, you wish me to take, upon condition that I make the degree before entering upon it, you may be sure I can do it next Aug.; and indeed any how if I take a position in America next fall, I shall probably pass the examination before returning home: though I could of course return the following summer and take it then. I am quite sure I can present a thesis that would be accepted, thought it would be nothing like as good, as I could prepare, if I finish the work I have planned. I could also next summer pass the examination in philosophy, though I would have to study a good deal—I know very little logic—for example. The "Nebenfäche" physics and biology or mathematics would be more doubtful,[3] as I would not have time to

---

[1] Again JMC's statements about overwork and the importance of physical activity are similar to those of S. Weir Mitchell. See 3.3 and Mitchell, *Fat and Blood: An Essay on the Treatment of Certain Forms of Neurasthenia and Hysteria* (Philadelphia: Lippincott, 1877).

[2] Robert Berber Youngman was a graduate of Lafayette (1860) who taught Latin and Greek there from his graduation through 1909 and served as clerk of the faculty throughout that time (Skillman, *Biography of a College*, vol. 2, pp. 178–180). "Moore" is James W. Moore (see 1.3).

[3] *Nebenfäche:* subsidiary subjects. A *Nebenfach* is equivalent to a required minor in twentieth-century American universities.

study them up, but it would be no great matter if I should fail in one of these (half the German students do)—the other examinations do not need to be repeated, and I could soon "cram" the one on which I had failed.[4] Still I consider it better not to try the examination next August, unless I am offerred a position in America.

The Ph.D. troubles me, as I have said, very little—the experimental work I have on hand is a larger factor in my plans. I sometimes regret that I began original research so soon, but any how it would be no great matter if I should give up what I have done. I have only spent on it about half my time for six months, and, disregarding the results, the training has repaid me for my time. Still I have results which Dr. Hall and Prof. Wundt consider important, and if I could complete and publish these, it would give me a certain standing, and make it easier for you to secure me that position. Therefore this Winter I shall try so far to complete the work, that I can print it in some good Magazine—Mind if possible—then I shall be given credit for what I have done, and need not worry about the work for the present. I should like to do this by the end of the winter semester, and think I can, especially if I can get Berger to help me. I should not however want to ask him to do it, unless I could offer to pay him well.

[4]Special schools known as *Presse* were established in Germany during this period to facilitate such cramming (see Paulsen, *German Universities*, p. 333).

4.16 From William Cattell, Glasgow, 7 September 1884

(One paragraph of gossip has been omitted.)

Yr mother has written to you today:[1] & I only add this supplement to urge what she said about yr employing Berger to help you in yr work. In one of my letters to you[2]—referring to the possibility of yr returning home & continuing yr studies—I suggested that Savitz, or some good man, could be got to aid you: —but of course Berger w^d be better. There must be many matters of detail which a careful man could finish & thus save you both time & labor; —which you c^d devote to something else—not forgetting those 5 or 6 hours of physical exercise wh., in yr last letter, you say w^d be a good thing for every one! Of course I do not mean for you to employ B. in those joint investigations & researches which w^d make it necessary for him to be associated with you in the "authorship"—I assume, of course, that he w^d simply carry our yr lines of investigations—doing the clerical work—for which he sh^d receive ample compensation.[3]

[1]Elizabeth Cattell's letter (not included here) was quite similar in tone to this one.
[2]See 4.14.
[3]Compare 3.12, especially nn. 12 and 18. The experiments discussed in this document formed the basis of both Berger's dissertation ("Über den Einfluss der Reizstärke auf die Dauer einfacher psychischer Vorgänge mit besonderer Rücksicht auf Lichtreize," *Philosophische Studien* 3 [1886]: 38–94) and an article by JMC ("The Influence of the Intensity of the Stimulus on the Length of the Reaction Time," *Brain* 8 [1886]: 512). Each author cited the other's help and encouragement, but in neither case was authorship shared.

I am very anxious that your work shd reach one of its natural fixed points—Complete qurad hoc[4]—and be given to the public in some Journal, such as Mind—& especially if Wundt would make some public reference to its value, such a publication wd be of great aid in securing you a position: but the "authorities" of our Universities & Colleges know too little about the subject to judge of such a paper—but they know Wundt—& in just his imprimatur w^d induce them to read, (& perhaps carefully) what you print. Such a position as you want is rare with us: You may be sure there will be many applicants—and some valuable original work from you w^d be a great "lever": —so that I sh^d regard five hundred dollars well spent that w^d secure from Berger such aid as will relieve you of yeomans work—& enable you, without injury to yr health to complete & print something valuable this winter in physicological psychology, the only delicate matter in the arrangement is that of the joint authorship—but you ought to be able to manage this: & you sh^d in some way connect this work at Leipzic with what you did at Baltimore—showing it to be a continuation &c[5]—I have written hurriedly but you will gather my meaning.

[4]*Complete qurad hoc:* as far as a particular point is concerned.
[5]JMC followed this advice in at least three articles, with sentences such as "These experiments, though begun in America, have been carried out in the psychological laboratory of the University of Leipzig" (see for example "The Inertia of the Eye and Brain," *Brain* 8 [1885]: 295–312).

4.17 To Parents, Prague, 26 September 1884

(Two paragraphs of gossip have been omitted.)

Yesterday morning I had rather an amusing adventure—though you Papa would have found it decidedly annoying. Before I was up in the morning a police officer came to examine my pass. Of course I had none, and so was taken to the police department, where I was examined in a most absurd way.[1] Every thing was taken away from me and three lists made, a pen-knife, 2 pair socks &c. All my letters &c were read—the one from Berger described my Kronoscope which they seemed to consider a very suspicious piece of apparatus. The prefect, or whoever he was, told me I might have to be kept in custody a week until they could find whether my story was true. Then he telegraphed or at least said he had to Leipzig. In the mean while I made them send to the American Embassy, the answer was they did not know me personally—which the prefect considered a very suspicious circumstance—but advised that I be let go, so I was—after having been kept there four hours. I kept in a fairly good humor, but gave them all plenty of plain advice. They had, I imagine, no right to detain me after I proved that I was a citizen of the U.S. I was going to make complaint at our embassy, but as I just had time to catch the train in which I wanted to come to Prague, I let the matter go.

[1]Baedeker (*Southern Germany*, p. xiv) assured the traveler that "passports are unnecessary in Austria" and that "custom house formali ties are now almost everywhere lenient."

5

Experimental Achievement: Leipzig, October 1884–September 1886

After the incident at the Austrian border (see 4.17), James McKeen Cattell was glad to return to Leipzig. In many ways it was like a homecoming for him—especially after he had sold the house in which he was raised—and he rejoiced at returning to a place he knew well.

5.1 To Parents, Leipzig, 1 October 1884

Here I am, not only at Leipzig, but sitting under my own vine and fig-tree.[1] The first thing I do in my new rooms shall be to write to you. You will get many letters written on this desk, but you will never know how often I think of you as I sit here or walk to and fro.

I arrived here Monday as I had proposed. I spent Friday at Prage, Sunday at Dresden, and Saturday between the two places—the ride down the Elbe through "Sächsische Schweiz" was very pretty.[2]

On arriving here I went first to the bankers, and was delighted to get your long letter, Papa.[3] I have not yet received a letter from you, Mama. There is perhaps one at the bankers—I was there at noon, and found the doors locked. The things I

[1] 1 Kings 4:25: "And Judah and Israel dwelt safely, every man under his vine and his fig tree."
[2] "Saxon Switzerland," a section of southern Saxony bordering on what is now Czecho-slovakia, starred by Baedeker (*Northern Germany*, pp. 232–237) as "very picturesque."
[3] JMC had asked his parents (22 September 1884) to write him in care of Frege and Company.

sent from Wiesbaden, as also the writing-machine, had come safely to hand.[4] From the bankers I went to the Hotel de Prusse where I spent two nights.[5]

I occupied myself in looking for rooms; as I had expected there was an embarras de richesse. I consider myself however very fortunate in the rooms I have taken. They are at the opposite side of the town from Arndtstr, a stone's throw from the Rosenthal.[6] They are not too far from the centre of the town, and yet on its edge where the air is better. I enclose a plan of Leipzig so that you can see exactly where I am.[7] You see I can easily get into the Rosenthal and the country. I have also marked the Gymnasium (where the Turnverein exercises) the winter lawntennis court, and the skating pond. They are all near enough to tempt me to take exercise. The rooms themselves are very nice—better than any rooms I have ever lived in before. They are tastefully and handsomely furnished—about like the parlors in Bailey's Hotel.[8] Hardwood floor, heavy stiff curtains, handsome desk, bookcase, sofa, stove &c. Both rooms are large—the study about twenty feet square. There are two large windows in the study, facing exactly south—the bed room adjoining also faces south. The street is broad and quiet. There is no family living in the flat. The family below have two rooms, and two more are rented to a merchant. It is thus very quiet. I seem to have the flat all to myself. There is a bath room. The sanitary arrangements seem to be perfect.[9] The W. C. will be but little used, is in good order, and is across the hall from me. It has a large window, and is shut off from the hall by two tight-closing doors. The kitchen is not used, and (as also the bathroom) is not near my room. Every thing seems beautifully clean. The Hausman and his wife take care of my rooms—they seem very nice—an electric bell enables me to call them without getting up from my chair or bed. The 'Frau' brings me my breakfast, and will cook for me what I wish, or brings things from the restaurant. I am on the third floor—there is however a flat and a half above me. I pay 45M, service, heating and light being extra.

[4] JMC had sent his heavy luggage on to Leipzig from Wiesbaden, where he had decided to take his tour of the Tyrol. The typewriter was a Remington No. 4, a model made in the United States during the early 1880s. Its design was basically that of Remington's well-known Model No. 2, which had introduced the shift key and a few minor improvements in 1878. JMC's typewriter is famous in the history of psychology, as Wundt, with his poor eyesight, was supposed to have been impressed with it and to have obtained one for himself. Many of Wundt's later American students remember him hunting and pecking on an American typewriter—either the one JMC had used and left for him or perhaps a Remington No. 5, designed for the Continental market, with a wider platen and additional keys. (JMC, "The Inertia of the Eye and Brain" *Brain* 8 [1886]: 295–312; G. Tilghman Richards, *The History and Development of Typewriters* [London: The Science Museum, 1964], pp. 27–30; JMC, Charles H. Judd, and Edward M. Weyer, "In Memory of Wilhelm Wundt," *Psychological Review* 28 [1921]: 155–159, 173–178, 181–183.) See also 6.17.

[5] The Hotel de Prusse was starred by Baedeker (*Northern Germany*, p. 243).

[6] JMC's rooms were at Humboldtstrasse 19 (JMC to parents, 4 October 1884).

[7] See figure 6.

[8] An excellent hotel in the West End of London, recommended by Baedeker (*London*, p. 7) with a caveat about its expense.

[9] JMC was greatly concerned about sanitation, and blamed in part the drainage at his old lodging for his illness during the previous summer (see 4.13).

It seems half natural, half strange, being in Leipzig once more. The streets and every thing seem familiar, but I know scarcely any one here. I trust you both and Harry are well. I hate to think that you will so soon leave me over here alone.

5.2 To Parents, Leipzig, 2 October 1884

(Two paragraphs of gossip have been omitted.)

You will be surprised at receiving a letter from me again so soon, and still more so when I tell you that I am going to write to you every day this winter. I will simply write you a line before going to bed, as I might come down stairs to say good-night if I were at home. Of course I shall write very briefly, and what I say will be common-place—just as if I were home I might say: it looks as if it might rain tomorrow.

I called on Prof. Wundt yesterday—he seemed glad to see me and was very cordial. He is willing that I should work in my own room[1] and offered to help me in any way he can. The day I came I went to see Berger—they told me he would not be back for a week or two—but he turned up the next day. We went rowing together yesterday evening. We shall soon begin work, though he is busy with his examinations. Next week he must pass an examination, which entitles him to teach gymnastics in the schools. I called at the Gerhard's Tuesday evening.[2] Things were as of yore—there were however no smells. They seemed glad to see me. Tuttiet— whom Harry knows is boarding there.

[1] See 5.6 and 5.8.   [2] See 3.7.

5.3 To Parents, Leipzig, 3 October 1884

(A postscript containing gossip has been omitted.)

I began work this morning—it is a long time since I have worked regularly. I shall try to get through with my hardest work in the morning from eight to half past one. I must hear some lectures in the afternoon, but I shall only subscribe to three courses Prof. Heinze (4 times) on the History of Modern Philosophy and Prof. Wundt; Psychology (4 times) and Ethics (twice). I heard the first two courses when I was here in '81–2,[1] and may not attend regularly—I shall see. I shall be out often in the evening. I think I shall exercise with the Turnverein twice a week, and shall of course go to the theatre and opera—perhaps an average of once a week to each. The gymnasium will only take half the evening however, and the theatre is often out by nine o'clock. To operas with which I am familiar, I often go to the top gallery for twelve cents, and only stay part of the time. In the evening I shall only read, and do work that does not require much mental effort.

[1] See 1.15.

Prof. Wundt's youngest child (four years old, he has two older) died Wed. at two o'clock.[2] I was there between twelve and one. He told me his child was sick, but the worst was over. It had been taken sick while they were at the sea-shore, but now that they were home, it would soon be all right. He is you know a physician. Poor man!

It began to rain yesterday for the winter.

[2]Wundt's third child was named Teresa.

5.4 To Parents, Leipzig, 5 October 1884

It has been raining most of the time for the past four days. As the reading room is usually closed, and I have not been at the theatre, I have had plenty of time to enjoy my room. It is fortunate it is so comfortable. It is in every way better than being at the Gerhards.

I have an excellent bed—the family hair matress on a spring matress. My Haus-frau calls me at half past seven and fifteen minutes after brings me my breakfast. She seems to be a model of punctuality. I have a very good breakfast. I buy my own oat-meal, which she cooks very nicely. Then I have half a pint of good cream; and a pint of new-milk from a "Molkerei" & bread and butter. In this latter however I experience the tribulations of house-keeping. I can't buy less than a pound, and that lasts longer than it will keep fresh. I have though a refrigerator in the hall. Dinner I eat at a restaurant at about two o'clock. I can get a good dinner soup, fish, meat, potatoes, and "compot" for about 35 cts. My breakfast costs 20 cts. I have been eating very little for supper. Fruit when I am out—in my room crackers. I shall buy a Bunsen burner—there is gas in my room and then I can cook any thing I want— the Frau will wash up the mess. The gas will be convenient. I do not use it to read by, but a Bunsen burner will be useful in my work, and I can heat my room when desirable by gas—it is not much dearer than coal. There are three argan burners, and pipe for a rubber tube.[1] I have a meter to myself.

I hate to think that you will soon be so far away. Still as I cannot be with you perhaps it will be less aggravating than having you so near.

[1]The Argand burner, a lamp invented by Aime Argand around 1783, was the standard laboratory source of heat until the invention of the Bunsen burner by Robert Bunsen in 1855.

5.5 To Parents, Leipzig, 6 October 1884

(This letter was typewritten, all in capitals. One paragraph of gossip has been omitted.)

This is my first letter written on the new machine. It seems to work very nicely. I have only practiced two or three hours and can already write quite rapidly, and with tolerable accuracy. I think it is easier than writing with the hand, and I shall soon be able to write about twice as fast. It is certainly a great saving of mental and physical energy, and would be a special relief after one is wearied writing with the hand. One does not use the eyes at all, which of cours is a special advantage to me.[1] It is further a great comfort to hav what one writes so throughly legible, especially if he is preparing manuscript for the pres.

It keeps on raining. I was at the opeara this evening—Gounods Margarethe.[2] It seems like old times to be in the Leipzig theatre, but it is scarcely as good as it was last winter and not nearly as good as it was two years ago.

[1] JMC always tried not to overuse his eyes (see   [2] See 1.6 and 2.41.
5.83 and 5.103).

5.6 To Parents, Leipzig, 8 October 1884

(One long paragraph about the weather has been omitted.)

I got all our notes, tables, curves &c. from Berger today. I shall study them over carefully, and see exactly what there is there, and what had better be done.[1] Of course this is only one part of my work—the other which I began at Baltimore and worked on some last winter in my room, I consider more important, and it is certainly more original.[2] Fortunately the two fit together nicely. In both cases I determine the time required by simple mental processes—how long it takes us to see, hear or feel something—to understand, to will, to think. You may not consider this so very interesting or important. But if we wish to describe the world—which is the end of science—surely an accurate knowledge of our own mind is more important than any thing else. Why should quite a stir be made over the matter when a new asteroid is discovered—what possible difference can it make to us whether there be one more or less? Why should men spend their lives making dictionaries of a language, that is only a hypothesis? What is the use of catalogueing every bug? You can just as easily explain to a man why music is pleasant, as why knowledge is good. But if one thinks that knowledge for its own sake is worth the pursuit, then surely a knowledge of mind is the best of all. Not only is the mind of man of infinitely more worth and importance than any thing else, but on its nature the whole world depends.

As to my special work—it is surely (apart from educational and practical bearings) in itself interesting to know how fast a man thinks—for on this, not on the number of years he lives depends the length of his life.[3]

[1] JMC abstracted this material for his own use; see 3.12.
[2] JMC published this work as part of "Über die Zeit der Erkennung und Benennung von Schriftzeichen, Bildern und Farben," Philosophische Studien 2 (1885): 635–650.
[3] The idea that the speed of thought determines the span of an individual's life is one

that Wundt had worked with in the 1860s (see Wundt, "Die Geschwindigkeit des Gedankens," Gartenlaube no. 17 [1862]: 263–265). JMC was to express it often in his later articles (see for example "Mental Measurement," Philosophical Review 2 [1893]: 316–332).

I received Harry's letter today. I shall see that he gets his ex-matriculation papers.[4] I bought my lamp from Mappin & Webb, Oxford St. W. (10% discount).[5]

[4] Henry Cattell was apparently concerned with obtaining a record of his study in Leipzig, as in the fall of 1884 he enrolled in the medical department of the University of Pennsylvania (see 6.72).

[5] Recommended by Baedeker (*London*, p. 20) as suppliers of cutlery and similar items. This comment probably refers to the Argand lamps mentioned in 5.4.

5.7 To Parents, 12 October 1884

(This letter was misdated "Sun. Oct 13th." A paragraph of gossip has been omitted.)

Berger has gone to Dresden today, where he is to try his "Staats-turn exam".[1] When he finishes his university studies he must teach for a year without any pay, but if he can secure a position, he can at the same time teach gymnastics or physical training, for which he would be paid $300. He returns on Thursday, and we shall then begin work together. He has now however to study himself, as he passes his Staats-exam, equivalent to a Ph.D. examination the end of next month.[2] After that he will have plenty of time, and will help me all he can. I have made such an arrangement with him as you advise.[3] It was not such an easy matter as he is very sensitive and a good friend of mine, whom I do not want to lose. He is however glad to earn the money, more especially as I had lent him $50 last winter, and offered to lend him more if he wished to continue studying and take his Ph.D. in the spring. It was of course I of my own accord who offered to lend him the $50 and what more he needed—at first he did not want to accept it. (If you kept any track of the money you gave me last winter, you see I spent $50 less than you had supposed.) I have offered to give him $250 this winter, and in case I obtain a position in America due to the success of my work $250 more. Thus he has a personal interest in its success. He will help me all he can—we shall make experiments together in the morning, and he will help me tabulate the results, calculate averages &c. I shall work hard this winter and in the direction most likely to secure me a position in the Univ. of Penna.—this wholly for your sake.

Please throw my letters in a box so that I can have them when I come home, they will serve as a journal.

[1] *Staats-turnexam*: state examination in gymnastics.
[2] That is, equivalent in difficulty (see Paulsen, *German Universities*, pp. 335–340).
[3] See 4.11.

5.8 To Parents, Leipzig, 16 October 1884

(Several paragraphs of gossip have been omitted.)

I got my largest piece of apparatus from the machinist's today.[1] I feel like an author with his first book. The machine is really a great advance on methods heretofore used. I imagine Prof. Wunt would give $500 if he had it four years ago—or indeed so much if he had himself thought of it now. I shall set up my elements tomorrow, and have things in working order. It is very much better having my apparatus in my room. In the first place every-thing is in good order, I shall never be disturbed, and it is far more convenient. But there are two greater advantages even than these. There will arise no difference of opinion between Prof. Wundt and myself, and Prof. Wundt will not be give credit for half the work.[2]

Berger came back from Dresden this evening. He received next to the highest possible grade, only one other out of the fifteen getting it. I "turned" this evening with the Verein.

[1] This instrument was the gravity chronometer (see 3.4, 3.10, and figure 8). Carl Krille was the man who built JMC's apparatus (see "Cerebral Operations," *Man of Science*, vol. 1 p. 43); little is known of him.

[2] JMC did most (but not all—see 5.16) of his experimental work in his rooms, primarily for the convenience of having everything at hand and always ready. But in all his papers he noted that he had begun his work in America and that at least some of it was "carried out in the psychological laboratory of the University of Leipzig," and he always made sure to thank Wundt for his help and encouragement. (See "Cerebral Operations," *Man of Science*, vol. 1, p. 52.)

5.9 To Parents, Leipzig, 17 October 1884

(Several paragraphs of gossip have been omitted.)

I had a fencing lesson and bath this afternoon. The weather has changed—it is now rainy and warm. I was at the theatre this evening—the first time I have seen a play for a long, long while. This evening I was attracted by the subject "Der Saloutyroler."[1] They play such pieces (comedy-farces, like "Krieg im Frieden"[2]) to perfection. I shall go to the Alt-theatre once in a while. It is right near by, and does not take all the evening. I only have to pay 25 cts. for the best seat in the house.

I got my tuning-fork through the custom-house free of duty[3]—I imagine I would have had to pay 10% ($2.50) if they had known what it was worth. I only had to pay $3 freight and duty on the writing-machine. So I really made $12 by bringing it to Europe, as they gave me 20% discount for exportation.

[1] This play has not been identified, at least in part because the transcription of its title from JMC's handwriting is uncertain.
[2] See 3.7.

[3] JMC, like many others, used tuning forks as time standards for the adjustment of the Hipp chronoscope (see "Cerebral Operations," *Man of Science*, vol. 1, p. 45).

5.10 To Parents, Leipzig, 18 October 1884

Berger was "bei mir" this morning and we made a few experiments. The apparatus is not yet in working order however—only those who work with electricity and complicated apparatus understand how difficult it is. For example I have a battery of three La Clanché elements—a new element supposed to be especially constant, yet at first it gave a strong current, and a few minutes afterwards would give none at all—it is apparently impossible to find out where the trouble lies.[1]

This afternoon I played foot-ball, and now am so stiff I can scarcely move. I enjoy foot-ball thoroughly. It seems physiologically good occasionally to take rather violent exercise—muscle cells &c are broken down and are replaced by new and healthier ones—to bring this about the blood is made richer and the appetite improved. I think I was hungrier this evening than any other evening since I have been in Leipzig. If the blood is richer not only are the cells which are broken down renewed, but the cells of the brain are nourished. If the brain is little used, it can perhaps be over-fed and become sluggish; but this is impossible with those who use their brain constantly. I think after severe mental work one's appetite is sharpened, but the brain being such a small part of the body, and the cells elsewhere not being able to take up nourishment (perhaps because they are unhealthy) the extra amount of food assimilated is not large, and the brain may not recover as much energy as it has used. Work seems to break down the brain cells, and if the blood is rich enough to build new ones, the brain is healthier than before—worry however seems to put them in a diseased condition. All of which should be kept constantly in mind by the man using his brain much. It does no harm to use up the brain, if the blood is healthy enough to renew it—therefore one who thinks should keep his blood rich in nourishment. He will not only be in better physical health, but his mental work will be more and better.[2] Because I believe the above, don't think I am a materialist—metaphysically I believe that the mind creates the brain.[3]

This evening I read one of Shakespeare's plays. It seems lonely to be here in Europe all by myself. I trust the voyage has had a good beginning, and will continue pleasant. Were there nice people on board?

[1] Georges Leclanche, a Parisian chemist and instrument maker, had developed in 1867 a battery with zinc and carbon plates in a solution of ammonium chloride, which as JMC noted was supposed to provide an especially constant current (*World Who's Who in Science*, p. 1014; W. R. Cooper, *Primary Batteries: The Theory, Construction and Use* [London: "The Electrician" Printing and Publishing Co., 1901?], pp. 179–192). JMC soon abandoned these cells, and "after considerable experiment . . . adapted a form on the zinc-copper gravity-battery," a modification of the Daniell cell (see "Cerebral Operations," *Man of Science*, vol. 1, p. 45; Cooper, *Primary Batteries*, pp. 207–208).

[2] JMC's ideas on mental physiology expressed here are similar to those set forth by Philadelphia neurologist Silas Weir Mitchell in *Fat and Blood: An Essay on the Treatment of Certain Forms of Neurasthenia and Hysteria* (Philadelphia: Lippincott, 1877). See 3.3.

[3] JMC expressed this opinion no place else, and if not a materialist in 1884 he was clearly one by 1886 (see 5.142).

5.11 To Parents, Leipzig, 19 October 1884

It continues to rain, while the barometer with most shameless assurance continues to mark "Schönes Wetter." My room numbers amongst its attractions the said hopeful barometer.

I was at the American Chapel this afternoon.[1] As I do not know who preached there is no harm in saying that it is a mental and moral injury to me to listen to such stuff. A clause like this: "Michael Angelo, a painter, a sculptor, a poet and an architect—unequalled in some of these departments, unparallelled in others, eminent in the rest" is only mentally hurtful, with a sentence like the following it is different: "A person looking at a flower may get some pleasure, but it is not a Eithe, no not a twentieth of the joy of the man who has studied botany and can understand the construction of the flower and knows the species which are related to it". I say it does me moral harm to hear such a sentence, for I listen with a mental sneer at the idea of such a man being put there to instruct and elevate me.

Our preacher quoted plentifully from Tennyson: he might in the above connection have quoted this

"What is it? a learned man
Could give it a clumsy name.
Let him name it who can,
The beauty would be the same".[2]

Or, from a greater than Tennyson (for he evidently owns a dictionary of poetic quotations)

"A Rose, if called By any other name would smell as sweet"[3]

It makes me utterly wretched to listen to the fluent words of such a man, who covers his lack of earnestness and faith, with a veneer of science and culture he does not understand.

The days go by very fast. I have an idea spring will come, before I am half ready with the work I want to do. There is no trouble however about the Doctor's thesis, as I shall have work enough to make two or three.

Some of the lectures begin this week; I shall not attend many. I have heard (and it may be added forgotten) the courses announced by Profs Wundt and Heinze, with the exception of Logik twice a week by Prof. Wundt.

[1] An American service was held each Sunday at 5 P.M. at one of the city's schools. In late 1884 the American Pastor at Leipzig was Eugene Luzette Mapes, a graduate of Union College and the Union Theological Seminary and a student at Leipzig from 1882 through 1885 (*Alumni Catalog of Union Theological Seminary*, p. 219.)

[2] From Tennyson's *Maud* (part 2, section 2, lines 9–12). (See 2.3.)

[3] From *Romeo and Juliet* (act 2, scene 2).

5.12 From William Cattell, Philadelphia, 19 October 1884

(Several paragraphs of gossip have been omitted.)

As I have already said to you, some good published work of yours will facilitate (if indeed it is not absolutely necessary) my getting a suitable place for you in America. I could easily get you a "Professorship" in some small college; —but I want you to have some position where you will have leisure for original work; & where yr teaching will be on the lines of yr work; also one that promises promotion. Then you can await the result, if such a position can be secured in the University at Philadª how happy we shall be! I believe it wᵈ be the very best place for you; —but added to this wᵈ be our joy to have you at home. You may be assured that dear Mama will have your rooms all you cᵈ wish—and we shall both not forget that you are now a man—our precious child indeed, but no longer a boy: and you can have all the freedom of yr bachelor's quarters at Leipzic! You may cook all the omelettes in yr room—& all the apples—that you fancy! and sit up as late or sleep as late as you please!!

5.13 To Parents, Leipzig, 20 October 1884

I enclose a note from a friend at Cornell.[1] If I become a college professor of course I should rather be with you at Philadelphia, or if this is impossible, in one of the large cities—New York or Boston. But for several reasons it would not be best at first to go where I want to settle for life. I would have to learn more than teach the first year or two, and might make mistakes tht would hurt me for a long time—as for example Charlie Scott.[2] If then I do not find at once a suitable position in Philadelphia, Cornell would be a good place. It is healthy & beautiful, the library &c is good, they have plenty of money. Mama thinks Cornell is rationalistic or something of the sort, but I imagine it is less so than other undenominational colleges. The present Prof of philosophy is a D.D,[3] as are many of the other professors. But in this matter I shall, at least for the present, do as you wish.

Berger and I worked together this morning and intend to continue it all through the winter.

[1] The note, from Henry Winchester Rolfe (see 1.19 and 4.7) and dated 26 September 1884, read as follows: "Some one has just given $50,000 to endow the department of 'moral philosophy.' That means a breaking up and recasting of the whole department of psychology, philosophy, and so on. Won't your father look the matter up? A visit or a letter to Pres. White would very likely be the opening wedge." Rolfe was referring to a plan of Henry Williams Sage, an old friend of Ezra Cornell and of Cornell University, to endow a chair of Christian ethics and moral philosophy. Its first holder would be Jacob Gould Schurman, a good friend of Sage and later president of Cornell. (See Morris Bishop, *Early Cornell, 1865–1900* [Ithaca, N.Y.: Cornell University Press, 1962], pp. 245, 267; *DAB* 16: 290–291; *DAB* Supplement 3: 696–699.)

[2] Charles P. G. Scott (see 3.7 and 4.6).

[3] William Dexter Wilson (see 4.7).

5.14 To Parents, Leipzig, 22 October 1884

I was made happy this morning by receiving your letters written at Queenstown. I am glad the voyage has begun pleasantly, and trust it will be safe, quiet and quick.

I enclose the signatures, but I think Brown Bros. send the letter to Frege and Co. and I sign it here.[1] I have just drawn all the ballance on the old letter as it soon expires. I have $210 in German money and £5 English. This is not so very much, as I must pay for my apparatus—I do not know how much. I suppose the above will last until New Year—but it depends on how much I give Berger. When we separated you proposed letting Mr. Wood send me the new letter on his return.[2] There is really no hurry, except that I feel more comfortable when I have a larger margin than I am likely to want. I do not think any extraordinary expenses will turn up from now on—except $200 for Berger. I shall come home in August, unless you secure me a position, you wish me to take, in which case I might stay until Sept. I cannot make my Ph.D. in Aug, Sept or Oct, so must make it in July, if at all before the opening of the college year in America. This would be possible, but I must know before Jan. 1st, the university authorities require about that long notice, and I would need six months to prepare myself. If you secure me the offer of a position, I can enter upon it immediately, and come over here next summer and pass the examination. I would however prefer to have a year's leave of absence—or at least until Christmas, in which case I would not come home in August, but would stay here and pass the examination in Nov.

[1] The letter referred to is a letter of credit sent from the American bank, Brown Bros., to the Leipzig bank, Frege and Company

(see also 1.1, 4.2, 5.1).
[2] On George C. Wood see 4.2.

5.15 To Parents, Leipzig, 23 October 1884

(One paragraph of gossip has been omitted.)

The sun came out this afternoon and the stars are shining now. I took a walk out past the Grosse Eiche and Leutch.[1] I always walk in the same direction—it give me more pleasure when I know the way well. Around the park the shorter way is two miles the longer three, further past the Schützenhaus is four,[2] still further past the Grosse Eiche &c perhaps nine. I fenced from six to seven, and "turned" from 7.45 to 9.15 so I have had plenty of exercise today.

I cannot however go out in the morning any more as Berger and I work together from eight to one.

James McKeen Cattell,
James McKeen Cattell,
James McKeen Cattell.[3]

[1] The *Grosse Eiche* was a monumental tree in the *Rosenthal* (see 3.11). Leutzsch was a suburban village beyond the *Rosenthal* (Baedeker, *Northern Germany*, pp. 242–243).
[2] *Schützenhaus:* guardhouse.
[3] See 4.12.

5.16 To Parents, Leipzig, 25 October 1884

Prof. Wundt met this afternoon those intending to work on psychology in his laboratory.[1] Berger and I were there, and took our old quarters for four afternoons in the week.[2] We shall work but little there[3]—I think however Prof. Wundt was pleased that we turned up. He naturally would rather that I should work there than in my room. He would be given a large part of the credit, for any work I might do, and if I correct any of his work, it would look more as though he were rectifying his own mistakes. I gave him this afternoon an account (written on the machine) of the work I want to do this winter.[4] When I spoke of it last semester he thought it would take ten years to finish it!

[1] The opening of the laboratory each term was a rather formal event for Wundt. See 3.2.
[2] That is, in the university's psychological laboratory.
[3] See 5.8.

[4] This account, an expansion of the statement that JMC probably prepared the previous spring, has not been found. However, a transcription of a handwritten draft of it is presented here as 5.17.

5.17 Draft of "Statement prepared at Prof. Wundt's request"

I should like to present as a thesis in application for the degree of doctor of philosophy an essay on Psychometry, or the time taken up by simple mental process.[1] I am well aware that this subject is too large and difficult for me to thoroughly investigate as a university student, but I trust I shall be able to prepare an acceptable thesis by giving (1) a brief summary of the work which has been done in this field. (2) a fuller account of experiments I myself have made and the results reached (3) the subjects which seem to me to need investigation.[2]

I give below an imperfect analysis, of the factors on which the reaction time depends,[3] underlinging with blue ink the subjects on which I have worked, and with red ink those which I should like to investigate this winter.[4]

Reaction Time

Analysis.

The <u>sense stimulus</u>. Sound, Light, Touch (touch proper, electric shock, temperature) Taste, Smell.

[1] See 3.7 for JMC's first use of the word "psychometry."
[2] JMC apparently expanded on "the subjects which seem to me to need investigation" in his final version of this statement.
[3] JMC added this clause above the rest of the line.
[4] There is no such underlining in the draft.

Sound—loudness, pitch, timbre[5]
Light—intensity, color (saturation)
Toutch proper—force, kind
Electric shock—intensity   } momentary
Temperature—degree (heat & cold)   continuous
Taste—intensity, variety[6]
Smell—intensity, variety

Sense organ, nerve, and muscle.[7] { normal
                                      abnormal

Sense organ & efferent nerve[8] used—toutch on lip and back, light image on yellow
or field of indirect vision, taste on lip of tongue or palate.
Muscle & efferent nerve[9] used, finger, wrist, foot.

The Subject[10]   (physiological)
                  (psychological)

Sex, Age, Temperament, Character,[11] Mental Acumen, Physical power, Avocation.

The Same Individual

Normal
Abnormal

Normal

Mental State— }   Fresh, wearied, interested, indifferent,
                   dull, excited
Physical State }   Eating, Sleep, exercise,[12] temperature,
                   weather (saturation & electric condition)

[5] JMC first wrote "quality" here, but replaced it with "timbre."

[6] In this line and the next, JMC first wrote "variety, intensity" but then changed the order of these words with arrows.

[7] JMC originally wrote this paragraph at the end of the outline, but marginal signs indicate that he wanted it placed in this position.

[8] JMC added "& efferent nerve" above the rest of the line.

[9] JMC added "& efferent nerve" above the rest of the line.

[10] Part of the mythology of psychology, based directly on JMC's own reminiscences ("In Memory of Wilhelm Wundt," *Psychological Review* 28 [1921]: 155–159), is that JMC met with Wundt in 1884 and "presented an outline of the work I wanted to undertake, which was the objective measurement of the time of reactions with special reference to individual differences. Wundt said that [the idea of measuring differences among reaction times of individual subjects] was 'ganz Amerikanisch'; that only psychologists could be the subjects in psychological experiments." This statement does call for the "objective measurement" of reaction times, without introspection of any kind (see also 3.12 and notes below). But this topic is one JMC did not investigate in the experiments that formed the basis of his dissertation, at least partially because he had "no time to spare" (see 5.96).

[11] JMC added "Character" above the rest of the line.

[12] JMC added "exercise" above the rest of the line.

Attention–Practice–Fatigue[13]

Attention $\begin{cases} voluntary \\ caused\ (signal, distraction[14]\ \&c) \end{cases}$

$\left.\begin{array}{l} Practice \\ Fatigue \end{array}\right\} \begin{cases} series \\ day \\ continuous \end{cases}$

Abnormal

$\begin{array}{l} Physical \\ Mental \end{array}$ $\begin{array}{l} (pain\ or\ disease[15]) \\ (distress, elation,[16]\ or\ disease \\ (insanity) \end{array}$ $\begin{cases} natural. \\ artificial. \end{cases}$

artificial, by use of drugs, alcohol, ether, caffein, morphine, &c.

The measurement of mental processes[17]

Prof Wundt in his Phys Psy[18] & Phil Studien[19] shows how we can determine the absolute time of certain mental processes U, W, & A.[20] The length of these times depends on the conditions mentioned above[21] and also on other factors—a thorough investigation of which would require many years of careful and thorough work. Last winter with Herrn c and. math.[22] Berger, I made a large number of experiments seeking to find the value of U for colors, but have come to doubt the possibility of determining with scientific accuracy[23] U (distinct from W) by any method with which I am acquainted.[24] We also made a number of series of W reactions. Before[25] coming to Leipzig I made a number of U and W reactions on a different & as far as I am aware new principle.[26] Printed letters (from Snell's Optotypi) were pasted

[13]JMC first ordered this heading and the following entries "Practice–Fatigue–Attention," but used an insertion and an arrow to change the order. He had experimented on these topics earlier (see 3.12).

[14]JMC added "distraction" above the rest of this line.

[15]JMC first wrote "sickness" here, but replaced it with "disease."

[16]JMC added "elation" above the rest of the line.

[17]This paragraph was written on the same sheet of paper as the first paragraphs of this statement.

[18]*Grundzüge der physiologischen Psychologie*, 2nd edition (Leipzig: Engelmann, 1880), vol. 2, pp. 247–279. See 3.12.

[19]"Über die Messung psychischer Vorgänge," *Philosophische Studien* 1 (1882): 251–260.

[20]*Unterscheidungzeit, Wahlzeit,* and *Apperception-zeit* (discrimination, will, and apperception times). See 3.12.

[21]JMC wrote "under reaction time" here, but then crossed this phrase out.

[22]JMC added "math." above the rest of the line.

[23]JMC added "with scientific accuracy" above the rest of the line.

[24]See 3.12. This point was of course a criticism of Wundt's views on introspection and, especially, on apperception. Wundt and a number of his later students attempted to get around JMC's objections after JMC had left Leipzig. See postscript.

[25]At this point JMC first began a new paragraph, starting with "This winter I should like to continue the W. reactions." However, he crossed this sentence out and continued the preceding paragraph with this sentence.

[26]JMC first wrote "an entirely different principle" here.

on the cylinder of a kymograph one c. from each other.[27] They were then read as they passed a slit in a screen, the slit being at first 1c. wide. The rate at which the cylinder revolved was adjusted so that the letters could just be read. When they were simply read U was measured.[28]

[27] In 1862, Herman Snellen (*World Who's Who in Science*, p. 1573), professor of ophthalmology at the University of Utrecht, published a "set of type of various sizes for testing central vision"—*Optotypi ad visum de terminandum*—which was made available in English almost immediately as *Test-Types for the Determination of the Acuteness of Vision*. JMC prob-

ably made use of the fourth edition of this set, published in 1868 in New York by Balliere Brothers and in London by Williams and Norgate (*Pre-1956 Imprints* 553: 30).

[28] See 2.41, and JMC, "Über die Zeit der Erkennung und Benennung von Schriftzeichen, Bildern und Farben," *Philosophische Studien* 2 (1885): 635–650.

5.18 To Parents, Leipzig, 27 October 1884

I received my papers from Dr. Hall today without a word of comment or explanation.[1] Yesterday I wrote a letter which I intended copying and sending to the man who continued my work[2]—he is an undergraduate, and as far as Dr. Hall seemed able to explain the matter had not added a single idea to what I had done—he only worked on minor points, following suggestions I had made. I enclose the letter, as I shall not send it now.

Dr. Hall has not acted honorably towards me. When I was at Baltimore he praised me highly, said there was no one he would so gladly see holding the fellowship, but unfortunately he had nothing to say in the matter and Dewey was a great favorite of Prof. Morris's and Pres. Gilman's. He added he hoped the university authorities would grant him an assistant, and if so he knew no one so well fitted for the part as me. Pres. Gilman showed me Dr. Hall's recommendation for the fellowship—Dewey stood first and I fourth.[3] Yet Dr. Hall always spoke to me rather slightingly of Dewey, and told me this year he was compelled to drop him. Dewey handed in his application for renewal of the fellowship, but at Pres. Gilman's advice withdrew it.

Dr. Hall knew that the fellowship would not be given to me (he had already recommended that it should not) when he got my papers from me. He therefore got them under false pretenses. He said my work ought to be printed immediately, less some one else should anticipate it, and promised to print it in Mind for Sept. '83. But he really wanted to get possession of the papers, knowing that I could not finish the work at Baltimore. He wanted half the credit for work to which he had contrib-

[1] See 3.4 and 3.6.
[2] Julius Friedenwald, who later graduated from Johns Hopkins and earned an M.D. from the Maryland College of Physicians and Surgeons, where he later taught gastroenterology (*Who Was Who in America*, 1897–1942, p. 427; *Baltimore: Its History and People*

[New York: Lewis Historical Publishing Co., 1912], vol. 3 pp. 517–519). JMC's letter detailed his relations with Hall, pointed out that Hall had said Friedenwald "had done much more work than" JMC and urged Friedenwald to publish his own results separately.

uted nothing, and in the end was not satisfied with this, but tried to get it all. In the paper he would have written, I would have been given as much credit as Frieden-wald and both of us precious little.

Dr. Hall treated me most kindly and was exceedingly flattering, when I saw him at Williamstown, until I said I could not consent to the use he proposed making of my work. Then he became very angry, said it was of no account, i.e., my work, &c.[4]

Dr. Hall is of course a man of ability, but his reputation is due to the fact that he has repeated in America what is a commonplace in Germany. He has no talent himself for doing scientific work, nor any accurate scientific knowledge. His suggestions to me when I was working at Baltimore were most silly, and his own work usually turned out worthless, not through lack of pains, but through sheer ignorance and stupidity.

I think I shall not write to Dr. Hall at all, and shall of course dismiss him from my mind—though if he improperly uses my notes, I will bring suit against him.

It is curious that scientific men should be fonder of praise (even undeserved praise) than of advancing science. It is the same with artists.

I am writing hurriedly & perhaps not very clearly.[5]

I think I said in a recent letter that I had $210. I really had $350—I multiplied the pounds wrong. So I have plenty of money, even though I give Berger $100. Another financial statement of mine was wrong, butter costs 40 cts a pound. I thought my half pound pieces weighed a pound.

[3] See 2.57.
[4] See 4.7.

[5] This letter contains many minor corrections.

5.19 To Parents, Leipzig, 28 October 1884

(Two paragraphs of gossip have been omitted.)

The lectures began yesterday. I have already heard the most important of them and the light is too bad to admit of taking notes. I am very busy, and as I want to work in the morning and evening, ought to take exercise in the afternoon. The air is so bad, as to be really injurious. In Prof. Wundt's lecture room there are packed some three hundred men—the ceiling is low and not a crack is left open.[1] For all of which reasons I shall attend but few lectures. German students in their last year scarcely ever hear any.

[1] For other similar descriptions of Wundt's lectures see "In Memory of Wilhelm Wundt," *Psychological Review* 28 (1921): 154, 160, 162–163, 164, 184.

5.20 To Parents, Leipzig, 29 October 1884

This has been the most tolerable attempt at a clear day since I have been at Leipzig. This afternoon I walked out the familiar Arndtstr. way and took a canoe for a little while. At four I heard Prof. Wundt lecture on Ethics.

After the lecture he talked to me for three quarters of an hour and was extremely pleasant. In the paper I gave him on Saturday, I classified the work I have done, and propose doing this winter under four heads.[1] Prof. said he would accept any one of these as a doctor's thesis, but recommended one as especially original.[2] He said he would accept this for his "Philosophische Studien." This offer is all the more of an honor from the fact that scarcely any of the work has been done—he must therefore have considerable confidence in me. He said he would let me use the set up type for printing my doctor's thesis, which would make it comparatively inexpensive. He also said he hoped when I printed work in English I would send him a translation for his "Studien". He is going to come to see me.

I went to a café this evening and waited until yesterday's Times came. I was made happy by the line "New York. Oct 27 The Servia, with the mail." Curious they should mention the mail, when there was something so much more precious on board.

[1] The draft of JMC's "Statement" (5.17) does not include "four heads," but see 5.22.
[2] The proposed work Wundt thought most original was JMC's plan to study association (see 5.22).

5.21 To Parents, Leipzig, 4 November 1884

(One paragraph of gossip has been omitted.)

We worked as usual this morning, but Berger gives another lesson tomorrow. This afternoon I was out past the Grosse Eiche. I have not been in the town today. I am not hearing many lectures.

I have paid my university bill, enriching the institution with $3.00. Neither was the machinist's bill so very large—$40—the apparatus would have cost at least twice as much in America, and not have been so good. I did not need to have a single thing altered—a compliment to both the machinist and me. Dr. Hall always had to send back his things half-a-dozen times, and then they were never right.

5.22 Journal, Leipzig, 5 November 1884

Except a brief account of my whereabouts, I have fallen into the habit of writing here, only when I have something fine to say, usually of a tragic-melodramatic turn. This is a pity, as it would be well for me once in a while to write down in simple English what I think and feel.[1] Things have been moving rather decently and in good

[1] See 1.1.

order for the past month. My life has been less of a mistake than usual. This is doubtless due to the fact that my health is better. The Tyrol weeks were time well spent.[2] I have been working on the whole regularly and just about enough. Then I take constant exercise. I row, swim, fence, play tennis & football and exercise in the gymnasium; what is even better than these I often take long walks of an afternoon. Mirabile dictu I am doing pretty well as to what I eat and the use of stimulant drugs. I could scarcely have hopes for this, as these are no external circumstances to keep me straight. No one knows what I do. I am mostly alone in the afternoon and evening. I eat nearly altogether in my room. The Hausmann's Frau brings my break-fast—oatmeal and cream, bread, butter and milk, she also cooks me apples and pears. I cook me mostly eggs. I have smoked about once a week, drunk two cups of coffee, and some chocolate and cocoa. These I have taken however almost exclusively in my room, and in a way scarcely objectionable. I have drunk beer once (seven glasses) at the Turnverein Angängskneipe,[3] and once in a while a Schmidt when I have eaten my dinner at a restaurant. As to my work: Berger comes to my room every morning at 8.30 and we work together until one. I have also spent a good many evenings writing our results &c. I have been more in my room of an evening than ever before. I am fortunate in having a definite work which it is important I should complete before I leave Leipzig, and want to leave Leipzig in May. This is my experimental psychometric work. I want to prepare papers for publication on four subjects all closely connected (1) The Reaction Time—with special relation to the influence of Attention, Practice and Fatigue (for Brain).[4] (2) The Time of simple mental processes (for Mind, perhaps also in German).[5] (3) The Association Time. (Prof. Wundt offers to print this in his Studien and recommends it for my doctors thesis)[6] (4) The Legibility of letters, words, and phrazes.[7] I write every evening to Mama and Papa—these letters will serve as a Journal. As to my emotional background, it is on the whole better than for a long time, perhaps than it has ever been. I feel rather strongly, and my emotions cover a fairly large circle. It is a good sign that I am able to be so constantly alone without undue restlessness. I have but few acquaintances here (whom I seldom see) and no friend except Berger.

[2] See 4.13 and 4.15.

[3] *Angängskneipe*: initiation drinking party.

[4] This work was later published as part of JMC's doctoral dissertation, in *Philosophische Studien*, and in *Mind*, but not in *Brain* (see below).

[5] This work later made up the largest part of JMC's doctoral dissertation, published separately and also as "Psychometrische Untersuchungen," *Philosophische Studien* 3 (1886): 305–335, 452–492, and as "The Time Taken Up by Cerebral Operations," *Mind* 11

(1888): 220–242, 377–392, 524–538.

[6] This work was later published as a supplement to JMC's dissertation as "Die Association unter wilkürlich begrentzten Bedingungen," *Philosophische Studien* 4 (1888): 241–250, and as "Experiments on the Association of Ideas," *Mind* 12 (1887): 68–74.

[7] This work, started in Baltimore, formed the basis of JMC's first published scientific paper, "Über die Zeit der Erkennung und Benennung von Schriftzeichen, Bildern und Farben," *Philosophische Studien* 2 (1885): 635–650.

I think there is no woman in the world, in whom I am "interested". I mean deeply of course—the order of interest perhaps stands [here Cattell listed in shorthand the names of about ten young women] *the latter seem mostly historical. Purely historical for* [here Cattell listed in shorthand about five more young women] *inter al. As for friends I might give some such list as this—Bailey, Rolfe, Berger, Leiper, March, Zinkeisen, Robinson, Armstrong, Hubner, Brown, Green, Smith, Wagener, Mackenzie, Scott, Thomas, Bright, shading off indefinetely.*[8]

[8]On Bailey, see 2.51; Rolfe, 1.19; Berger, 3.2; Zinkeisen, 1.19; Robinson, 1.19; Armstrong, 2.64; Brown, 1.5; Green, 1.5; Scott, 3.7; Thomas, 2.9; Bright, 2.1. "March" is Francis Andrew March, Jr., the son of JMC's professor at Lafayette, himself a Lafayette graduate and later was his father's successor (*DAB* 12: 270). David Douglas Wagener was a Lafayette classmate of JMC who later entered business in Easton (*Men of Lafayette*, p. 232). Joseph M. Leiper was an employee of Dodd, Mead & Company, the New York bookseller, who later became a farmer (Cattell papers). The others listed have not been identified.

5.23 To Parents, Leipzig, 6 November 1884

Prof. Wundt came to see me this morning. He stayed three quarters of an hour and was very cordial, as he has always been recently. He has treated me very nicely, considering that I have called his attention to mistakes in his work, and have left his laboratory, where he would much rather have had me stay.[1] *That he has acted this way is not only plesanter for me, but also better for him. If, as many would have done, he had insisted that my corrections were wrong and my work of no account, and had put obstacles in the way of my setting up my own laboratory,*[2] *the facts would not have been changed; but I would have published sometime or other my work in a way not pleasant to him. As it is he offers to print my work in his own magazine and is in every way pleasant and obliging. I shall therefore probably correct his work in a way that will be almost a compliment, and I shall be considered a pupil of his, and he will be given credit for what work I print now or later.*

He praised today my arrangement of apparatus highly, and will copy it in some respects for his institute. He also offered to accept as my doctor's thesis and print in his "Studien" the work I began in Baltimore. This I have been working on and can easily finish by Christmas. So I could probably get my degree the first of August. In six months I could study up enough Philosophy, physics, and biology or mathematics, I imagine. Still unless there is some special reason for this I would rather finish up all my experimental work by August, then come home to see you, and return and pass the examination in the Winter. Besides the division of my work above mentioned—which is on the Time of Simple Mental Processes—there are three others.[3] *The Association time, which Prof. Wundt recommended for a doctor thesis and offered to accept for his Journal. This is scarcely begun. Then there is a good deal of work already done on the Reaction Time—this is what Berger and I worked on last*

[1]See 3.12 and 5.8. JMC did continue to do some work in Wundt's laboratory; see 5.16 and 5.27.

[2]That is, in his own room.
[3]See 5.22.

winter. Forthly there is work on the Legibility of letters, words and phrases, practically the most interesting and valuable of all. I have quite a good deal of work on this, but there is any amount more to be done. As I have said I can leave the last three of these for the present, but in a recent letter I gave reasons why I should like to finish it all.[4]

[4]See 5.18.

5.24 To Parents, Leipzig, 13 November 1884

We have stopped our experimental work now for nearly two weeks. I hope Berger passes a good examination—it is of great importance to him. It is the Staats-Exam and the grade of school to which he is appointed depends on the result of the examination.

This morning I was at the library. It is very badly managed. One cannot consult a book, unless he has ordered it the day before, and then it does not usually turn up. The other day I did not for one reason and another get a single one of five books I had ordered.

It has been a beautiful day. This afternoon I was out past the Grosse Eiche. This evening I attended the third Schiller performance—Kabale und Liebe.[1] I missed an act to attend a meeting of the Tennis Club in Mrs Sheperds rooms. There were twenty members there. I wanted to be present to see that we engage the skating rink again this winter. Muirhead[2] was rather inclined to give it up. You know it is very convenient for me.

[1]JMC attended a cycle of Schiller plays during this period (JMC to parents, 6 November 1884).
[2]Findlay Muirhead (Who Was Who [Br.], 1929–1940, p. 980) was a Scotsman who had been educated at Edinburgh and who, in 1884, was a student at Leipzig. His brother, James Fullerton Muirhead (see introduction to section 1), was employed by Baedeker, the Leipzig publisher, and JMC was a good friend of both brothers. See also 5.143.

5.25 To Parents, Leipzig, 15 November 1884

(Two paragraphs of gossip have been omitted.)

I received your second letter today, Mama, it nearly overtaking the first. I am glad things seem to be moving smoothly. I hope you will take a house or flat, rather than live in a hotel or boarding-house. You must both be very careful of yourselves, I would gladly have given the $3000 I have if you could have stayed in Europe another year, but I think we decided wisely according to our knowledge, and trust it will prove to have been as best.

I have found a new psychologist—a very good man, named Wolfe from Ne-braska.[1] *He wants to pass his examination here in psychology and afterwards to teach it. He is coming to me two evenings a week to be experimented on, and would come oftener if I needed him; but Berger can come of an evening after he is done with his examination.*

[1]Harry Kirke Wolfe (*DAB* 20: 450–451) was a graduate of the University of Nebraska who later taught psychology there after earning a Ph.D. at Leipzig (1886). Like JMC, he was a charter member of the American Psychologi- cal Association. See also L. T. Benjamin, Jr., and Amy D. Bertelson, "The Early Ne- braska Psychology Laboratory, 1859–1930," *JHBS* 11 (1975): 142–148.

5.26 To Parents, Leipzig, 26 November 1884

(Two paragraphs of gossip have been omitted.)

Berger began work with me again this morning. I do not think I shall hurt myself in working, if I keep taking such constant exercise. It is undoubtedly true that making experiments on one's self is trying, but I do not do this continuously. If I spend six hours a day at this work, perhaps two must be given to looking after apparatus, preparing things &c. This is very easy work indeed. Then in two of the other four hours the other man is the subject and my work is not especially difficult. So you see I only spend two hours in work that strains. It were a pity if at twenty five I could not stand that.

This afternoon I skated and afterwards with Berger and two other Germans bowled.

5.27 To Parents, Leipzig, 27 November 1884

So this is Thanksgiving day—the day of family reunions, and I am way off here and can only think of you. I celebrated the day by eating boiled eggs for dinner.

We worked together this morning, and shall every morning except Sunday. We shall tomorrow afternoon set up the apparatus we used last year in Prof. Wundt's laboratory. Berger is going to use the thesis in psychology, which he presented in his Staats-Exam, as a Doctor's thesis, and must enlarge it. I shall help him, and so we must work some of an afternoon. We have "engaged" the laboratory for Mon. Tues. Fri and Sat. afternoon. I shall also work usually in the evening. So my work ought to make some progress.

This evening I heard "Tell," the last and in many respects best of Schiller's plays.[1] *They give however one other—a translation from Corneille.*

[1]*Wilhelm Tell* (see 5.24).

After the theatre I was for an hour at a Thanksgiving celebration, given by the church. It seemed to be quite a success—there must have been nearly three hundred people there. I knew very few of them. I met a Mrs Sheperd, a sister of Dr. Nixon's, who saw you once, Papa, when you were a boy.[2] I like Mr & Mrs Mapes.[3] Frl. Olga Gerhard was there[4]—I must call on them. It is quite late—which is against psychological law.

[2] On Dr. Nixon see 4.1. Mrs. Sheperd was involved with the Leipzig Tennis Club (see 5.24).

[3] On Eugene L. Mapes see 5.11.
[4] On the Gerhards see 3.7.

5.28 To Parents, Leipzig, 30 November 1884

(This letter was misdated "Dec 1st (Sun)" [see 5.29]. One paragraph of gossip has been omitted.)

Housekeeping seems much simpler in Germany than in America. There is but little trouble over servants. Living in rented flats is very easy. Then things for the table and house can be bought to such advantage—if one can afford to pay for them. There is no scolding the butcher, grocer and milkman—you get exactly what you pay for. You can get beef for 12 cts the pound, or pay 40 cts. At the same price the quality will be always the same. You can get butter for 20 or 40 cts the pound. I pay the higher price—the butter is excellent—and does not vary a jot from week to week. You can get perfectly fresh eggs in the middle of winter if you pay the current price. Then you can buy things ready for the table—there is no need of baking bread, cake, or pie—it is indeed cheaper to buy them. You can buy jelly preserves &c as good as you can make them and about as cheap. Then you can buy cold meat as good as you can cook it and no dearer. In a butcher shop that I always pass one block off I can buy for seven cents cold tongue or martbeef—enough for a meal. The excellent cream I get every morning is alway exactly the same. If I paid two cents less it would be thinner, if two cents more it would be thicker. You see how very easy housekeeping is. I dare say it is easy to get extra help or waiters. Washing is done as well and cheaply by the washwoman as at home. You can order from the restaurants whole meals. I don't find keeping house at all hard work.

5.29 To Parents, Leipzig, 1 December 1884

I have no almanac or newspapers and one day is so much like another, and they all go so fast, that I am usually uncertain as to the day of the month, and must sometimes even count up to find the day of the week. I believe I dated my letter yesterday Dec. 1st.[1] At all events it is here now, and a cold winter day, with fine skating and sleighing.

[1] He did (see 5.28). This error exemplifies the problems involved in dating JMC's correspondence.

I was skating this afternoon and afterwards had a fencing lesson. Berger and I also fence for fifteen minutes in the middle of the morning. I fence very well now— almost as well as the fencing master we had last Spring. I struck him six or eight times this afternoon and disarmed him once. I exercise the entire afternoon, and am very careful of my health. Did Harry tell you the very surprising figures I gave him? My breast increased in circumference 4¼ inches in three months, and the rest of my body in proportion. I had not supposed this to be possible. I am not fatter—my stomach measures only 31½ inches, whereas my hipps are 38¼.[2] I also sleep better than for some time. Last night for example ice froze ¼ in thick right by me, yet under two covers I slept very comfortably. Last winter I would have suffered from cold. I am glad to say I have no ear-ache which troubled me constantly last winter—as indeed also all summer.

[2] Of course, JMC was working daily with the Hipp chronoscope at this time.

5.30 To Parents, Leipzig, 3 December 1884

It has been psychology all day. We worked as usual this morning. This afternoon I went out to the "Physikalisches Institut" to examine some apparatus—telephones, microphones &c.[1] It would be of considerable importance if I could make a piece of apparatus which would register the instant a word is spoken.[2] I must go out there again tomorrow. This evening both Berger and Wolfe were working with me until after eleven.

The nice winter weather with snow and skating is gone. It is raining today.

[1] See Otto Wiener, "Das Physikalische Institut," *Festschrift zur Feir des 500 Jährigen bestehens der Universität Leipzig* (Leipzig: S. Hirzel, 1909), vol. 4, part 2, pp. 24–60.
[2] JMC accomplished this goal with his lip key, which broke an electric circuit the moment the subject's lips were opened, and his voice key, in which the sound waves produced by the subject's voice broke a circuit. He developed these instruments later in his stay in Leipzig. (*Man of Science*, vol. 1, pp. 47–48.) See 5.65.

5.31 To Parents, Leipzig, 4 December 1884

One day with me is a repetition of the preceeding, so that after I have once told you what I do, for the following days I can simply write under it ditto, ditto, ditto. We made experiments of course this morning: this afternoon I was at the Physical Laboratory. I introduced myself to the Professor of physics—Prof. Hankel[1]—he was very

[1] Wilhelm Gottlieb Hankel (*DSB* 6: 96–97), best known for his work on the electrical properties of crystals, was professor of physics at Leipzig from 1849 through 1887.

kind—spent about an hour with me looking over apparatus. I am glad he knows me, as he will examine me in physics. He wrote about twenty years ago an article on psychology, which I fortunately had read[2]—which pleased him. This evening I did not feel like "turning" and shall go to bed early.

[2] Possibly "Über . . . der ober Fläche des Menschlichen Körpers beobachteten elec- trischen Erscheinungen," *Annalen der Physik* 106 (1865): 299–306.

5.32 From William Cattell, Philadelphia, 7 December 1884

(Several paragraphs of gossip have been omitted.)

I am always specially interested in the accts you give about yr work. You must make all the use of Berger you can: that is in everything that will expedite yr work or render it more satisfactory to yrself. You need not hesitate about increasing his compensation. And get whatever help besides you may need. I quite agree with you as to the importance of yr printing some piece of good work:—by all means in Wundt's Journal if, as you intimate, he is willing to insert it: or in "Mind—and if Prof. W. would only commend it! I must have some such point d'appui as this,[1] to secure fixed attention when I speak to my parental "color-blindness"—& want some proof of what there is in you. And the average "Trustee" w^d probably not be able to judge of the real value of yr work, yet he knows that what Wundt commends must have value. It w^d of course be comparatively easy for me to secure you a position as "Professor" in some feeble College—but this is not desirable. You will do better to take a subordinate place among scholars—where there is a chance for you to win a good position. And as our longing to have you with us at home w^d be gratified by yr having some position in the University here, that is of course my objective point. I need not give details:—but scarcely a day passes that I do not lay some anchor to the windward with a view to this. But I fear that I shall not get on much further without something from you in print—or from Wundt about you.

[1] *Point d'appui*: fulcrum.

5.33 To Parents, Leipzig, 8 December 1884

(One paragraph of gossip has been omitted.)

My own plans depend somewhat on yours, and are generally uncertain—there are so many contingencies. I had originally intended going to Italy or Greece in the Spring vacation, March and April, and coming home in the summer vacation. However Berger hopes to get a position on the 16th of April, so it would be better to finish up my experiments by then. I find further that it makes no great difference to me whether the university is in session or not. So if I keep reasonably well I should rather work right on with perhaps a week's vacation at Christmas and at Easter. Suppose I finish my experimental work the middle of April—it would perhaps take *six weeks to get the results ready for publication. Then I could leave Leipzig the first of June—my doctor's thesis being ready. On the supposition that by that time I should not be very well, I have thought of walking in Norway for a while and then returning home. But I might return to America immediately and exercise there. If I were to take position in America in the Fall, I should only do what experimental work I need for my thesis, and could easily get ready to pass the examination at the beginning of August. I expect to go home with Berger and spend Christmas with him.*

5.34 To Parents, Leipzig, 11 December 1884

(Brackets indicate a word lost because of a rip in the page.)

It is a pity if I spend my days badly as they are all spent pretty much alike. Berger comes every morning at eight and we [   ] experiments which will turn the world upside down until one. I usually have some remnants of the morning's feast (I see my metaphor is sadly confused by the sentence following—I mean feast of reason)[1] to gather up, and afterwards cook my dinner, once in a while eating it at a restaurant. In the afternoon I work helping Berger with his thesis, or travel out in the country. After it gets dark I usually fence, swim or go to the gymnasium—one or all of them. In the evening I work twice a week with Wolfe—and twice with Berger, and ought to work the other three alone—but of course don't always.

[1]JMC added the parenthetical sentence as a footnote.

5.35 To Parents, Leipzig, 12 December 1884

(One paragraph of gossip has been omitted.)

As I have told you I expect to spend Christmas with Berger—his father is postmaster at Sangerhausen two or three hours away by train. We shall probably go Wed. and on Fri. if the weather is favorable walk for several days in the Harz mountains. Sangerhausen borders on these mountains. If the weather is bad I may go to Berlin for a couple of days. We will not begin experimental work until the Monday after New Year's.

5.36 To Parents, Leipzig, 17 December 1884

The psychologic laboratory was running this evening with Berger and Wolfe[1]—I have good reason to think that I am doing more (= quantity $\times$ quality) work than is being done in Prof. Wundts laboratory, and probably twice as much as is being done in Baltimore.

[1]JMC may have meant that he worked that evening in the laboratory of the university, as he did on other afternoons and evenings (see JMC to parents, 5 December 1884).

Imagine, the student who is going to do some averaging for me only asked 15 pf. 3½ cents the hour.[2] I give him 25, but it is enough to make me uncomfortable to think that I have often played billiards for 60 cts an hour—representing twenty hours of that man's work. I am also made nervous by a tailor, who sits at a window in the top story opposite me—he works all day long—beginning by lamp light long before I am up, and working till after I am in bed at eleven or twelve. I hope however in this instance it is only a Christmas celebration. But people must work very hard for very litte here in Germany. I imagine I should be a socialist if the company was not so bad.[3]

[2]JMC later identified this student (see 5.147) as G. Heise, who first enrolled at the university in 1884 as a student of theology (*Personal-Verzeichniss der Universität Leipzig*, no. 107, summer semester, 1885, p. 62).

[3]JMC later (see 6.108) got to know Wilhelm Liebknecht, the leader of the German Socialist Party and the father of Karl Liebknecht, who headed the *Spartakist* uprising in Germany at the end of World War I (JMC, "In Memory of Wilhelm Wundt," *Psychological Review* 28 [1921]: 155–159; *Encyclopedia of the Social Sciences* 5: 454–456). Still later JMC proposed "A Program of Radical Democracy" (*Popular Science Monthly* 80 [1912]: 606–615) that was socialist in many of its provisions.

5.37 To Parents, Leipzig, 18 December 1884

(One paragraph of gossip has been omitted.)

I am tolerably well satisfied with the Fall's work. My negative results are at all events very valuable—that is I have found the work done by others in this department to be mostly wrong.[1] The positive work is good as far as it goes, but it goes slowly as I am very careful, trying to avoid the errors into which others have fallen. If I worked more rapidly my results would be more showy—but of course less reliable.

[1]JMC is probably referring to his discovery of problems in the use of the Hipp chrono-scope and to Wundt's ignorance of these problems (see 5.47).

5.38 To Parents, Leipzig, 22 December 1884

(One paragraph of gossip has been omitted.)

Berger has just gone. I thought we would end up this evening, but we have decided to work for a while tomorrow. We leave for Sangerhausen at 12.40.

I ate one piece of candy today—the first since Harry used to treat us last winter. I shall send a lb. box to Pauline Gerhard, and take some to Berger's "Geschwister."[1] I imagine I shall not eat any, or at least not much this Christmas—for the first time—it seems almost a pity.

[1]On the Gerhards see 3.7. *Geschwister*: brothers and sisters.

I wish that tailor over there would go to bed. Saturday night he worked until after one. We have been having a great trial here—of anarchists who tried to explode dynamite at Niederwald—the Emperor and King of Saxony being in the procession. The accused (or rather convicted) are men of the working class—a cobler &c. Three of them were today sentenced to death.[2]

My letter of credit came as you know safely. I must draw money tomorrow—I shall just about begin the new year with it. I have given Berger in all $125.

[2]The details of this conspiracy and trial are discussed by Andrew R. Carlson, *Anarchy in Germany*, vol. 1, *The Early Movement* (Me-

tuchen, N.J.: Scarecrow, 1972), pp. 288–301.

5.39 To Parents, Sangerhausen, 23 December 1884

I came on here with Berger today—we leaving Leipzig at 12.40 and arriving here at five. You would be interested if you could see me. I am sitting now in the general room and of course can't think or write, as I am so used to being alone. Berger's father receives a very small salary and they live very plainly. You can best picture this to yourself when I tell you that the father, Berger, the little boy and I are to sleep in the same room. But there is somehow a difference from what one would see under similar circumstances in America. I cannot explain exactly what it is, but there is a certain ease—I might almost say culture in the way they live, and—for example treat me, which we would scarcely find at home. Beside the father and mother, and the boy of nine above mentioned, there are two girls—of twelve and fourteen. The children are bright and handsome—like Berger.

As I say I cant write under the present conditions. I would find it rather hard to write every day any how if I were knocking about. At Leipzig, it is very easy.

5.40 To Parents, Sangerhausen, 25 December 1884

(Two paragraphs of gossip have been omitted. Brackets indicate wormholes.)

This has been the most curious Christmas I have ever spent.[1] *I have been with happy children—which makes the best Christmas now that we cant be children ourselves—but while it seems Christmas for them, somehow [   ] not seem like Christmas at all. This morning Berger and I made a long excursion—fifteen miles—into the mountains. It was very pretty and interesting, but by no means easy walking as the snow was [   ] foot deep. We had for Christmas dinner [   ] and potatoes—apple sauce being [   ] me. We help ourselves [   ] own fork and to butter [   ]. I am not spoiled any more as I used to be at home, so dare say I wont be so particular—until you spoil me again.*

[1]See also 3.4.

This letter many not be worse than usual, but it is curiously difficult for me to write when surrounded by people. At all events this is not written for a biographer as Carlyle's wife suggests he wrote his letters to her.[2] It takes me five minutes to think of a sentence, then after I begin it, I forget how it was to end.

[2] Though Jane Welsh Carlyle's specific comment has not been found, it is clear that, as the editor of his correspondence noted, Thomas Carlyle "clearly wished the letters preserved as a record of his life that would be as full as possible" (Charles Richard Sanders, ed., *The Collected Letters of Thomas and Jane Welsh Carlyle*, Duke-Edinburgh edition [Durham, N.C.: Duke University Press, 1970–1977], vol. 1, p. xi).

5.41 To Parents, Leipzig, 28 December 1884

(Two paragraphs of gossip have been omitted.)

I received your letter today, Papa. Your letters come with great regularity two weeks after they are written. I don't know why mine turn up all at once. Herr Berger does not think that letters are sent by way of England, but says there are two mails a week sent by way of Bremen and two by way of Hamburg.

I don't know what you mean, Papa, when you say "I want the ipsissima verba of Prof. Morris about you." "Send them in your next letter."[1] If you mean Prof. Wundt's letter, I was about to ask you to let Harry send me a copy of it. I gave it to you the night before I left Keswick and Harry took a copy before I left Leipzig.[2] It is a pity if it is lost—if you have it plese ask Harry to copy it carefully for me.

[1] *Ipsissima verba*: identical (or exact) words. JMC is quoting here from his father's letter of 14 December 1884.
[2] Apparently Wundt had written a letter about JMC and had given it to him the previous Spring (see JMC to parents, 9 December 1884.) No copy of this letter has been found. (But see 5.88.)

5.42 To Parents, Leipzig, 30 December 1884

(Brackets indicate wormholes.)

All my plans have been upset today. I had intended to finish my thesis and see it printed in Prof. Wundt's "Philisophische Studien" before I left Leipzig, and also to get any other experimental work I may have ready for the press—then when this was ready I intended coming home, studying some there for the oral examination and the rest here so that I could pass a [   ] in the winter or early [   ] Prof. Wundt tells me my doctor's thesis can not be printed any where until I have received the degree. My other work I cannot well print before that contained in the thesis—so what am I to do? I must either wait and take my degree next winter as I had proposed and put off printing my work until then, or try to pass the examination this summer. Even if I pass the examination this summer my thesis wont be printed immediately, as Prof. Wundt wont issue a number of his "Studien" for my benefit. (It is issued when he has enough manuscript to fill a number—three or four times a year). If I pass the examination this summer my oral examination will be poor, my thesis will not be completed as I would like to see it, and I cannot finish up the other experimental work—three serious objections.

There is a third alternative. I can [   ] work that has been printed [   ] to advise me. On the whole I [   ] in this whole matter for your [   ] rather than mine. If you were not living I should probably never have done this experimental work—if I had I should print it, without special regard to the doctor degree. Should go to Greece and read Plato and the Dramatists, should probably spend a year in the East, and then another year in Germany. At the end of this time, I should it is true want a doctor's degree, but should not concern myself about the grade.

But I really will do what you wish if you are decided as to what you think best. I don't think I can print anything soon. I wanted to write a paper for Brain, but this was to contain some of the work Berger and I [   ], and Berger is to use this in his thesis, and ought to print it.[1]

[1] Berger did publish this work first as his dissertation; then JMC published it in *Brain* (see 3.12).

5.43 To Parents, Leipzig, 31 December 1884

This is the last thing I shall do this year—it will soon be gone. I have been celebrating New Year's Eve by writing on my doctor's thesis. I have written fiften pages since I have been back—there must be in all about seventy-five to make forty or fifty in print. I can write it easily enough, as fast as I find out what I have to describe. I shall write it in English, and then dictate it to Berger in German, and he must set the German right. Prof. Wundt says I can do this, indeed wants me to do it.[1]

I have further celebrated the end of the old year by deciding to try to pass my examination about the first of July. It is Sylla & Charybdis, but unless my mythology is at fault Ulysses got through. I risk a good deal, but that may be the best way to win. I shall get my thesis done the end of April, and if Prof. Wundt will print it in the next number of his Journal (which he proposes issuing in June or July) I shall let him. Then just before the Journal appears I shall try the oral examination. If I fail in this I must write another thesis, but shall have almost enough work ready for this and can finish it in the Fall. I shall try to make my thesis as good as possible, and shall devote myself to this during the next four months. The other experimental work I must let go mostly, but can perhaps get it in shape after the examination is over—supposing it is over. After I have handed in my thesis I will of course "cram" until the examination. I shall not of course concern myself about the grade—if I only get through! The grade of the thesis is (or can be) printed on it, but no one knows anything about the oral examination. I must be examined in philosophy, physics and

[1] JMC's German was never very good (see Cattell "Autobiography," p. 631).

zoology. I can pass in philosophy, and can probably get ready to pass in physics. It would be wise to take English as third subject, but I don't intend to. I called on Prof. Leuckhart (Zoology) yesterday.[2] He said he would examine me mostly on biology and comparative physiology and anatomy, and I know something about these.

You see I have decided to do all I can to help you to get me a position. If you cannot however I shall look to spend next winter studying over here.

[2] Karl Georg Friedrich Rudolf Leuckart (*DSB* 8: 268–271), professor of zoology at Leipzig, was best known for his work in parasitology.

5.44 From William Cattell, Philadelphia, 1 January 1885

(Several paragraphs of gossip have been omitted.)

I was glad to read that you had employed help to relieve Berger & yrself from computing averages &c—such clerkly work is a waste of time for either of you. What money may be necessary to have it taken off yr hands is wisely expended.

And I noted with special interest what you said about correcting Prof. Wundt's tables.[1] It is not in human nature to be thankful for such services. Prof. W. may appreciate the work you have done, but it will require skill & prudence for you to retain his kind personal feelings towards you and this goes a great way with such men if they are "applied to" for a recommendation &c—They w^d not deny—probably w^d admit—the value of certain work, but w^d couple their opinion of <u>this</u> with some unfavorable expression about <u>you</u>, that might be very damaging to yr prospects. I incline to think that Prof. W. w^d do for you all that could be expected: and I have great confidence in your tact, and also know that you see the great importance of my receiving a favorable opinion from him not only of yr work, but of <u>yrself</u>. I expect to spend a day in Princeton with Dr. Taylor soon; & shall ask him (as a Trustee of Princeton College) to write to Prof. W. about you.[2] I wish the college w^d offer you a place: —but I shall not be satisfied until you get in the line of promotion here

[1] See 5.37. However, the specific letter to which William Cattell refers earlier in this letter—one dated 15 December 1884—has not been found.

[2] No Dr. Taylor was listed among the trustees in any Princeton catalog of the early 1880s. William Cattell inserted the phrase "with Dr. Taylor" above the rest of the line.

5.45 To Parents, Leipzig, 3 January 1885

(One paragraph of gossip has been omitted.)

I have written about one third of my thesis, this is all I can write for the present. Of course it must be thoroughly revised and translated into German. Berger is to return tomorrow and we begin work once more Monday. We can work for about three months—I hope in that time to get the thesis in good shape. I shall not trouble myself about the examination, until the thesis is out of the way.

5.46 To Parents, Leipzig, 4 January 1885

(One paragraph of gossip has been omitted.)

I received yesterday a letter from Surg. Billings.[1] I think he is going to give me a copy of a catalogue issued by our government and edited by him[2]—if so he will send it to you. It is a valuable work and would be useful to me. I think he will set up in the museum at Washington the psychometric appartus used and to a large extent invented by me.[3] He is Professor at Johns Hopkins.[4] You remember I met him on the steamer. I must have been almost as good as you, Papa, on that steamer. I got quite well acquainted with Mr Depew and his family—also with Christine Nillson, Geo. Jones (editor of the Times) Prof. Wm. Darling, and a great lot of people.[5]

[1] John Shaw Billings (see 4.10).
[2] *The Index-Catalogue of the Library of the Surgeon-General's Office*, edited under Billings' direction from 1880.
[3] Billings was curator of the Army Medical Museum, and later noted that the museum "fitted up one large room with instruments and apparatus for anthropometry in its widest sense, including psychophysical investigation" (Billings, "On Medical Museums, with Special Reference to the Army Medical Museum at Washington," *Medical News* 53 [1888]: 309–316). Billings wrote to JMC in the letter referred to here (19 December

1884) that he had just "purchased for the Museum a set of the Anthropometric instruments of Francis Galton as exhibited by him at the Health Exhibition in London" and that "I should be glad to add to those instruments one or two instruments for recording reaction time provided I can obtain something that is simple and can be used with very ignorant people."
[4] Billings was medical advisor to the trustees of the estate of Johns Hopkins, and played an important role in the establishment of the university hospital and medical school.
[5] See 4.10.

5.47 To Parents, Leipzig, 5 January 1885

Berger turned up early this morning and we started work. By way of variety not one but both of the electric batteries were out of order. You have no idea how much one must fuss over apparatus. The trouble is not that one must know physics, but that he must be an original investigator in physics. For example Prof. Wundt thought that when a magnet was made by passing a current around a piece of soft iron, it was made instantaneously. I find with the current he used it takes over one tenth of a second.[1] All the times he measured were that much too long.[2] Now the time re-

[1] JMC is probably oversimplifying the situation in an attempt to make it clear to his parents. He gave a more reasonable description of Wundt's assumptions in a letter to John Shaw Billings (3 January 1885, Accession Files, Medical Museum of the Armed Forces Institute of Pathology): "The times obtained by Prof. Wundt and others using the chronoscope are wrong because the times required for the magnet to attract the armature and let it go were not the same." See also 3.12.
[2] The corrections JMC thus made in Wundt's data are probably those referred to earlier. (see 5.37). It has been suggested to the editor (by Solomon Diamond) that the pages on

which JMC adjusted his data to compensate for this error were later found by Lincoln Steffens, the American journalist, who studied with Wundt in 1890 and 1891. Diamond further suggested that these pages led Steffens to charge that one of Wundt's best-known American students—left unnamed—had purposely falsified his data so as not to arouse Wundt's ire at having his theories challenged. See Lincoln Steffens, *Autobiography* (New York: Harcourt, Brace, 1931), pp. 150–151; Alfred C. Raphelson, "Lincoln Steffens at the Leipzig Psychological Institute, 1890–1891," *JHBS* 3 (1967): 38–42.

quired for magnetism to be developed in soft iron has nothing on earth to do with psychology, yet if I had not spent a great deal of time on this subject all my work would have been wrong. Then it is hard to get apparatus made properly. The machinist thought he would give me a pleasant surprise by finishing up what I had ordered while I was away, but it must mostly be done over. I go there every day. He has been making a piece of apparatus for Prof. Wundt for the past six months—it was to have been done Oct. 1st, but I imagine it wont be ready for a long time yet. All this, the getting the apparatus and running it, is very aggravating when one is in a hurry.[3]

[3]Portions of this letter—dated incorrectly— appear in Michael M. Sokal, Audrey B. Davis, and Uta C. Merzbach, "Laboratory Instruments in the History of Psychology," *JHBS* 12 (1976): 59–64).

5.48 To Parents, Leipzig, 9 January 1885

Berger and Wolfe have just gone—the psychological institute meeting Wednesdays and Fridays. We of course worked this morning from eight to one. We must begin before it is fairly light. That tailor over there and I are martyrs to advancing progress and civilization.

This afternoon however I "cut" laboratory—skated, exercised in the gymnasium and fenced. I disarmed the fencing master with my left hand, he using his right. My left arm is stronger than my right was at this time last year. The skating is good— we having genuine winter weather. The water in my bed-room is already frozen.

5.49 To Parents, Leipzig, 10 January 1885

After the lecture this afternoon I told Prof. Wundt what I want to do about the examination &c. I said I would give him the thesis the end of April for his "Studien", and a second copy to the Rector as a doctor's thesis. Supposing the thesis is accepted in both places, I shall make application to be examined in philosophy, physics and zoology just before the "Studien" comes out (in June or July). It will be a great pity if I fail in the examination—likely enough—as I must then prepare a new thesis. I could make sure of getting the degree, if I should not work very much more on the thesis, and should take English instead of zoology. But it is of the greatest importance that the thesis should be good. Almost any thesis I hand in (I imagine) would be accepted—but of course with a low "Censur," and Prof Wundt would not want it for his "Studien".[1] It is really more of an honor to have my paper printed by Wundt that to get the Ph.D. so I had better make sure of that, and if I decide that I had better not try the examination, or fail in it, I can easily enough get the degree next year.

[1]The term *Censur* meant more than a grade; it carried something of the connotation of an imprimatur of the university faculty (see George M. Stratton to George H. Howison, 12 May 1896, Howison papers).

5.50 To Parents, Leipzig, 12 January 1885

Nothing much happens to me—my chief excitement being when something gets a matter with my apparatus. I got my last machine on Saturday, and am glad my daily visit to the machinist is at an end for the present.[1] It was hard work—however stupid he might be I had to keep in good humor (outwardly at least) and keep him in a good humor. Prof. Wundt was in the laboratory this afternoon. I see him often and we are good friends. Give me credit for that, as well as for the way I get along with Berger. My relations to both are very delicate—could only be steered straight by one with my great suaviter in modo.[2]

I don't know what these two pieces of apparatus cost—perhaps $25 and $5, but I imagine I wont get any more—I haven't much time now.

[1] The machine mentioned is probably the voice key (see 5.30 and 5.65).

[2] This phrase is from the Latin expression *"suaviter in modo, fortiter in re,"* meaning "gentle in manner, firm in deed."

5.51 To Parents, Leipzig, 14 January 1885

(Several paragraphs of gossip have been omitted.)

The psychological laboratory was running this evening—which is not especially good for me as it keeps on running through my head all night. But perhaps subconscious cerebration will evolve some original hypothesis.[1]

[1] The concept of subconscious cerebration was current throughout the nineteenth century, and William Benjamin Carpenter had used it in the text JMC had read while at Baltimore (see 2.23; Carpenter, *Principles of Mental Physiology* [London: Kegan Paul, 1877], pp. viii–ix; *DSB* 3: 87–89; Lancelot Law White, *The Unconscious Before Freud* [New York: Basic Books, 1960]). Wundt's own work relating to apperception was based on the assumption of two levels of consciousness, one closer to immediate awareness than the other (see 3.12), and Wundt had earlier developed the concept of what he called "unconscious conclusions" (see *DSB* 14: 526–529). JMC did not use this idea explicitly in any of his published work. However, in his criticisms of Wundt's method for measuring discrimination time (see 3.12) he implied that discriminations were made without conscious action, and hence were unconscious. In the same way, in his analysis of reading, he wrote that it was "a completely automatic act," not requiring a "conscious act of choice" (*Man of Science*, vol. 1, pp. 15, 23).

5.52 To Parents, Leipzig, 15 January 1885

(Two paragraphs of gossip have been omitted.)

For your sakes I hope you will find me a position at Philadelphia or Princeton or someplace. I'm doing all I can to help you. A paper in German fifty pages long full of plates and tables with some such title as Die Apperceptionzeit bei Lichtvorstellungen,[1] ought to be imposing on people who know neither psychology nor German. I

[1] "The Apperception Time of Light Images."

shall write the paper in German, only the grammer being corrected by Berger. Then if I get the Ph.D. that ought to be useful. Not many Americans have taken the degree at Leipzig or Berlin (it being far easier at other universities, Göttingen, Heidelberg &c) and perhaps none as young as I.

5.53 To Parents, Leipzig, 16 January 1885

(One paragraph of gossip has been omitted.)

Prof. Wundt lectured yesterday and today on my subject[1]—I suppose you won't consider it egotistical when I say that I know a great deal more about it than he does, but you will be surprised when I say that half of the statements he made were wrong. I cannot understand how he is willing to give as positive scientific facts the results of experiments which he knows were not properly made.[2] I could write a paper on these two lectures most damaging to Prof. Wundt. It is to be hoped for his sake as well as mine that he passes me in the examination on philosophy.

The psychological laboratory has just adjorned. It is cold again. I skated for an hour at dinner time.

[1] See *Grundzüge der physiologischen Psychologie*, 2nd edition, vol. 2, pp. 219–291 for an approximation of Wundt's views on reaction-time work during this period.
[2] See 5.47.

5.54 To Parents, Leipzig, 17 January 1885

This afternoon I had the "usual half holiday"—but I forgot; your acquaintance with Gilbert and Sullivan does not extend beyond Pinafore. I played tennis—I would rather have skated, but had made the engagement to play before the ice came. Do you think it wrong to skate on Sunday here in Germany? At Göttingen the clergy-man's daughters used to bring their skates with them to church.

This evening I have been fussing over my thesis—I'm sure I shant be able to write it to suit me. After doing such a mass of experimental work one ought to wait a long time before writing it out, so that he can see things in their proper perspective. Then think of preparing fifty pages of printed German.

5.55 To Parents, Leipzig, 18 January 1885

Sunday comes as a welcome interruption in my work. I got up good and late this morning—ten o'clock—and it seemed as though I might have slept on indefinitely. Some people wonder why mosquitos were ever created, I wonder why things are so arranged that we dont want to go to bed at night, and don't want to get up in the morning. My menage (I must get me a French dictionary—I have no idea where to put ` ´) moves along smoothly. I eat nearly altogether in my room. I do not think however that I have added any dishes to my bill of fare. I have plenty of my old friends apple sauce and cream. The Hausmann's Frau makes me a big dish of apple sauce twice week, and I get a pint of cream a day. I am not, and never of course was a vegetarian but I dont each much meat—three or four times a week perhaps. If I were home I think I should eat it once a day, but then there would be so many good things, I should not know where to begin. I wish you had my Frau for a servant. My family is quiet and keeps in good order—at least as long as I let it alone. I only fed it zink, mercury, copper sulphate, and salts of ammonia.[1]

[1]*Ménage*: household. JMC's "family" is his batteries, which required regular additions of solutions to keep them working (see 5.10).

5.56 From William Cattell, Philadelphia, 18 January 1885

(This letter was dated "Sunday, 18th." Several paragraphs of gossip have been omitted.)

I made my first "presentation of the cause" before a Presbytery, in N.Y., on last Monday afternoon; had a very pleasant visit at Mr. Woods where I spent the night[1]—the next day stopped at Princeton, as I told you in my last letter was in the programme. I had quite a long talk with Dr. McCosh at his house, Tuesday evening.[2]—inter alia—learned that they had two young graduates of the College in the new department of Philosophy—both of them had studied in Germany as "Fellows" of the College; & the Dr. was much pleased with them both: —and no addition to this teaching force was contemplated. So the way does not seem clear for an invitation even to a tutorship there! And the fact is, such a position as we want for you is sure; —and the applicants many with strong "backing" too. One of these young men is a grandson of Dr. Hodge; the other, the son of a very liberal benefactor of the College![3] I met however in N.Y., a Prof. Stevenson of the Univer-

[1]On George C. Wood see 4.2.
[2]On James McCosh see 4.14.
[3]William Cattell's account of the situation at Princeton was confused. Henry Woodhull Green, (*DAB* 7: 546–547), an alumnus of the College of New Jersey and longtime Chief Justice of the New Jersey Supreme Court, had established the "Chancellor Green Fellowships in Mental Science," awarded each year to a recent graduate of the college. For 1883–1884 the Chancellor Green Fellow was John Gormley Murdoch (*Who Was Who in America*, 1897–1942, p. 881), who spent his fellowship year reading at Princeton under McCosh's direction, and who later taught English at Rensselaer Polytechnic Institute. The fellow for 1884–1885 was James Mark Baldwin (*DAB* Supplement 1: 49–50), who studied for a while at Leipzig (see 5.67n.1)

Berlin, and Freiburg and who later was a major figure in the psychological community as professor at Toronto, Princeton, and Johns Hopkins (Baldwin, *Between Two Wars, 1861–1921: Being Memories, Opinions and Letters Received*, 2 vols. [Boston: Stratford, 1926]). Neither Murdoch nor Baldwin was related to Charles Hodge (*DAB* 9: 98–99), longtime member of the faculty at the Princeton Theological Seminary, and neither was from a family that had contributed much to the college. Baldwin, however, returned to Princeton from Europe in 1886 as instructor in French and German and read for a Ph.D. in philosophy with McCosh. He then taught at Lake Forest University and Toronto before returning to Princeton in 1893 as Professor.

sity there:[4] he told me that Dr. McCracken, the new Professor of Psychology with whom I had quite a long talk in Edinburg,[5] had spoken to him about you & wished the University could give you some position. So my talk with him was, at least, not forgotten. But the University is very poor.

[4]Probably the geologist John James Stevenson   [5]See 4.14.
(*DAB* 17: 632–633).

5.57 To Parents, Leipzig, 22 January 1885

(One paragraph of gossip has been omitted.)

I worked in Wundt's laboratory this afternoon probably for the last time. Berger has enough material for his thesis—he is to hand it in next month. I have no doubt but that it will be accepted, and if so mine will be as a matter of course. Wundt's laboratory has a reputation greater than it deserves—the work done in it is decidedly amateurish. Work has only been done in two departments—the relation of the internal stimulus to the sensation,[1] and the time of mental processes. The latter is my subject—I started working on it at Baltimore before I had read a word written by Wundt—what I did there was decidedly original. I'm quite sure my work is worth more than all done by Wundt & his pupils in this department, and as I have said it is one of the two department on which they have worked. Mind I do not consider my work of any special importance—I only consider Wundt's of still less. The subject was first taken up by Exner, and Wundt's continuation of it has no originality at all; and being mostly wrong has done more harm than good.[2]

[1]That is, classical psychophysics. For example, see Gustav Lorenz, "Die Methode der richtigen und falschen Fälle in ihrer Anwendung auf Schallempfindungen," *Philosophische Studien* 2 (1885): 394–575.

[2]Sigmund Exner (*World Who's Who in Science*, pp. 538–539) was a physician who would long be connected with the University of Vienna. See also 5.68. In several articles on "Experimentelle Untersuchungen der einfachsten psychischen Processe" (*Archive für die gesamte Physiologie des Menschen und die Tiere* 7 [1873], 601–660; 8 [1874]: 526–537; 11 [1875]: 403–432, 581–602) Exner coined the phrase "reaction time" and interpreted all reactions as "prepared reflexes" involving no mental activity. Wundt argued that varying the conditions of the reaction-time experiment introduced such mental processes as discrimination or choice into the reaction. JMC, whose experiments led him to doubt that these processes took place during reaction, therefore thought more highly of Exner's work in this area than of Wundt's, and he cited the articles noted above in his dissertation. See also 3.12; JMC, "Cerebral Operations," in *Man of Science*, vol. 1, p. 53; Robert S. Woodworth, *Experimental Psychology* (New York: Holt, 1938), pp. 305–306.

5.58 To Parents, Leipzig, 27 January 1885

(Two paragraphs of gossip have been omitted.)

I talked to Prof. Wundt this afternoon about my arrangements, and have not been in an exactly comfortable state of mind this evening. One great trouble is Prof. Wundt does not know when the next number of his magazine is to be published. I am to give him my paper the end of April. Berger must leave Leipzig about the 8th of April. I think I would give $100 to know when the next number of Wundt's Studien and the one after the next is to appear. If the one after this should appear in the Fall I think it would be better for me to print my thesis in it. I could write it better if I had more time—if for example I could consult the libraries of Berlin or London. I could also complete the experimental work I have on hand, which is impossible between this and April. Then the examination. If I fail in physics or biology, as I have explained, I have forfeited my doctor's thesis. It would it is true appear all the same in Wundt's magazine, but I would have to prepare a new thesis, and as I intend to include all the experimental work I have in the first thesis, the preparation of a second would be no easy task. It would take me a year to prepare another as good as this one.

There are only two objections of consequence to my putting off the examination until Fall or Winter. The first is that I cannot publish any thing till then. Still if what I print then is made better by the delay perhaps it would be better to put it off. Second my coming home is somewhat involved. Still I might work on here until the first of August, and go to America in the regular university vacation. If you secure me a position I would pass the examination the first of August. Berger has not been able to get a position as Turnlehrer, he must therefore teach a year without any salary.[1] He cannot even do this here in Leipzig, and must go to Mühlhausen at Easter. But if I should pay his expenses he might be willing to stay here and help me. At all events he has vacation during July, and would come here then. We could then make a few more experiments, and put the thesis into German.

[1] *Turnlehrer*: gymnastics teacher (see 5.7).

5.59 To Parents, Leipzig, 28 January 1885

I talked to Prof. Wundt again today about my thesis and examination. His Studien are published in this way—as soon as he has a manuscript to make up a number of from 150–200 pages he publishes it. He has now a paper of his own and two others, and says he can at any time receive four or five papers—but does not know exactly when they will turn up. If I should give him mine, he would be ready to print the next number. Berger's thesis is also to be printed, but was to appear in the number after this.

It seems to me best (and also to Prof. Wundt) that I should wait for the second number. He is very kind about it—says I need not decide now, but can wait until April, and if my paper is ready he will print it in the first number, and if it is not ready I need not trouble myself. But I can decide now as well as then. I can have the paper ready, but it will not be as good as if I wait. You see the experiments, the arrangements, the writing out, and the translation must all be going on at the same time, and this only two months. It would be far better if I could do the things one after the other—I think it would be worth waiting six months longer. Berger and I can devote all our time to the experiments until the sixth of April. Then I can take rather a long vacation, and on returning to Leipzig, carefully review the work and write the thesis. Berger is willing to spend his vacation (July) helping me (I would give him $50). We could then do any further experimental work I had found desirable, correcting any mistakes I might have found, and put the thesis into good German. I could give it to Prof. Wundt the first of August and go to America, being in Leipzig again at the opening of the semester at the end of October, and would pass the examination as soon as ready. I need not wait for the "Studien" to print my thesis. Prof. Wundt will get the publisher to set up the paper (without expense to me) whenever I wish, and I can print the thesis. This seems to me the wiser course. But if Prof. Wundt delays the publication of his Studien until August, or if I were to take a position, I would finish up everything next semester. Prof Wundt thinks I need not worry about the examination.

In your letter just received, Mama, you seem to indicate that I am staying longer here than you had expected. But when we came to Europe in '83 we intended to stay two years and I proposed passing my examination in the summer or fall of '85. Last summer my work was interrupted for six or eight months. Yet I am ready to pass the examination this summer, if it is desirable. I am considerably younger than the average age at which Germans take the degree, and the fact that Prof. Wundt is going to print my thesis, proves it to be better than the average.

Of course you understand the trouble is the oral examination. I could have the thesis ready, but if the Studien were issued in May I could scarcely be ready for the examination. I might pass the examination the first of August and print my paper in England Oct. 1st, but this would displease Prof. Wundt.

5.60 To Parents, Leipzig, 30 January 1885

(One paragraph of gossip has been omitted.)

The psychological institute has just adjourned. The third man however is not of special use. Berger and I work very well together. We undoubtedly do more than all of Prof. Wundt's laboratory. But of course the work we do is only understood by specialists—the ground of science is so cut up now that each one must cultivate a very little plot. Still I suppose I might write a semi-popular article on the time it takes to see, to will and to think.[1]

[1] JMC did so later: "The Time It Takes to Think," *Nineteenth Century* 22 (1887): 827–830.

5.61 From William Cattell, Philadelphia, 1 February 1885

(Several paragraphs of gossip have been omitted.)

Now as to yr thesis & Prof. W's Journal—An article from yr pen in the Journal w^d be of much more use in calling attention to you here, then even yr thesis: especially if the Professor w^d make some reference to it. I should want a no. of copies of the No. of the Journal also a good English translation. Can you form any notion as to when it will be printed?

I need not impress upon you the importance however of its being good work: and I want to say again, don't waste yr time & strength by doing the <u>clerkly</u> work. Get Berger for all the time he can give you: —in such subordinate work as does not, of course, give him a "claim" as joint author with you &c—

I go to N.Y. this week and shall take the original copy of Prof. W's letter with me to show to Prof. MacCracken (I referred to him in a recent letter) and this will give me an opportunity to reopen the subject with him.[1] The Princeton door seems closed:[2] and it does not look very promising here:[3]—so as the next best thing I shall keep the subject before the authorities of the N.Y. University—I hope they may offer you some position: but we cannot give up the thought of having you <u>at home</u>.

[1]Wundt's letter has not been found; see 5.41. On Henry Mitchell McCracken see 4.14 and 5.56.   [2]See 5.56.

[3]That is, at the University of Pennsylvania.

5.62 To Parents, Leipzig, 4 February 1885

(One paragraph of gossip has been omitted.)

My last piece of apparatus is today in good order. Now all the apparatus I use I have invented myself. It is true I bought the clock,[1] but it had never been properly used before, and can only be made to measure correctly by aid of apparatus I have made. You know what I want to do is to find how fast we think—how long it takes to see, to will &c. No one can directly measure the time of mental processes, and it is perhaps the most difficult (certainly as important as any other) subject science tries to investigate. Science has always solved the easiest questions first. Astronomy and physics were well advanced before Chemistry and geology were thought of, then came physiology and biology, last of all psychology.[2] Psychology is of course the most important of the sciences—all science and all knowledge depends on the nature of the human mind.

[1]"Clock": the Hipp chronoscope.   [2]These ideas are similar to Comte's (see introduction and 2.36).

5.63 To Parents, Leipzig, 11 February 1885

I received a telegram from Berger last night, saying he would be detained today in Berlin, so our work has been put off another day. However I shall have materials for a good thesis by the end of next month, so I need not really worry about that.

I am invited to supper at Prof. Wundt's tomorrow evening. I am also invited to a Mask-ball at Frau Goedecke's on Saturday.[1] I called intending to excuse myself, but she said I need not wear a costume, and as I would rather like to see it (it will be very large and fashionable—some thirty or forty officers being invited) I accepted. Now I scarcely know what to do. I may go very late without a mask, or may wear a cricket (tennis) suit.[2] I have ordered what I need for this, as I want it here any how. I neglected to bring my flannel trousers with me, and the poor blue coat has had added to its already varied experience six weeks in the Tyrol and all winter devoted to psychology.

It is a bit cold again, and I skated this afternoon though the ice was bad.

[1] Possibly Elise Louisa Henrietta Gödecke (*Leipzig Adress-Buch für 1901* [Leipzig: Albergander Edelmann, 1901], p. 278).
[2] JMC did go to the ball in such an outfit, "which looked very nicely and did very well, as no one knew what it was, and they naturally thought it was gotten up especially for the occasion" (JMC to parents, 14 February 1885).

5.64 To Parents, Leipzig, 13 February 1885

As I told you I was invited by Prof. Wundt to supper with other members of the laboratory.[1] I cant say that I enjoy such things, I have no special reverence for any one I know personally, and it gives me no special delight to hear Wundt talk about the opera and such like. Mrs Wundt is however nice and Prof. Wundt seems to like me and to appreciate my phenomenal genius.

I received a letter from Washington last night which pleases me.[2] In the first place Surg. Billings is going to give me the very valuable dictionary or catalogue he edits.[3] A new volume appears every year and this is also to be sent to me. Then he sends me an order for $250 for which I am to get a set of the instruments for measuring the reaction-time which I use and have for the first time.[4] He also asks me to set up the apparatus in the museum at Washington. If I wish I can give the government the apparatus I am using, and so would have spent nothing on apparatus, but it would be better perhaps to keep my own things even though I do not use them. They have cost me about $175. Of course I'm not to make any money out of

[1] Wundt played host to his students fairly often, at least in later years (see "In Memory of Wilhelm Wundt," *Psychological Review* 28 [1921]: 164, 171, 176).
[2] From John Shaw Billings, 28 January 1885, in answer to a letter from JMC, 3 January 1885.
[3] See 5.46.
[4] See 5.80.

this arrangement, but it is no trouble at all, as I give the things to the agent of our government here in Leipzig. I am also to get for the library at Washington a copy of Wundt's "Studien"—this pleases him, as does also the order for the apparatus.

I enclose a most valuable communication which I find in the pocket of a coat I have not worn for two weeks.[5]

[5] A letter JMC had written earlier but forgotten to mail.

5.65 To Parents, Leipzig, 13 February 1885

Berger and I worked as usual today and yesterday. The last piece of apparatus opens up work which would keep us busy for a long time if we had the long time. The apparatus registers the instant at which one speaks.[1] A membrane is set in vibration by the sound and an electric current is broken, when this current is broken the magnetism in an electric magnet disappears and a spring draws away the armature and breaks a second current which stops the hands of the chronoscope. I have indeed invented an imperfect telephone—quite different from (as also inferior to) the telephones in use. I have found some interesting things—for example you can see the picture of a hat in less time than you can see the printed word "hat," but it takes much longer to name the picture than the word.[2]

I was skating this afternoon on very bad ice.

[1] This instrument was the voice key (see 5.30 and 5.50).
[2] See "Uber die Zeit der Erkennung und Be-nennung von Schriftzeichen, Bildern und Farben," *Philosophische Studien* 2 (1885): 635–650.

5.66 To Parents, Leipzig, 19 February 1885

It has been a beautiful day—bracing air and a warm sun. I found a Schneeglöckchen in the woods this afternoon.[1] The papers say it has been very cold in America this week. This evening I have been arranging the work—for my thesis. I am glad I do not have it all ready in German the first of April. Berger "gymnasticizes" with the Verein Monday and Thursday. I should like to too, but so seldom have an evening to myself. I do not get to the opera as often as I should like to. I should also like to attend the Gewandhaus concerts Thursday evenings.[2] You can get seats in the new building. I was in it for the first time last week—I heard the Probe the day Berger was in Berlin.[3]

[1] *Schneeglöckchen*: snowdrop (a flowering herb).
[2] See introduction to section 3.
[3] *Probe*: test. This clause refers to the final testing of the new hall.

5.67 To Parents, Leipzig, 25 February 1885

I worked with Berger this morning and this evening. This afternoon I played tennis with Muirhead, Tuttiet and Baldwin;[1] afterwards took a bath and read the papers. When I add that I had scrambled eggs for dinner and a sandwich for supper, I have given you a circumstantial history of the day. The constant excitement and novelty here will I fear lead you to think that I shall not be happy when I settle down at home. Think of the mad emotion, the wild longing, the vibration between triumph and dispair,—think how the heart beat fast and loud, how the blood flowed fire, how the brain whirled—when there was a deuce game at tennis!

[1]On Findlay Muirhead see 5.24n.2. On James Mark Baldwin see 5.56n.3. Baldwin stayed only one term at Leipzig and "did no more than familiarize myself with the methods and principal researches of the laboratory, acting as subject (*Versuchsthier* [experimental animal]) for the more advanced students" (Baldwin, *Between Two Wars*, vol. 1, p, 32). Less than a year after the date of this entry (on 3 January 1886) Baldwin wrote to JMC from Princeton asking for details about experimental procedures and apparatus.

5.68 To Parents, Leipzig, 26 February 1885

This afternoon I walked out into the country, took a bath and read medical magazines. This evening I have been busy over physiology—for about seven hours in fact. I must be thoroughly familiar with nervous physiology in order to write my thesis properly. That this is no easy task you can see when I tell you that over 2000 pages are devoted to it in Hermann's physiology.[1] I have just bought the two volumes treating it.[2] Besides physiology there are a number of attractions Thursday evening. There are the theatres and the Gewandhaus concert.[3] Then the Turnverein exercises, and the English I know best bowl Thursday evenings. But I cannot be omnipresent.

[1]Ludimar Hermann edited the standard German *Handbuch der Physiologie* (Leipzig: F.C.W. Vogel, 1879–1883).
[2]JMC was especially interested in the section by Sigmund Exner on reaction-time experiments (vol. 2, part 2, chap. 4, pp. 252–283), and cited this section and several other articles by Exner in his dissertation (see 5.57).
[3]See introduction to section 3 and 5.66.

5.69 To Parents, Leipzig, 8 March 1885

(Two paragraphs of gossip have been omitted.)

Just as I have decided to pass my examination this summer, I receive your letters advising me to put it off until the winter semester. In this case you do not seem to think it best that I should come home in the summer. I do not look on crossing as such a great matter, and consider it a healthful, perhaps the most healthful way of spending a vacation, so only the expense were to be taken into account.

However now I think it better to get through with the oral examination. I don't believe it will be so terribly hard, but it troubles me to think about it, and the longer I put it off and the more I study for it, the more formidable will it become. I shall now take it very easily—read through a physics and biology, and review philosophy a bit. If I fail in some subject I shall either make it up or choose an easier one. I suppose I shall have to correct the proof sheets, and get the thesis printed in September, but I cannot do this until the thesis has been officially accepted, so it might be desirable for me to be here at the opening of the semester. Then it might be possible to write something an English review would print—if so it had better be done here and at once.

5.70 To Parents, Leipzig, 12 March 1885

I am going to have a paper in the next number of Wundt's "Philosophische Studien." [1] It ocurred to me a couple of days ago that it would be better to print two papers than one. I think they will let me change this a bit and include it in my doctor's thesis to make that as formidable looking as possible. At all events the thesis will be accepted whether it contains this special work or not. [2] You remember it is not allowed to print the thesis or part of it before the oral examination has been passed.

This paper contains about the work they did not give me my fellowship on, and which Dr Hall wanted to appropriate. [3] I wrote the paper nearly all at one sitting, finishing it this evening—it will make eight pages or so, including elaborate looking tables. [4] It was very good of Prof. Wundt to accept it, as I told him of it this morning for the first time, and only what it is to contain in a general way. He is to give in the manuscript for the next number of his journal in one week, but lets me take two weeks if I need it. I must translate it into German and complete a few tables. Of course Prof. Wundt will not publish the paper unless he finds it good, but it is kind of him to accept it in this way without seeing it, and after the number has been made up.

The number will be published in May. I can send you as many copies of my paper as you wish, the whole Journal costs too much ($1, I could get them for about 75 cts). Of course it is printed on the paper that it is an abstract from such a number of Wundt's Philosophische Studien. I suppose they would cost about five cents each. English scientific journals give contributors forty to two hundred copies of their paper, and these are sent around, it being a highly acceptable method of advertising.

[1] "Über die Zeit der Erkennung und Benennung von Schriftzeichen, Bildern und Farben," *Philosophische Studien* 2 (1885): 635–650.
[2] JMC's dissertation omitted all material published in this article.
[3] See 5.18.

[4] JMC may have emphasized his tables because Hall and E. M. Hartwell had recently published a paper ("Bilateral Asymmetry of Function," *Mind* 9 [1884]: 93–109) in which they had cited "tables far too extensive to print here."

5.71 To Parents, Leipzig, 17 March 1885

(One paragraph of gossip has been omitted.)

I have arrived at the interesting stage of my investigation, where I must make experiments on four women.[1] I showed my good sense this evening by making them on a Miss Hopkins, about the nicest young lady here.[2] They are to be made tomorrow at Inselstr, 11.[3] Frl. Kuhn wanted to make them, and they have invited me to dinner.[4]

[1]These experiments were reported in JMC's first paper in *Philosophische Studien* (see 5.70n.1).
[2]Myra L. Hopkins, from San Francisco, and Alma Kühn, of Leipzig, both mentioned in this letter, were students at the Leipzig Conservatory (see introduction to Section 3 and *Königliche Conservatorium*, pp. 57, 64).
[3]JMC had lived at Inselstrasse 11 during his first year at Leipzig (see 1.14).
[4]See also JMC to parents, 19 March 1885.

5.72 To Parents, Leipzig, 21 March 1885

(One paragraph of gossip has been omitted.)

We did not work this morning, but began again this afternoon. We start now in the morning at seven, so I must get up pretty early. There are only about two weeks more now—of course there is any amount of work I might continue at indefinitely, but I have pretty well finished every thing I had planned, and here worked at a good many new things, which have turned up. It will be hard to get things into order, but I shall have time enough.

5.73 To Parents, Leipzig, 22 March 1885

(One paragraph of gossip has been omitted.)

I was very busy all last week. Besides the regular experimental work which we tried to keep up—there was this special work to be finished. I had to make a few simple experiments—what you did last year when you read as fast as you could—on ten persons, and this was a great bother.[1] Then the translation into German has not been an easy thing. It has taken us about twenty hours. The translation of the doctor's thesis will take a hundred. Besides this translation, there was a great deal of work to be done in arranging the figures and tables. It is to be hoped that after all my trouble Wundt will print the paper. I shall take it to him Tuesday or Wednesday.

[1]These experiments were reported in JMC's first paper in *Philosophische Studien* (see 5.70n.1).

5.74 To Parents, Leipzig, 24 March 1885

(One paragraph on the weather has been omitted.)

Not much happens now except work in psychology. For example we began this morning at seven and worked until twelve, I continuing after Berger had left for an hour. Then we have been working this evening from four on until now—after ten. Of course this wont last very long, but writing up things will be still worse. Berger thinks I cannot write the thesis in less than six weeks, and translating it and getting the tables &c. ready, will take quite as long. I'm sure ! dont see where I'm to find any time to prepare for the oral examination. Prof. Leuckhart's specialty is human parasites, and they say he expects candidates to know all about this attractive subject—indeed he has printed a book on it especially for university students.[1]

[1]Karl Friedrich Rudolf Leuckart, *Die menschlichen Parasiten und die von ihnen herrührenden Krankheiten* 2 vols. (Leipzig: Winter, 1863–1876) See also 5.43.

5.75 To Parents, Leipzig, 25 March 1885

We worked as usual this morning and this evening. This afternoon I played tennis, and ran about six miles in the skating rink. Afterwards Muirhead and a friend of his were in my room.[1] The latter is fellow & tutor at Oxford. There was also another man here a Dr. Lorenz whose thesis was published in the last number of Wundt's Studien.[2] I mention him because if I have time to work after Berger leaves I shall try to get him to help me. He has been working in psychology for several years, and is a very good man. He is not well off, and must teach this year without any salary. If one could afford it, it would be pleasant to help such a man any how. I am very glad that Berger has been able through me to take his doctor's degree, he could not otherwise have done it.[3]

For completeness sake I enclose a valuable manuscript which I find in one of my pockets. Have any of my letters failed to turn up?[4]

[1]On Findlay Muirhead see 5.24. The friend was Henry Tresawna Gerrans, a graduate of Christ Church College, Oxford, and a tutor in mathematics at Worcester College, Oxford (see JMC to parents, 29 March 1885; John Foster, *Alumni Oxoniensis, 1715–1886*, reprint edition [Nendeln: Kraus, 1968], vol. 1, p. 517).

[2]Gustav Hermann Lorenz's dissertation was "Die Methode der richtigen und falschen Fälle in ihrer Anwendung auf Schallempfindugen," *Philosophische Studien* 2 (1885): 394–474, 655–657.

[3]Here JMC is probably referring more to his financial help of Berger than his intellectual help.

[4]See 5.64.

5.76 To Parents, Leipzig, 26 March 1885

You will perhaps be horrified when I tell you that we have been making experiments on me pretty much all day, from seven this morning until now half-past one. My object was to examine the effects of fatigue. It has been really quite as hard for Berger to run the apparatus & write down the times, as for me to make the reactions.[1] We do not begin tomorrow however until nine. We made a pause at dinner & supper time. In the former I went to see Prof. Wundt. I had given him my manuscript yesterday.[2] He was much more reasonable than I shall be likely to find most editors, asking for no omissions, and only suggesting one or two verbal changes. The fact that I announce in this paper that in the next number of the Studien I shall have another paper, is the most substantial compliment Prof. Wundt could pay me, as he thereby pledges himself to print a paper not a word of which is written. This paper would have been accepted as a doctor's thesis, and I shall print enough this summer for five more doctor's theses.

[1] See "Cerebral Operations," in *Man of Science*, vol. 1, pp. 90–92.   [2] See 5.70.

5.77 To Parents, Leipzig, 27 March 1885

I dare say my letters will be no more interesting after Easter than they are now, but at all events they won't contain the monotonous repetition, "we made experiments." But that is about all I do, we working this morning and again from four o'clock on this evening.

I can really however tell you something in this letter. Prof. Wundt yesterday kindly agreed not to issue the number of his review to follow this one until after the opening of the semester next fall. This is exactly what I have been wanting all along. I could have passed my examination this summer, but it would have pressed me very much. Now I can pass it with some comfort at the opening of the semester. Of course I will be safer as to getting through but the chief advantage (besides the great relief) is that I can devote all my time to my thesis, and can make it far better than if I had to prepare for the examination. I shall get the degree at exactly the same time, as if I should have passed the examination in the summer, as my thesis would not have been accepted until the fall, and I could not have printed it till then.

I suppose you will scarcely accuse me of changeableness, though I have changed often. The trouble was largely with Wundt's Studien. He thought he would issue this number in June, and I supposed that meant July or August. I wanted to get my paper into that fearing he might not issue another number for a year. I at first as you know intended to get my thesis ready for that number and to pass the examination next winter. Then about Xmas I learned that I could not print my thesis until after I had passed the examination. After much worry I decided to try to pass the examination before the number of the Studien was issued. That was because I did not want to put off printing something for a year or more. However Wundt decided to issue this number in May, I would therefore have had to give up my thesis now the end of March and pass the examination the end of April—both almost impossible—that is if my thesis was to be as good as possible, and the examination reasonable safe. Of course when Wundt told me he would issue another number in the Fall I was delighted to wait till then. I feared however he would issue it before the opening of the semester, in which case I should have had to pass the examination this summer. This would have been possible but would have sadly interfered with the thesis. I was much pleased yesterday when Wundt promised to give me time at the opening of the semester to pass the examination.

I must therefore be here the first of November, though of course if it is absolutely necessary for me to be in America at that time, I can pass the examination this summer. I am awfully tired tonight so shall say

(to be continued)

5.78 To Parents, Leipzig, 28 March 1885

I want to continue my letter of last night, by saying a few words about the future. It seems to me best that I should come home this summer as soon as I finish my Arbeit. Unless therefore you do not advise it, or something new turns up I shall look to arrive the the end of August and stay about six weeks. During August, Sept and Oct. I must take some vacation, and prepare for the examination. Quite putting aside the pleasure and advantage of our being together, a trip to America is the best way I can accomplish these two things. The expense (about $200) will not be greater than if I stay here.

You seem to think it best that I should stay here until say next Feb. and then come home for always. I scarcely understand why I had not better come home this summer, supposing I am to return for good next winter. I suppose you think that if I see you this summer I shall be inclined to stay longer than I otherwise would. That is undoubtedly the case. But is not a separation for eighteen months more serious than two shorter ones even though their sum may be two or three months longer?

There is no harm in saying that my work this winter has been more successful than I had hoped. I have a paper in press, and if I wish Prof. Wundt will print papers by me in the next two number of his journal. I shall also possibly be able to print something in England. So I have done all I can towards securing me a position, having worked hard and exactly in the direction most likely to accomplish this. After I have printed three or four papers, one or two more would be of but little service towards securing me a position. I have of course this year been compelled altogether to give up those studies and that line of thought, which are more to me than experiments in psychology.[1] It is right that I should take these up again this winter,

---

[1] The meaning here is unclear.

and I can do it better here than in America. I am not staying longer nere than I had intended. Last April I wrote to Dr. Hall "I intend to stay here (i.e. in Leipzig) for about two years, after that I will probably try to obtain a position in the Univ. of Penna, though I would not object to carrying on my studies even longer free from the responsibility of teaching".[2] You know you intended to stay in Europe two years. I shall finish my work here more quickly and I think better than seemed probably a year ago. I have therefore in some sort earned a right to time, which I trust will not be wasted. As to the expense—I shall accept the money I can draw with my letter of credit, after which I shall try to take care of myself, using of course as far as necessary the money you have given me and which I have with Bailey.

The life of each of us is a little circle, which does not cover the circles of other lives—even those who love each other most cannot fully understand each other. I must tell you frankly that if you were not interested in the matter, I should not make the slightest effort to secure me a university position, and should hesitate about accepting one if it were offerred me. I cannot sympathize with those whose aim in life is peacable respectability, a comfortable living and a good position in society. I cannot very well explain what I want in life, nor do I exactly know want you want for me, but I imagine I am more likely to follow your way than mine. We are carried along by the strong current, but it may be better to resist it for a while.

[2] This letter has not been found, but see 5.127.

5.79 To Parents, Leipzig, 30 March 1885

Our work must be brought to an end now in nine days. I scarcely know whether I am sorry or glad. I imagine quiet, uneventful, regular work makes the happiest life. Still when at it we always want the end. Lamb was always longing to be free from his desk at the India House, but when at last released was disconsolate.[1] However I shall not make a sudden end. We shall work again in July. In the mean while there is some work I can do alone, and it would be desirable to verify the experiments I have made on Berger and me on some one else. I have given so much time and money to these experiments, that I had better make them as good as possible "I am in blood,

Stepped in so far, that, should I wade no more,
Returning were as tedious as go o'er."[2]

I suppose the money spent is not a bad speculation, and the work has been the best possible training whether I become a professor or lettercarrier.

[1] Charles Lamb was an English essayist who worked as a clerk at the East India House from 1792 through 1825 (*Oxford Companion to English Literature*, pp. 459–460).
[2] From *Macbeth* (act 3, scene 4).

5.80 To Parents, Leipzig, 31 March 1885

We have been making fatigue experiments today on Berger, like those we made on me last week. To think that 2184 times I have said "jetz" and started the clock,[1] and made a noise or something, and stopped the clock, and written down the time. I imagine I shall dream that I am saying "jetz" etc. all the rest of the night.

The second chronoscope came today.[2] I shall pay for it, so it is to be hoped our government wont fail before I get any money back. It is in good order. If there is any thing I understand better than any one else in the world it is this chronoscope. I should not like to count up how much time I have spent on it.

It is beautiful spring weather. My window has been open all day. I'm going to live out of doors half the time after next week. Even Leipzig is beautiful in Spring.

[1] Before starting each experimental trial, the experimenter signaled the subject that he was about to begin with the word "*jetzt*," or "now."

[2] This comment refers to the gravity chronometer JMC had ordered for the Army Medical Museum (see 5.64).

5.81 To Parents, Leipzig, 1 April 1885

I am sure you will be surprised at the contents of this letter, and indeed I scarcely know how to write what I want to. I cannot myself realize that I am nearly twenty-five years old—the age at which it seems to me a man should marry. Unfortunately under our present social condition most men cannot support a wife and increasing family at that age, still it seems to me better that a man should choose a wife while young, even though the marriage cannot take place immediately. Several years between the engagement and marriage bring many advantages, indeed the marriage sometimes does not take place at all. I need scarcely say that it is the duty of every one to marry, for if no one should marry the worst possible example would be set to the coming generation. I know that as a rule you think it better that Americans should not marry foreigners, still you

April 1st!

After which elaborate effort I have doubtless earned a right to go to bed.

5.82 To Parents, 2 April 1885

We have been working pretty much all day—from seven until eleven. It is lovely weather. I suppose when Berger goes and I have more time it will be cold and wet. I do not know exactly what to do with myself. I shall have an immense amount of work to do, but it is just that kind of work which one can do more of sometimes in one day, than at other times in a week. It were therefore well to be in the best possible health. First because when real well I am more likely to be able to write to advantage, and secondly because when feeling like writing I can write all day or all night without hurting myself. It would therefore pay, even though we do not look beyond the first of August, to take a vacation. But I do not know where to go. I should rather walk, or be out of doors somehow, but it is too early for that here, and the places within reach are not especially attractive. The Harz and Saxon Switzerland, the best regions in North Germany are not so attractive as the country around Easton. If I could only be at the Italian Lakes or in the mountains about Florence, that would be the best of all, but it would cost nearly $50 to get there and back. I may go there, or may stay here until May and walk through some of the mountains (hills) near headquarters. I can keep my room while away for two thirds the regular rent. I shall not decide until I see how I feel next week after Berger is gone. I should like to make some sort of a start on my work at once.

5.83 To Parents, Leipzig, 7 April 1885

This has been the last day of experiments. We work again tomorrow until eleven, and then adjourn sine die. Berger leaves for Muhlhausen the morning after at five. Except perhaps Freshman year at college I have never worked so hard as during this winter, especially the last three months. It is surprising that I should have kept so well. I am sure that I am in better health than when I came here last September. Just now with Spring weather and all I am feeling a bit tired, but that would go away with a week out of doors. You can see how well I am from the fact that I intend to continue working, whereas there is nothing pressing to keep me from taking a vacation if I needed it. Perhaps the most surprising thing of all is that my eyes seem stronger than they have ever been in my life.[1] I could not have thought out a more continuous and consistent way of hurting them. Either my eyes are much better or the work was not so bad for them, as one would suppose. Indeed it seems to me the injury of making experiments on one's self, has been overstated. I think I am so well owing to the walk in Tyrol last year and to the constant exercise I have taken all year,[2] also to the fact that worry is so much worse than work, that work may be a pastime.

[1] See 5.5.

[2] On JMC's walk in the Tyrol see 4.13 and 4.15.

5.84 To Parents, Leipzig, 8 April 1885

Schluss der Vorstellung this morning at eleven![1] *It seems decidedly queer to have absolutely nothing to do. That is to say nothing that must be done at a fixed time, though six months work must be done in the next three months. The writing out of my work is difficult chiefly because there is so much of it—though I ought first to study a great lot of psychology and physiology. There is no unreasonable conceit in saying that my work is better than I had thought it would be. Last winter I hoped to do enough to have it accepted in a doctor's thesis here, this winter I hoped it would be good enough for Prof. Wundt to accept it for his Studien, now things are quite changed and it is Prof. Wundt who hoped that I will give him all my work for his Journal. He is willing, indeed anxious, to publish papers by me in the next two numbers.*

While praising myself I must speak of the way I have gotten along with Berger. He is a strong man—the best German I have ever met. A good proof of this is that for the past four years he has been absolute dictator of the Turnverein. Two facts made our relations extremely delicate. Money matters, and the fact that he wrote his thesis on the same subject that we are working on. Our friendship has not however at any time been for a moment in danger.

[1]*Schluss der Vorstellung*: end of the performance, or finale.

5.85 To Parents, Leipzig, 10 April 1885

(One paragraph of gossip has been omitted.)

I worked over my notes this morning. There is a tremendous amount of work there, before I can begin to write to advantage. I am thankful that I dont have to reckon all the differences and averages. I paid my man the other day for fifty hours work, and he does it faster than I could. I have at least 150 hours more of work for him—I would be utterly discouraged if I had so much purely mechanical work before me. About that many hours work must be done arranging and tabulating, but that no one else can do, and it is important and not uninteresting. Berger will copy the thesis and tables for me.

5.86 To Parents, Leipzig, 14 April 1885

I rode again this afternoon. I had a first rate horse and rode over two hours. I am consequently pretty stiff now. I have just gotten back from the theatre where I heard Lohengrin—so I have heard three of Wagner's operas in five days. How proud he would be if he were alive and knew it! Except Fidelio and several of Mozart's, I have scarcely heard any beside Wagner's operas this year. I have not spent so very much money on the theatre this year, but I have had to draw a good deal. Even deducting Berger and apparatus my expenses seem rather large. I paid $25 for books—"school books"—the other day, and had to order a suit of clothes today. I know lots of people here now, and unfortunately cannot continue to wear my old clothes indefinitely. Three pairs of trousers have come through at the seat simultaneously—that however is one of my psychological results.

5.87 To Parents, Leipzig, 17 April 1885

I took Prof. Wundt this morning an analysis of my proposed thesis, and spent over an hour with him. He was extremely cordial and made a most surprising offer. I said I feared the thesis would be long; he said I should not try to shorten it, that he would put at my disposal half of the two numbers of his Journal next to be issued (almost 200 pages!) (for me)[1] That is probably the most remarkable compliment I shall ever receive. He also said my paper would make <u>six</u> doctor's theses. He will also let me print my whole work as a doctor's thesis before the second number of his Studien appears. Of course all this is extremely satisfactory—all my work appears in three succeeding numbers of Wundt's Journal, and together as a separate treatise. It will fill perhaps 100 pages. I save of course the large expense (perhaps $200) of having it printed, and I understood Prof. Wundt to say the publisher would give me 200 copies free. It contains engravings & tables which cost a great deal.

Prof. Wundt also let me have all the reviews & books he has on my subject.

[1]The intended place of "(for me)" is uncertain, as JMC added it between lines.

5.88 To Parents, Leipzig, 22 April 1885

(One paragraph of gossip has been omitted.)

I send a letter from Prof. Wundt with which I think you will be pleased.[1] He offered himself to write it. I am sure you will give me credit for having worked hard this winter in the direction most likely to secure me a position—now I can do nothing more. Though it might be considered my duty to accept a so called "position of usefulness," it is scarcely my duty to hunt around for it. If we try to find me a position such as you have in view, it is with the same selfish motives we would have if we were looking to get me a government clerkship. The advantages of a university position are three (1) salary (2) quiet & settled way of living (3) a certain position in society. The salary is an important matter, but I imagine you consider it secondary to the other two. For example you would advise me to accept a position at Columbia with a salary of $500, yet my expenses in N.Y. would be more than $500 above what they would be if I were living with you at home. I do not trouble myself about money matters, if I have money I shall spend it, and get I think my money's worth, if I have but little, I can get along with little. I shan't starve or freeze while I am young and strong, and should rather spend the last five years of my life in the poorhouse, then devote the whole of my life to make it possible to die a rich man. As to the second advantage—that is a matter of age and character. I am sure it would not have been—from the common sense point of view—an advantage if immediately

[1] On 28 September 1886 William Cattell sent to William Pepper, provost of the University of Pennsylvania, a copy of this letter from Wundt, dated 15 April 1885, together with a translation into English (faculty applications, 1886, University of Pennsylvania Archives). As translated (see 5.108), this letter reads as follows. "Honored Sir: I gladly embrace the opportunity to communicate to you my opinion concerning the studies and the success of your son, Mr. James M. Cattell, during his sojourn in Germany. Mr. J. M. Cattell had made already in America solid acquisitions in metaphysics and natural sciences. Through it and through the zeal and the enthusiasm with which he devoted himself to his work during his two years' stay at Leipzig, he was enabled by his own investigations and experiments partly to solve, partly to bring nearer to a solution, a large number of important psychological phenomena (occurrences). In his investigations, Mr. Cattell showed especially individual inventive power and great self-dependence of judgment, not only in stating, formulating and solving the problems, but also in the technical execution of his experiments and in the management of the requisite apparatus. The extensive work, which he is about to complete, and with the results of which I had an opportunity to get acquainted, will appear in the next number of the *Philosophische Studien* (a periodical edited by me). From what I know about it, I am confident it will be one of the best works in this department. It solves for the first time a series of questions, which are of theoretical, and some of them of practical, value and interest; in respect of other questions, it contains valuable corrections and amendations of former works. Mr. Cattell has proved by this investigation, that he is eminently able not only to give into himself, independent of another, problems, but also to invent the methods and means for their solution. Moreover, Mr. Cattell possesses in other branches of philosophy manifold learning and an independent judgment; especially has he made successful effort in familiarizing himself with the more recent German Psychology and Philosophy. (Signed) With highest respect, PROF. W. WUNDT"

after graduating I had settled down at Lafayette or Princeton to be tutor, assistant, professor—and so died. You perhaps think the time has now come when it would be better for me to do this—I do not know. I can only say that without wife and children such a life seems rather hopeless and dreary. I can make my home with you—but I am apt to live twenty or thirty years the longer. I do not mean that I do not want to be a college professor—only that there are many long years in life. As to the third point above mentioned—a good or leading position in society, this has perhaps the most influence with you and the least with me. If I am not worthy to hold such a position I had better not have it, if I am it makes but little matter whether I have it or not. "A respected citizen, appreciated by the general public", an ideal epitaph!

5.89 To Parents, Leipzig, 23 April 1885

(One paragraph of gossip has been omitted.)

I enclose a note I have just received from Prof. Wundt.[1] You can see (from the signature for example) that he likes me. I shall be glad to meet Prof Bain who is one of the most celebrated philosophers living, as you know.[2]

[1] See figure 11. Wundt's note to JMC has been translated as follows: "Dear Mr. Cattell: Prof. Alex Bain of Aberdeen is here and wants to see the psychological laboratory tomorrow afternoon. It would certainly be of interest for him to meet you and get acquainted with you (and vice-versa). Also, as you know, I speak English poorly, and you can serve as an Interpreter better than I can, so it would please me if you could at all possible come. I have made an appointment with Bain, whom I will fetch, and we hope to be at the laboratory at 4 o'clock . . .With best greetings, W. Wundt"

[2] Alexander Bain (*DSB* 1: 403–404) was professor of logic at Aberdeen for many years and a celebrated philosopher whose work represented the culmination of nineteenth-century British psychological thought.

5.90 To Parents, Leipzig, 24 April 1885

There has been quite a philosophical event today. Prof. Wundt wrote to me again asking me to come for him, and we called together on Prof. and Mrs Bain. They went with us to see Prof. Wundt's laboratory, and afterwards came home with me to see my things, and were in my room an hour and a half. This evening I was invited to meet them at Prof Wundt's, no one else being there.[1] Bain and Wundt are, I suppose, the two greatest psychologist that have ever lived, and perhaps the greatest philosophers now living. I do not think them philosophers of the very first rank

[1] In his *Autobiography* (London: Longmans, Green and Co., 1904), p. 378, Bain remembered this dinner with Wundt and the young American who interpreted for the two professors.

(Lotze was perhaps) but there is no one better living. It seems curious to be treated as an equal by such men. They are both exceedingly "nice", Prof. Bain is one of the nicest men I have ever met—a genuine Scotchman, but not at all what I had expected. Mrs Bain is also very nice, she knows a little German, (but not as much as you, Mama) he does not know any.

5.91 To Parents, Leipzig, 25 April 1885

I went to the station with Prof. and Mrs Bain this morning.[1] *It is very pleasant for me to have met them, and from a philosophical and worldly point of view highly profitable. He will give me letters of introduction to any philosopher or scientist in England. They invited me to stay with them during the meeting of the British Association for the Advancement of Science to be held in Aberdeen next Sept. Prof. Burdon Sanderson of Oxford the greatest English physiologist is to be their guest.*[2] *I would really like to attend the meeting, though of course I would not stay at their house. Prof. Bain is rector of the university of Aberdeen. He said Pres. McCosh was the other candidate for the professorship of philosophy there, at the time he was elected.*[3]

I am making an experiment tonight. I slept from five until ten, and shall go to bed again towards morning. We sleep much the most soundly the first two hours—it is probable that they are worth at least four later hours. It might be highly profitable to sleep from seven till ten, then work five or six hours, then sleep three hours more.[4]

[1] JMC apparently helped the Bains get on the proper train for Jena and Berlin, and impressed them greatly. (Frances Anne Bain to JMC, 30 April 1885: "I shall always remember you amidst my good student friends.") Bain added a postscript to his wife's letter: "I intended to enclose some remarks on a point connected with your experiments; but the hustle and fatigue of Berlin and the noise of life in a hotel are too much for my nerves . . . I will write you as soon as I feel settled."
[2] John Scott Burdon-Sanderson (*World Who's Who in Science*, p. 1473; *DSB* 2: 598–599).
[3] On James McCosh see 4.14.
[4] One of JMC's sons later carried out an experiment based on reasoning similar to this, with disastrous results (Anna D. Burgess to Michael M. Sokal, 23 September 1978).

5.92 To Parents, Leipzig, 27 April 1885

I haven't written a word of my thesis—I wrote twenty pages or so last Xmas; but that is not of much use. If it is 100 pages, and it takes two hours to write each page and two hours to translate it into German, I have 400 hours work before me—and the writing out is scarcely half the work. I must read a great deal—general psychology and psychology, and papers on my special subject. For example today & yesterday I read three such papers containing about 150 pages. Then there is the getting of the experiments in shape, calculating averages &c. Lastly there is the making of the general plan of the thesis. German do this very badly—most of the scientific papers I read are much longer than necessary and not at all clear. I told you Prof. Wundt said he thought I would need 200 pages to describe my work—a German (with honorable exceptions) would.[1]

I rode this evening for over two hours. It continues hot, but is raining now.

[1] JMC's dissertation, as published separately, was 72 pages long; few, if any, of Wundt's articles in the first three volumes of *Philosophische Studien* were so long.

5.93 To Parents, Leipzig, 29 April 1885

(One paragraph of gossip has been omitted.)

I spent this afternoon in the psychological laboratory, where Prof. Wundt met those who are to work this year in experimental psychology. I don't imagine science will be especially advanced by students working to get their doctor's degree. I think I have made it impossible for any student beginning at the beginning and working two or three times a week say for a year to do anything in my department, and that is about a half of the laboratory. I really pity the poor fellows—I could make them their doctor's theses so easily. Prof. Wundt started two of them this afternoon on work I am sure can give no good results (except the theses)!

5.94 To Parents, Leipzig, 2 May 1885

It has been raining all day. I have been in my room most of the day trying to write something towards my thesis. The result has been two or three pages I cant use. I can describe my experiments I think readily enough, but it is extremely difficult to come ones self to a conclusion as to their meaning, and to give them their place in the whole of science. I should know perfectly psychology and physiology, and a great part of physics, mathematics, &c. I could of course simply state the results of my experiments, and that was what I had originally intended to do, but it would be better to get myself the credit for various ideas, which can be deduced from the work. The worst of it all is that I must not only know for example all that is known on the physiology of the brain, but must for myself make some order out of the areas of confusion and contradiction there confounded.

5.95 To Parents, Leipzig, 3 May 1885

(One paragraph of gossip has been omitted.)

I received my proof today.[1] *I cannot realize that it is really mine. It does not excite me as it might if it were a poem or a novel, for I can judge quite objectively of its value—indeed better than any one else. It seems to be quite accurately set up—it*

[1] This comment refers to the proofs of JMC's first *Philosophische Studien* paper (see 5.70 and 5.76).

is of course hard to correct the German, but Prof. Wundt has offered to do this. I think I shall not trouble myself with changing the form, even where it seems as though I might improve it. They have written on it asking me how many copies I want—I think they give me some free.

5.96 To Parents, Leipzig, 4 May 1885

I have taken scarcely any exercise for the past 3 days and (consequently?) have headache this evening. A Hungarian Herr Papai who works in the Laboratory came to be experimented on this afternoon.[1] I should like to verify my work on different people, but have no time to spare.[2] I heard Prof. Wundt lecture on Logik at five, as I wanted to speak to him. I shall not attend any lectures this semester. I have heard Prof. Wundt's Logik, so can subscribe to that for M 1.50. Prof. Heinze lectures on the History of Philosophy. I have paid for his courses on Ancient & Modern philosophy, but suppose this is different. It will probably be wise to invest the $4 in that direction, as the money goes directly to the professor, and he knows who pays. The same holds good of Prof. Leuckardt's lecture's on biology. He examines me, you know.[3] It would be well if I could hear these lectures, but I cant take time from 12 to 1, losing two hours to hear what I can read in ten minutes. I think I shall let Prof. Hankel and physics go.[4]

[1]C. Papai had enrolled the previous term at Leipzig as a student of philosophy (*Personal-Verzeichniss der Universität Leipzig*, no. 107, summer semester 1885, p. 86).

[2]See 5.17.
[3]On Karl Leuckart see 5.43.
[4]On Wilhelm Hankel see 5.31.

5.97 To Parents, Leipzig, 5 May 1885

(Two paragraphs of gossip have been omitted.)

I have received a letter from Mrs. Bain and a note from Prof. Bain.[1] You see we are quite good friends. You will perhaps smile when I say that I feel, not exactly wiser, but somehow older than Prof. Bain and Prof. Wundt. Yet it is really true—they seem to me like children playing with their psychological toys. And like other children they build their house of blocks, not to realize their sense of truth & beauty, but to get it a bit higher than some one else's—and they are quite ready to give up the whole enterprise, if they see Mama going towards the pantry. This last remark was suggested by Prof. Bain's wanting to be returned for parliament. The oracle was right and Socrates was the wisest man in Athens if he was the only man there who knew that he knew nothing. But as conclusion of the whole matter we may grant that it is better children should play with their toys than that they should sulk in a corner.

[1]See document 5.91.

5.98 To Parents, Leipzig, 7 May 1885

I worked as usual this morning. I received the second half of my proof. One table is not set up—either Prof. Wundt or the printers lost the manuscript—so I must make that again. It is a great trouble and nuisance to prepare manuscript for the printer, and the proof for publication. After this year I dont want to print anything for five years.

This afternoon and evening I was with Aunt Nemie and Lizzie.[1] They came to my rooms, and I took them to the Prüfung of the students of the conservatory.[2] This is really interesting—I had never been there before. In many cases they play their own composition. Nearly half the students are American and English.[3]

[1] See introduction to section 3.
[2] *Prüfung*: examination (in this case, a final recital).
[3] From 1843 through 1893, 1,808 of a total of 6,166 students at the Leipzig conservatory were from Great Britain or America, and most of these were there after 1870 (*Königliche Conservatorium*, pp. 113–114).

5.99 From William Cattell, Baltimore, 8 May 1885

(Several paragraphs of gossip have been omitted.)

I came here yesterday, by invitation of the Presbytery, to deliver "the charge to the Pastor", my friend Mr. Proudfit,[1] who was installed over the 2^d Church:—Bright & Emerson saw the announcement in the Papers and were present:[2]—had pleasant chat with them after the services. Bright was in London last summer while we were there. Emerson hopes to secure a position in Toulane University Louisiana:—& I think I can help him. I have met many friends here—but have not seen my old friend Gilman. Both Mr. & Mrs. Garrett have died since I was here.[3] The house (on the corner) is I see being thouroughly overhauled. Young Mr. Garrett is completing the house near by,[4] commenced by his father:—& now has a law suit on his hands brought by his neighbor who complains of a "projection" in the front.

I hear no University news,—thought I w^d call on Gildersleeve but concluded not to do so for fear something w^d be said by him on me that had better be "unsaid:"[5]—Baltimore looks lively this beautiful May morning. Vegetation is far ahead of that in Philad^a.

[1] On Alexander Proudfit see Joseph T. Smith, *Eighty Years: Embracing a History of Presbyterianism in Baltimore* (Philadelphia: Westminister, 1899), pp. 47–48.
[2] On James Wilson Bright see 2.1. On Alfred Emerson see W. Norman Brown, *Johns Hopkins Half-Century Directory* (Baltimore: Johns Hopkins University, 1926), p. 104.
[3] On the John Work Garretts see 1.17.
[4] On Robert Garrett see Hugh Hawkins, *Pioneer: A History of the Johns Hopkins University, 1874–1889* (Ithaca, N.Y.: Cornell University Press, 1960), p. 105.
[5] On Basil Lanneau Gildersleeve see 1.17.

We were delighted to hear of Wundts offer—to place at yr disposal the half of the next two numbers of his Journal. It is, as you say, a great compliment to you. I have no fears but what the article will do you credit—and I want some copies of the Journal: —at least twenty. I will send you the money so that it will not be a draft on yr funds! —How much will they cost? —And I want a translation. Will it be printed also in English?

5.100 To Parents, Leipzig, 20 May 1885

I saw Prof. Wundt this afternoon. He is always as cordial to me as it is possible for any one to be. He is going to get the apparatus I have invented for his laboratory— so the machinist has orders for two sets. It costs about $50. He says he will issue the next number of his Journal as soon as I am ready. On account of my examination it cannot be issued before the opening of the semester, but will be issued shortly after, so I must pass the examination the end of October. I shall probably have to sail on the 7th of Oct. so shall only be home a little over six weeks, but I'm sure it is worth crossing to spend that much time with you. Of course if it is advisable for me to be home this winter I can stay here and study through the vacation and return to America after my thesis is printed in November. I must study pretty hard even if at home.

5.101 To Parents, Leipzig, 23 May 1885

The days have no special individuality, and so I scarcely realize how fast they come and go. Nor is the amount of work I get through with so large as to seem to represent much time. The six days since I have been back from Berlin seem at the most like two or three.[1] I have written about twelve pages (of print), but this, describing apparatus and methods, is the easiest part of the whole paper, yet it seems I have only written it at the rate of two pages a day. Berger is coming to Leipzig again the middle of June. He must serve as a soldier for two weeks. Whatever work Germans may be at they must give it up for this.[2] He will probably stay here then and help me translate the thesis, and make more experiments, if there is any time.

[1] JMC had taken a short trip to Dresden and Berlin early in May (see JMC to parents, 13 May 1885).

[2] See Paulsen, *German Universities*, p. 128.

5.102 To Parents, Leipzig, 27 May 1885

I have not yet engaged my passage though it is probable that I shall sail on the Etruria Aug. 15th.[1] I will not however come, if you would rather have me stay home this winter, or if my work gets into such a way that I cannot finish it. I really havn't been settled in my plans all winter, but one must have some sort of a working theory, and I have frankly told you all my temporary compromises. It is the old story. The longer I keep at the work the better it will be, and the less I shall be pressed. The paper already printed by me (the Philosophische Studien has just been issued) would have done for a doctor's thesis, and I could have passed the examination by this time and be a Ph.D. Now however as Wundt wants to issue his Studien again in October and I would have to hurry back and pass the oral examination before then, as well as hand in the entire thesis the first of August, I sometimes think it might be well to give Wundt a paper for the coming number of his Journal, and save the larger part of my work for doctor's thesis and the following number or numbers of the Journal. I could then pass the examination say in December, and should have time to make the thesis better.

Sometimes too I think it might be better not to come home until Nov. and to finish up everything by that time. I should indeed do this, if I thought I could work on steadily until then without hurting myself, but I do not feel especially strong now, and doubt if I could work on through the summer to advantage.

[1]The Cunard steamship *Etruria*, launched in 1884, is generally regarded as the most luxurious ship of the period (Harry Parker, *Mail and Passenger Steamships of the Nineteenth Century* [London: Sampson, 1928], p. 106).

5.103 To Parents, Leipzig, 30 May 1885

It is the last straw that breaks the camel's back, you know, and I really believe an insignificant accident that happened this afternoon has decided me not to try to write out all my work between this and the first of August. I swam for the first time out of doors, and without noticing it must have gone in with my glasses on, and likewise without noticing it must have lost them. I dived for them for a good while, but the water was so muddy I could not see six inches. The man will dredge the bath, but I fear he will not find them. Instead of ordering a new pair I would rather consult an occulist here. I have wanted to all winter (as also last year) but have never had time. I am not sure it is necessary for me to wear glasses all the time. A normal eye is astigmatic, and I doubt whether the defect in my eye was great enough to warrant Dr. Thomson in giving me glasses to correct it. The trouble is now my left eye is more shortsighted than the right. I think I owe this to Dr. Thomson's glasses. I was scarcely shortsighted at all when I went to him first, he gave me glasses (to be worn only when I read) to correct the astigmatism, and then I grew shortsighted very rapidly, then he made me wear the glasses all the time, and since then my eyes have been doing pretty well. But I imagine I should never have worn glasses at all. When a boy I think the trouble was more in the head than the eye. You, Papa, and I dare say half the people in the world have as much astigmatism as I, weak eyes can best be cured by, first, not overstraining them, and, second, by looking after the general health.[1] I find the other pair of spectacles has somehow been broken, so I have none at all.

As I said above I have almost decided to write now a paper for the next number of Prof. Wundt's "Studien", but to keep most of my work back, and not hand it in as a doctor's thesis until the following number of the Journal is to be published. The only reason I can see for finishing it all up in the next two months is the innate restlessness which always leads us to hurry up with what we are at, in the foolish expectation of being happier afterwards. I have always said that five hours mental work is enough, with perhaps two hours of mentally exciting reading or conversation.[2] Now exactly why I should work ten hours a day as hard as possible for the next two months, and continue the same thing when at home on my vacation, in order that I may afterwards take a vacation I don't exactly see. Yet I have always a strong impulse to push forward as fast as possible. My reason tells me it is silly and probably impossible to do eight months work in four, yet somehow I want to try it, and find it hard to give it up. Yet I have really done enough work for one year—I think I told you when I showed Prof. Wundt last fall what I intended to do this winter, he said it would take me ten years! but I have done it all and more.

When I reason about the matter I am absolutely certain that it is better not to hand in the thesis in July. The chief point is that if I wait I can make it much better. I must in the first case write it hurriedly and translate it hurriedly, besides sins of omission, there would doubtless be absolute errors, which I should be the first to correct if I had time. Further if Berger and I do not spend our entire time translating the paper, we can make more experiments, which would be of great value. In the second place I am not very strong and might seriously hurt myself, if I work under pressure through two months of this enervating weather. You see I could not take a vacation afterwards either, as I would be studying (even on the steamer) for the examination. Further the consequences would be terrible if I should fail in my oral examination, the number of the Studien would be printed before I could make up the subject on which I had failed, and I should have to prepare an entire new thesis. As to minor points—the English reviews Mind and Brain appear quarterly. If I wished to print an article in one of them, I should have no time to prepare it for either the Oct. or Jan. number, and later after the results have been published in Germany, they would not care to reprint them in England.[3] Further it seems scarcely fair for me to compel Prof. Wundt to postpone the issueing of his Journal until I pass my examination, and I could not in any case correct the proof of my paper. I think of other points but need not trouble to write them, only that if I pass the examination in Oct. I can only stay at home six (perhaps five) (and should be extremely busy) weeks, whereas if I postpone it I can stay ten weeks or longer.

[1] See 5.5.
[2] See 3.3 and 4.15.

[3] Yet JMC's *Philosophische Studien* papers were regularly published, in translation, in *Mind* and *Brain* (see 5.110 and 5.134).

The present "working hypothesis" would then be, to write now a paper—twenty pages perhaps—for the next number of the Journal, and to translate it with Berger;[4] and to make further experiments for the doctor's thesis. What extra time I have between this and Aug 1st I will of course spend in writing or preparing to write the thesis. Then when home this fall I will complete the thesis, and returning to Germany will translate it during December, and hand it in the first of Jan, passing the examination the first of Feb. I suppose the Journal containing the thesis would be printed in Feb. or March.

Supposing I should write a paper in the time it takes to think, would you think it better to have it appear in "Mind" or the "Princeton Review."? Of course beggars are not choosers, but the two reviews would be about equally likely to accept such a paper.

[4]This paper was published as "Über die Trägheit der Netzhaut und des Sehcentrums," *Philosophische Studien* 3 (1885): 94–127, and later as "The Inertia of the Eye and Brain," *Brain* 8 (1886): 295–312.

5.104 To Parents, Leipzig, I June 1885

I spent an hour at Prof. Wundt's today. He has been very kind all winter concerning the difficulties of publishing my work and passing the examination. He says he is willing to put off the issuing of the Journal a month on my account, but thinks the plan I last wrote you to be the best. I shall not therefore include the paper already printed in the doctor's thesis, nor the one to be written for the next number of the Journal; but the thesis will be printed in the following number. If I wish Prof. Wundt will have my paper set up at any time, and let me publish it as a doctor's thesis, but for reasons I have given you I do not really care to have it printed before the first of next year.

I went to see an oculist today—Dr. Küster. He agrees with me in thinking that the astigmatism in my eyes was not sufficient to warrant Dr. Thomson in giving me spectacles, and that I might get along without them, so I shall try for a while. I have not been working very hard for the past two or three days, but have not missed the glasses especially.

5.105 To Parents, Leipzig, 3 June 1885

(Two paragraphs of gossip have been omitted.)

I sent you today twenty copies of my paper.[1] I have plenty of them. They gave me 25 and I ordered 25 extra. Then when I decided not to reprint it in my doctor's thesis I ordered 50 more—when I get these I shall send you 20 more—though I'm sure I don't know whether you will want them or not. I was going to send you one copy of the Journal, but in your last letter, Papa, you seem to want more, though you refer perhaps to the number only, which is to contain the thesis. I shall send you however several, though they are expensive, and scarcely better than the reprint. I shall not print anything better than this—only more of the same sort. I will send copies to Profs. March and Bloomberg²—and perhaps Annie Pardee³—I think of no one else to whom you might send one—you would not of course send any to Baltimore.

[1]"Über die Zeit der Erkennung und Benennung von Schriftzeichen, Bildern und Farben," *Philosophische Studien* 2 (1885): 635–650.
[2]For more on Professor March see introduction. Augustus Alexis Bloomberg was professor of languages at Lafayette from 1867 through 1906 (*Men of Lafayette*, pp. 333–334).
[3]The daughter of Ario Pardee (see introduction).

5.106 To Parents, Leipzig, 13 June 1885

I called on Mrs. Wundt this morning—she is very nice, as I have told you.[1] The professor also came in to see me—I am going to row them up to Connewitz some day next week. There are two bright children—a girl of eight, and boy of six.[2] Prof. Wundt is coming to see me tomorrow. I am invited to dinner at Prof. Heinze's the day after[3]—so you see I am on excellent terms with my professors. Indeed I imagine—though I have often been told the contrary—that I am quite clever in getting along with all sorts and conditions of people. The only question is how far it is manly—or worth the while—to spend ones life playing a part or rather a dozen different parts.

[1]Frau Wundt (according to Wolfgang Bringmann, who is preparing a biography of her husband) had studied in England and spoke English well.
[2]Max Wundt would publish several books on Greek philosophy (*Pre-1956 Imprints* 676: 273), and Eleanore (nicknamed Lolo) would publish a bibliography of her father's writings: *Wilhelm Wundts Werke: Ein Verzeichniss seiner sämtlichen Schriften* (Munich: C. H. Beck, 1927). Eleanore Wundt also later did much of her father's actual writing, as his eyesight failed, and acted as hostess after her mother's death (Charles H. Judd, "In Memory of Wilhelm Wundt," *Psychological Review* 28 [1921]: 173–178).
[3]On Max Heinze see 1.15.

5.107 From Elizabeth Cattell, Philadelphia, 16 June 1885

(Several paragraphs of gossip have been omitted.)

Come home take a long vacation, go back in the Fall to Leipzig, and stay as long as you think it necessary to accomplish what you desire. Our income is such that you can go on with your studies as long as you think best, but dear Jim do of all things take care of your health.

5.108 From William Cattell, Cape May Point, N.J., 29 June 1885

(Several paragraphs of gossip have been omitted.)

I have been trying to read yr article in the "Studien";—but its too much for my German. Prof. W's letter however I could read.[1] I sent in to Prof. Bloomberg to get a good translation.[2] He made quite a pleasant notice of it wh appeared editorially (local) in the Free Press—Mamma will send you a copy; & a little paragraph, taken from it, is going the rounds of the "Personals" in the Papers.[3] At Cincinnati, before recᵍ an ansʳ from Prof. B., I met Dr Schaff[4]—we were at the same Hotel—& he made for me a translation wh. I think better than Bloomberg's. The letter is one of unusual warmth for a German & perhaps gratifies your father more than it does you. My first impulse was to send a copy of it to Gildersleeve—with the expectation of course that he wᵈ show it to Gilman; —I mean the letter as well as the article— and I am at a loss to understand why you did not want it "sent to Balt." G. wrote to me kindly;[5] and stated the opinion of his "Colleagues" that you had not fulfilled their expectations. I want him to know that you have "fulfilled the expectations" of men immeasurbly superior to his "Colleagues"—but this is one of the subjects I find myself yet unable to write about or think about with composure.

[1] See 5.88.
[2] See 5.105.
[3] These notices have not been found.
[4] David Schley Schaff (*Who Was Who in America*, 1897–1942, p. 1085) was a distinguished Presbyterian minister best known as co-editor of the Schaff-Herzog Encyclopedia (see *Pre-1956 Imprints* 523: 691).
[5] On Basil Lanneau Gildersleeve and Daniel Coit Gilman see 1.17.

5.109 To Parents, Leipzig, 20 July 1885

(One paragraph of gossip has been omitted.)

I have something to tell you that will please you, though for a moment it may frighten you. I am to be Assistent in the university here next year, having under Prof. Wundt charge of the psychological laboratory.[1] Such a position has, I suppose, been but rarely held by a foreigner, and is seldom given to any one before he has passed the Ph.D. examination. Though of course I get no salary the position is very nice. I only need to be at the laboratory about two hours a day, and during this time can continue my private work, having a room to myself. There are usually about a dozen men working in the laboratory, and I dare say I shall have pretty complete control of what work they take up and how they do it. New men for their first semester must

[1] It is probable that Wundt gave JMC this responsibility in order to free himself from directing students, so as to be able to devote more time to his own philosophical writing (see introduction and 5.111). *Assistent* in a German university laboratory was an important position. Wundt, for example, had qualified as a *Docent* before becoming Helmholz's *Assistent*. See also 3.12n.12 for some indication of the *Assistent's* importance.

help me in my private investigations if I want them, and I can get the apparatus I need from the university appropriation. My name will of course be printed in the next catalogue.[2]

I think I should have no trouble in passing the privat-docent examination, and what I mean by suggesting that you may be frightened is, that you may imagine I am inclined to settle down here like Gregory.[3] But you need not worry about that. I told Prof. Wundt that I could not stay longer than next summer, as I hoped to find a position in American next Fall. I should think my holding this position here, would help you to secure me a position in America. There is more room to fear that I shall not much care to hold a position anywhere, than that I shall want to stick to this one here.

[2] See *Personal-Verzeichniss der Universität Leipzig*, no. 108, winter semester, 1885–1886, p. 31.

[3] On Caspar Rene Gregory see 1.17.

5.110 To Parents, Leipzig, 29 July 1885

I feel quite lighthearted or at least "lightminded" tonight as I gave Prof. Wundt my paper this noon.[1] I have been working on it for over two years, and quite incessantly for the past month. Berger read half of it aloud to Prof. Wundt today, the rest to be continued tomorrow. He seems much pleased with it, and only suggested one or two verbal changes.

When one thing is finished however, something new always turns up. I received today a nice letter from the Editor of "Brain" whom I had offered to give an abridgement of this paper.[2] Mind, Brain and the Journal of Physiology are the great English Journals. Mind is devoted to philosophy & psychology, the Journal of Phys. to physiology, and Brain comes between treating especially nervous diseases &c. Bucknill, Ferrier, and Crichton-Browne, the three most eminent men in England in this department are associate Editors.[3] Well, the "acting editor" writes "We shall always

[1] This is the paper referred to in 5.103.
[2] See 5.103.
[3] John Charles Bucknill (*DNB* 22:331) was the medical superintendent of the Devon County Asylum for many years, and was also long associated with the University College (London) Hospital. David Ferrier (*DSB* 4:593) was a Scottish neurologist, educated at Edinburgh and Heidelberg, and long-time physician at King's College Hospital in London; his best-known work was *The Functions of the Brain* (London: Smith, Elder, 1876). James Crichton-Browne (*DNB*, 1931–1940: 106–107) was director of several British asylums and, for many years, the Lord Chancellor's Visitor in Lunacy (inspector of asylums).

be glad to have any article on psycho-physics you wish to publish in English. Perhaps you might let us have one on some other subject, as well as the proposed 'Résumé' of your very interesting experiments on Vision".[4]

I wanted to go to bed early tonight, but I see it is twelve. I have not been in bed until after twelve for a long time, but shall sleep all I can for the next three months.

[4] Armand de Watteville, a graduate of University College, London, and a physician specializing in neurology at several London hospitals, served as Acting Editor while *Brain* was officially controlled by an Editorial Committee (1883–1887), and then took the formal title Editor (*Brain* 1 [1878]: 224; George Watson, ed., *The New Cambridge Bibliography of English Literature*, vol. 3, *1800–1900* [Cambridge University Press, 1969], col. 1858). The experiments mentioned are probably those eventually reported in "The Inertia of the Eye and Brain," *Brain* 8 [1886]: 295–315.

5.111 Journal, Philadelphia, 5 September 1885

(Several lines of gossip have been omitted.)

πάντα ρεῖ οὐδὲν μένει,[1] All things go by. I am only I by the mystery which links change to change. Easton is gone, the past is gone. Forgotten or to be forgotten, dead or dying, write it on the front banner. With the exception of a couple of days (June 23–7) in the Harz I stayed at Leipzig until the end of July. I do not think I did much in the way of work during June; and was in rather a bad way part of the time—quite upset in the Harz. I played tennis and rowed some, and [several undecipherable symbols in Pitman shorthand] Menzies.[2] Berger was with me during July and I worked hard. We made a lot of experiments—mostly on the sensitiveness of the retina and limits of consciousness, and I wrote a paper on this subject which we translated into German for the Studien.[3] We often worked half through the night. I am to be made assistant in the university—having charge of the psychological laboratory.[4] My plans were changed during the last month or two of my stay at Leipzig. I concluded not to hand in my doctor thesis until next winter, and the offer of the assistantship decided me to spend the whole year at Leipzig.[5] I had intended to pass the examination soon and go to Greece & Italy. On the night of Fri. Jul. 31st I went with Fine to Frankfort.[6] Then alone down the Rhine, and on Sat. night to Antwerp.

[1] See 3.1.

[2] Alice Marian Menzies was a student at the Leipzig Conservatory from London (*Königliche Conservatorium*, p. 58).

[3] This work was reported in "Über die Tragheit der Netzheit und des Sehcentrums," *Philosophische Studien* 3 (1885): 94–127. See 5.103.

[4] See 5.109.

[5] Part of the mythology of psychology, derived directly from Wundt's memoirs (*Erlebtes und Erkanntes* [Stuttgart: Alfred Kröner, 1921], p. 312), is that JMC approached Wundt and declared, "Herr Professor, you need an *Assistent*, and I will be your *Assistent*." This entry, however, written soon after the event rather than thirty years later, and in a journal not meant for other readers, makes it clear that JMC was offered the job. This point is reinforced by the view that Wundt wanted to free himself from laboratory drudgery.

[6] Henry Burchard Fine, a graduate of Princeton was studying mathematics at Leipzig. He later taught at Princeton and became dean of the faculty there (*World Who's Who in Science*, p. 565).

There was a world's exhibition there. Sun. night I went via Flushing to London. I called on Dr. Ferrier—was also there to dinner—he was glad to accept the paper above mentioned for "Brain", so I had to rewrite it which took up a good deal of my time while at London. I ate chocolate constantly and was in a queer way. My sight-seeing was mostly confined to Inventors Exhibition and National Gallery.[7] The Academy exhibition was worse than usual. I called on Miss Menzies. Not much was going on at the theatre. I had a not very ideal "romance" with Louise Smith. Fine turned up Thurs. (13th) and we spent a day together. I went to Liverpool Fri. morning and spent the afternoon and evening with Prof. Robertson at Formby.[8] Arrived in N.Y. Sat. night, making the shortest voyage so far 6 days 2 hours. Papa missed meeting me. I came on to our new Phila. home Sunday. Both Mama and Papa are in better health than when I left them a year ago. Harry is growing. We went to Spring Lake on Wednesday (26th) and spent a week there.[9] Its a queer world.

[7] The Inventors' Exhibition JMC mentions was the third of an annual series of major exhibitions held at South Kensington in celebration of aspects of late-nineteenth-century British civilization. (*Engineering* 39 [1 May 1885]: 437–441). The 1884 exhibition was the International Health Exhibition (see introduction to section 6).

[8] George Croom Robertson (*DNB* 16: 1294–1295) was professor of mental philosophy at University College. He had been Alexander Bain's student and protégé at Marischal College. Robertson edited *Mind* from 1876, while Bain supported the journal financially. JMC's meeting with Robertson was probably a result of his previous meeting with Bain at Leipzig. (see 5.89, 5.90, and 5.91). See also Anthony Quinton, "George Croom Robertson: Editor 1876–1891," *Mind* 85 (1976): 6–16. Formby: a suburb of Liverpool (Baedeker, *Great Britain*, p. 339).

[9] A resort on the Atlantic coast of New Jersey.

Cattell spent the month of September with his parents, and talked with them of his plans for the future. He left for Leipzig on 3 October 1885 on the *Servia*.

5.112 To Parents, aboard the S.S. *Servia* off Queenstown, 10 October 1885

(This letter was damaged, apparently by water staining. Brackets indicate unreadable sections. Two paragraphs of gossip have been omitted.)

There seem to be plenty of nice people on board. I have seen most of Pres. and Mrs White, who are here for all.[1] They are both very nice, but I do not think he has much strength of character or is in any way a very remarkable man. Mr. Hughes seems well satisfied with himself and the world.[2] I guess he was something of a prig when he was a boy and is yet. He has been writing a life of Peter Cooper all the way over.[3]

[1] On the Andrew Dickson Whites see 2.29.
[2] Thomas Hughes was an English writer, best known for *Tom Brown's School Days*, pub-lished in 1887 (*Oxford Companion to English Literature*, p. 403).
[3] *The Life and Times of Peter Cooper* (London: Macmillan, 1886).

I scarcely know what I have been doing—you know how time goes on shipboard. I have read half "The Adventures of [   ] and that is about all. We had the usual concert Sunday, but worse than usual—though Mr. Hughes recited a humorous poem. If I had written to you yesterday my letter might have been quite optimistic, tonight I am not likely to do any thing agreeable, neither internal nor external circumstances tending to make [   ] better.

5.113 To Parents, Leipzig, 27 October 1885

(This letter was misdated "Tues. Oct. 26th.")

I have been working on my thesis today and have written eight pages. I had already twenty-four pages written last summer, and it is now a third or at all events a fourth done. It looks as if I had a good deal to do yet, but I think if it were necessary I could write eight pages a day—and so have it all done in ten days, whereas I have fifty days before I need to begin to translate. I suppose it will take up about 75 pages in Wundt's Studien. My saving in not having to print it myself will be considerable.

I have received a letter from Berger. He will come on for a week or ten days at Xmas, but if we need more time than that to translate, I shall have to go to Mühlhausen.

5.114 To Parents, Leipzig, 28 October 1885

(This letter was misdated "Wed. Oct. 27th.")

The psychological laboratory opened officially today. Ten new men were admitted. I have two men to help me so far as I need them in my own experiments—a Dr Lehmann, privat-docent at Copenhagen and a Dr. Steinmetz.[1] I shall not however make many new experiments, until I write out accounts of those I have already made.

I do not think you appreciate the advantage of the position I hold here. I consider it much more to my credit than having held a fellowship at Baltimore, and if I want a position in America or England nothing could be of more help to me in securing it.

[1] Alfred Lehmann (1858–1921) was to be a pioneer in the establishment of experimental psychology in Denmark (Ingemar Nilsson, "Alfred Lehmann and Psychology as a Physical Science," *Abstracts of Scientific Section Papers: Fifteenth International Congress of the History of Science*, Edinburgh, 1977, p. 292). "Dr. Steinmetz" is probably Sebald Rudolf Steinmetz (*International Who's Who*, p. 1077), a Dutch ethnologist educated at the universities of Leiden and Leipzig.

5.115 To Parents, Leipzig, 5 November 1885

I was invited to supper at Prof. Wundt's this evening. He leads quite a model life. He likes to work, and works with the greatest regularity, so in the course of the year accomplishes a good deal. His life is completely and finally settled—he has nothing to hope for, nothing to fear. He has a good income, and does not need to worry about saving up or looking after his investments. He will receive his salary from the University all his life, and his wife a pension in case of his death. He is very happy in his family. He has a solid reputation, which he knows will last. Who would not envy him?

5.116 Journal, Leipzig, 11 November 1885

I had rather a good voyage. Pres & Mrs White were on board, and I saw a great deal of them. Saw something also of Thomas Hughes. Besides these Dr. Waldstein (Cambridge) and Dr. Weber, (Cleaveland) should be mentioned and remembered.[1] We were just in dock Sun. night, and I went to London Mon. morning.

I dined with Prof. Robertson and Dr. Ferrier—saw Mr. Galton, Dr. de Wattevill, Dr. Hughlings-Jackson and the Menzies.[2] Came to the continent Fri. night, spending Sat night at Hanover, stopped at Brunswick and Magdeburg and arrived here Sun. night (the 18th). I could not keep my old quarters—Humboltstr 7—but have been fortunate in finding others equally good—Liebigstr 9.[3] I have done a reasonable amount of work since my return. Sent a short paper to "Brain" and the first section of my doctor's thesis to "Mind".[4] This latter is about half done, and I suppose I shall have no trouble in finishing it in the next six weeks. I am reading some little in the direction of psyco-physiology. I find a number of acquaintances here, but there is no one I care much for—my good friends Fine, Muirhead and Berger being gone. I am mostly alone. I exercise quite a good deal—taking often a six-mile walk, over the fields to Connewitz and through the woods to the town. I fence four times a week and swim nearly every day. This exercise is all that keeps me straight, as my mental course is sadly crooked. I drug my brain too much, albeit only with theobromine and

[1]Charles Waldstein (later Walston), an important member of the Cambridge University community, was a fellow of King's College, reader in classical archaeology, director of the Fitzwilliam Museum, and the author of many books on Greek art (Venn, *Alumni Cantabrigiensis*, vol. 4, p. 311; Baedeker, *Great Britain*, pp. 440–441). Henry Adam Weber (*DAB* 19: 582–583) was a German-educated chemist who had worked with the Ohio Geological Survey and taught at the University of Illinois and who later played an important role in the passage of the Federal Pure Food and Drug Act.

[2]On George Croom Robertson and Alice Marian Menzies see 5.111. John Hughlings Jackson (*DSB* 7:46–50) was a neurologist connected with the National Hospital for the Paralyzed and Epileptic.

[3]JMC's Humboldstrasse address had been changed from 19 to 7 when the city of Leipzig renumbered its streets. Liebigstrasse was on the other side of town.

[4]"The Influence of the Intensity of the Stimulus on the Length of the Reaction Time," *Brain* 8 (1886): 512–515; "The Time Taken Up by Cerebral Operations," *Mind* 11 (1886): 305–335. See 3.12.

nicotine.[5] I lead a curious life. I live rather nearer to the central realities (or phantasms) than most people. I think constantly about life and death and such great— relatively great—matters. I imagine I am an able man, but do not think it will come to any thing—what should it come to? I wrote a clever sketch, "At the Crossways" three years ago, and stay by that.[6] Tonight and a good deal recently, I am inclined to accept ideal No. I. It seems to me on the whole the most possible and tolerable. On this hypothesis I only need a position in an American college and a wife: both, I take it, easily within my reach. Still it may be better to put them off for a couple of years—a year is gone, a year is come, says the new church built in memory of God. I have a dream of the past—a boy in the dark, looking for the light, stumbling. aching. I have a dream of the present—a man holding a candle in his hand, lighting ten feet around him, likely to be blown out with every breath. I have a dream of the future, a man sitting by his own fireside, watching his wife, his children, thinking of tomorrows work—tomorrow's duty. The dream changes as dreams will change. A man stands in a crowded room sought after, envied, disliked, busy, successful, proud. It changes again, a man sits at his desk, alone, weary, hopeless, sick-at-heart. Which dream will last—in the end they all fade into the same darkness. It is God's world, not mine: His will be done.

[5]Theobromine is the principal alkaloid of the cacao bean, and a constituent of cola and tea. (*Merck Index*, p. 8996).

[6]This "sketch" has not been found.

5.117 To Parents, Leipzig, 14 November 1885

I received today both "Brain" and the "Studien" with my paper.[1] My work is of considerable importance, and some parts of it are very easy to understand and of general interest (especially Sec IV.)[2] (which my next paper will not be). I think it would be well if you get it noticed in the papers, quoting possibly some paragraphs. Indeed it might not be impossible and might be worth the while to get it reprinted in one of our papers—the Sunday Tribune for example. I shall myself send copies to some journals (mostly European) and to some people—I shall not however send any to Philadelphia, Princeton or Bethlehem.

A paper like this gives me a very secure place in the scientific world, makes me equal with any American living. One likes however to be given credit for what one has done even by people who know nothing about the work. I scarcely know why we like to be praised by fools, but we do. Still there is some reason in my case. I may want a position or a wife, and certainly do want to be able to pick out the people I associate with.

[1]"Über die Tragheit der Netzheit und des Sehcentrums," *Philosophische Studien* 3 (1885): 94–127; "The Inertia of the Eye and Brain," *Brain*, 8 (1886), 295–312. (The *Brain* paper was published in the issue that appeared in October 1885; however, the title page of volume 8 is dated 1886.)

[2]Section IV was on "The Relative Legibility of the Letters of the Alphabet."

I shall send you 20 reprints of the German and 50 of the English—and can send more if you wish them. I have 250 reprints, (that is have ordered them) of the English. I first ordered 100, but the editor told me it would scarcely cost more to get 250—that he always got 250 reprints of his papers. I send you a copy of the "Studien." The first paper is by Fechner, perhaps the greatest philosopher living (over eighty years old.)³ The second by Berger, my name being often mentioned.⁴ If you really want it I can send you other copies, and you can get "Brain," but I think the reprints quite as good.

You will see the German version is much the longer. I give there full tables and details—I myself like it the better, but perhaps the English contains every thing of interest to others than specialists.

³Gustav Theodor Fechner (*EoP* 3: 184–185) was a German pantheist best known in the history of psychology for his development of the psychophysical law. See also 3.12. The paper to which JMC refers, "In Sachen des Zeitsinns und der Methode der richtigen und falschen Fälle, gagen Estel und Lorenz," *Philosophische Studien* 3 (1885): 1–37, discusses psychophysical techniques.

⁴"Über den Einfluss der Reizstärke auf die Dauer einfacher psychischer Vorgänge mit besonderer Rücksicht auf Lichtreize," *Philosophische Studien* 3 (885): 38–93.

5.118 To Parents, Leipzig, 16 November 1885

I was at the opera tonight, the first time for quite a litlle while. They gave Nicolai's "Lustige Weiber von Windsor"—I consider this the best of all comic operas, except Mozart's. I think I ought to go to the theatre quite a good deal—I shall scarcely have such another chance as here at Leipzig. It costs however something—I have gotten into the way of sitting in the parquet, and somehow feel as though I were deceiving my acquaintances if go elsewhere. Then anyhow the increase in price is so gradual—the cheapest seat one can engage is M.1.25 the side of the third gallery; but when one is there it seems better to pay 50 pf. more and sit in the middle where one can see. But when one has once paid M.1.75 it is certainly better to pay 25 pf. more and sit down stairs, where one can see and hear much better and it is tolerably cool. Then paraquet right near the stage and in respectable company only costs about 50 pf. more (tonight it only cost M2) so by degrees one has come to pay 50 to 75 cts for the seat.

This is the second week, Papa, I have not heard from you.

5.119 To Parents, Leipzig, 25 November 1885

I go to the laboratory Wed. and Sat. afternoons for about an hour, and meet Wundt there. This is all my work, corresponding with my salary. It has been very pleasant for me to know about and help arrange all the work that is being done. One of our members has gone home to fight against the Servians. During the past year we have had in the laboratory, two Russians, two Finns, a Norwegian, a Dane, a Hungarian, a Bulgarian and two Americans! It shows the attraction of the subject, for Wundt himself is scarcely a great man.

5.120 To Parents, Leipzig, 28 November 1885

(This letter was misdated "Sat. Nov. 30th.")

I got my letter of credit today. £300 would have been plenty, even though I make the doctor's examination before returning to America. This is said to cost $200. I give Berger $200. I may use the $300 interest of my money out west for apparatus, books and pictures. I think my general expenses here would be from $800 to 1000 including a reasonable amount of travelling. I should like to go to Italy or Greece this Spring, but if I am well, shall not have the time. As I have always said I would rather not try the examination until Fall or Winter—I never thought of taking it much sooner, as I expected we would all stay over here two years. Then my work was seriously interrupted this year. I can however without doubt manage to get the degree even this spring if it is desirable. Prof. Wundt conducts the oral examination in philosophy and examines the thesis. There is not the slightest doubt, but that he will except the latter, and I know much more about philosophy than Berger, and he passed, though not well. If I should fail in physics or zoology, I can study it up in a couple of weeks, or take English in its place. I must take Zoology or Mathematics, because Physiology (which I know pretty well) is in the medical department. There is no Biology.

5.121 To Parents, Leipzig, 4 December 1885

I have about made up my mind that it would be well for me to become doctor of medicine as well as of philosophy—led to this partly by the thought that you too think this would be my best course.[1] If I am to teach psychology I ought to spend at least a year studying the physiology and pathology of the nervous system, with one year in addition to this I should be ready to pass the M.D. examination. What I should learn this second year would to a great extent be useful to me, though I should not practice medicine.

The knowledge and the training a study of medicine would give me, and even the title, would be of service to me as a professor of psychology, and would enable me to fill other professorships. Then if for any reason I could not get, or could not or would not keep a college professorship, I could practice in a branch of medicine connected with my studies, and thus support myself, and occupy the position of useful and respected citizen, which you consider so vital.

Of course if I am offered a good college position next year, I can postpone these medical studies or pursue them in addition to my other work—otherwise I look to study chiefly with a doctor examination in view.

[1]Apparently JMC had discussed this possibility with his parents the previous September while in America. He had been talking and corresponding with such distinguished medically trained neurologists as John Hughlings Jackson and David Ferrier before making this decision (see 5.110 and 5.116).

I think the adoption of such a course by me, will relieve you of any anxiety you may feel as to securing me a position. As a student or practicioner of medicine I could afford to wait until a position was offered to me; indeed from your point of view I should be more more of success as a physician (treating paralysis or insanity) than as a professor.

5.122 To Parents, Leipzig, 6 December 1885

I received yesterday the proof of the dear little paper to be in the Jan, "Brain;[1] the Jan. "Mind" summary is also only a couple of pages.[2] I shall however have a long paper in the Apr. and probably July numbers of "Mind";[3] and my thesis (nearly 100 pages) in the next number of Wundt's "Studien".

I have certainly done a great deal in just one year. I have printed and in press five papers, I have three more ready, and after I pass the examination shall write three or four more, most of the work being ready. I could not have done all this if I had had a position, nor, I imagine, under any circumstances in America.

[1] "The Influence of the Intensity of the Stimulus on the Reaction Time," *Brain* 8 (1886): 512–515. (See 3.12 and 5.116.)
[2] "The Time It Takes to See and Name Objects," *Mind* 11 (1886): 63–65. This short article was really a summary of JMC's first paper in *Philosophische Studien*.
[3] These papers were the English version of JMC's thesis (see 5.116).

5.123 To Parents, Leipzig, 10 December 1885

The days go round so fast I scarcely know what I do. I have one man making tables for me, and another drawings, the which cost money, and still more patience. I have nearly finished a book by Bastian "The Brain as an Organ of Mind"[1] but do not somehow find time to read much or to prepare for the examination. After Xmas I must spend two months doing something else. Even in Philosophy I am badly prepared. I know psychology, but that is only one branch with history of philosophy, logic, metaphysics, ethics and aesthetics, some of these, logic for example, I have never studied, and I have forgotten a good deal I once knew about the history of philosophy &c.

[1] Henry Charlton Bastian, *The Brain as an Organ of Mind* (London: C. A. Paul, 1880). This book was especially important in the development of JMC's physicalist perspective on psychology (see 5.142n.1). Bastian (*DSB* 1: 495–498) was educated at University College, London, and practiced at the National Hospital for the Paralyzed and Epileptic. He also taught at University College, first as professor of pathological anatomy and later as professor of the principles and practice of medicine.

5.124 From William Cattell, Philadelphia, 10 December 1885

(One paragraph of gossip has been omitted.)

I hope this letter will reach you in time for Christmas; it bears to you greetings from our hearts: and when our Christmas comes <u>here</u>, we shall think of you often during the day: and we shall rejoice & give thanks we have such a son to think of and to love.

I have thought it w^d be better—so as not to appear in yr accounts—to send you a draft for yr Christmas present; the usual sum of $25 in money and $10 for the present.

The remaining $15 I send so as to cover the expense of some copies of the Studien which I am anxious to have;—& dont want you to pay for them out of yr bill of credit. I would like to have <u>two</u> more of each of the numbers in which yr articles have appeared: and 6 or 8 of the no. wh contains yr thesis. For my purpose, the "reprints" are not worth much; a much more <u>vivid</u> impression is made by the entire review;—& then the art. is more likely to be preserved. This may seem to you unimportant:—but my experience has always shown that you must attend to these matters. And I much preferred the form of cover &c in which yr first art. appeared. This last looks like a pamphlet. The announcement on the second page of the cover (separatabdruck aus &c) does not attract attention. The art. in Brain appeared in the best dress of all—the regular cover of the magazine. I wonder if you could not have yr <u>thesis</u> art. in this way:—the regular cover of <u>The Studien</u>, list of contributors &c and yr article marked (on the cover) by a pencil. This tells the whole story at a glance;—but if not this, to have it in such covers as the first;—not as the last which is too much like a pamphlet.

I read the art. in Brain with great interest. Could you have the english of yr thesis carefully transcribed for me & on one side of the paper? I w^d like to make some use of it. In fact I wish I had a carefully written copy of all three. I suppose the art. in Brain is similar to the last one in the Ph. Studien. If these copies can be made, without troubling <u>you</u>, please let me have them. The expense I do not mind.

I am hopeless as to any opening in the University here.—but Princeton, New York (University) & Lehigh are worth <u>working</u>:—Cornell also. What is yr real preference? <u>Our</u> heart's desire is to have you near us. But there is a great pressure for these positions. We must be satisfied with what we can get.

5.125 To Parents, Leipzig, 13 December 1885

(Brackets indicate wormholes.)

I was at dinner at Prof. Heinze's[1] [   ]   It was very kind in him to invite [   ] it was his fiftieth birthday. We had an elaborate (and very good) dinner with champagne &c. Prof. Heinze sent greetings to you, Papa: as he always does when I see him. He is one of the nicest men I have ever met, and has a strong individuality of his own. Their daughter is a remarkable child, she is [   ]

years old and acts quite like a [   ] I'm sure I never saw [   ]
child quite so grown up. Mrs. Heinze too is nice and an interesting study. She is
very much down on womans rights, woman's education, woman's ability &c. yet has
educated herself and is educating her daughter ([   ] go to school) elabo-
rately well. There are two sons (one in Berlin, and one studying here)
[   ] and good students—they are studying to be [   ] which
means something here.

[1] Max Heinze (see 1.15).

5.126 To Parents, Leipzig, 24 December 1885

I have had as pleasant a Xmas Eve as I could have looked for away from home. I
went to Prof. Wundts at six, where the children, nine and six years old, are just of
the age to be as happy as possible on Xmas. They are very good friends of mine—
Lolo and Max. At seven I went to Prof. Heinze's and stayed until after eleven. As I
wrote recently Prof. Heinze's family is very interesting—the sons are about the
nicest Germans I know. They gave me presents at both places—which seems sort of
a curious idea. I gave the children bonboniéres and the illustrated English papers. I
am to take dinner at Prof. Wundts tomorrow.

5.127 To Parents, Leipzig, 25 December 1885

So I have spent another Xmas apart from you—far away, of course, for otherwise I
would be at home. As I wrote Harry this morning I can scarcely understand that this
is the sixth Xmas I have spent in Germany. It is however only the third that I have
been away and you at home. I trust next year our family circle may be unbroken—it
is really a family square, and so one corner counts for a good deal.

It seems a pity that I have been away so much, but I am glad you sent me to
Germany. I am sure it is far better so, than if I had stayed at home and studied and
practiced law. If I had gone into business, I do not think it would have been possible
for me to have kept on and I should have had to begin over again. After all you
have not done more than most parents do in sending their sons to college. I have
been away just four years (including this) first separated from you 10 months, then
eights months, then a year, and now it will, I fear, be another year. It is true I have
been farther away than most sons, but on the other hand I think I have been more
with you than most sons during their college course. I suppose the expense has been
about the same as at one of the more expensive colleges.

That I have not changed my plans recently is evident from a letter I wrote Dr.
Hall from here Apr. 5th '84.[1] I say "I intend to stay here for about two years. After
that I will probably try to obtain a position, though I should not object to carrying on

[1] Not found. See 5.78.

my studies even longer free from the responsibility of teaching." I shall pass my examination within a week of the time I then planned (Mar. 1st), though at one time when anxious to publish my results, and not knowing that I would have more than enough for the thesis, I thought of passing the examination sooner.

This leads me to turn to what you wrote of in a recent letter, Papa. I am not anxious about securing a position. I think psychology is likely to be the science of the next thirty years—at all events the science in which most progress will be made. I have, little as it is, done more for psychology than any American, and have no reason to doubt that I can easily stand among the first in the future. It is further likely enough that I shall do better if looking for a position, than when finally settled. If I had my choice of a good position either now or three years hence, I should choose the latter. As to the where, we all think that as you are at Phila., Phila. would be the best place. As to my own inclination, I should prefer to live in the country, Cornell for example, if I had a family, if not in a city. Lehigh would for the present seem to be the most "likely" place. The $50000 given to Cornell was to found a chair of "Christian Ethics" "whatever that may mean" as Pres. White said, who did not seem pleased with the condition.[2] It would be curious if I could not get a position later, if I really tried. I shall have printed about ten papers in a year or so. I am assistant in the university of Leipzig. Your personal influence is great, and my own can be made considerable. I am not in the slightest hurry (money being the only and not very serious question) and certainly shall not worry unless I fall in love and want to marry her.

I had a pleasant afternoon at Prof. Wundts talking psychology, and playing with the children. I trust it has been a merry Xmas for you.

[2] See 5.13. JMC's quotation of White is probably from a shipboard conversation (see 5.116).

5.128 From William Cattell, Philadelphia, 27 December 1885

(Several paragraphs of gossip have been omitted.)

Mama & I read with great interest yr views as to studying Medicine. I have no doubt this w^d be a great advantage even if you do not practice;—tho I fear, to get a degree, you w^d need to study many technical things of use only to a practitioner. Positions here in colleges are difficult to secure. And indeed the professions are all over crowded. But you w^d be sure to suceed as a specialist in nervous diseases—the role of Dr Weir Mitchell.[1] However this is a matter to be thought over very seriously. And there is yet time.

Thanks to the Warren Foundry extra dividends we have got through the year without debt,[2] and we have had unusual expenses, in fitting up the house, new

[1] See 3.3.
[2] On the Warren Foundry see the introduction.

furniture &c. I have been working away, as usual, tabulating the expenses! —and have at least the satisfaction of knowing where all the money has gone. We can hardly believe it as we look at the "total"—yet in reviewing the "items" scarcely know which we c^d have avoided. But we are going to try next year to live on less; in fact this will be a necessity. But expenditures for health must not be reduced— unless they are unwise—God bless you my dear son

5.129 To Parents, Leipzig, 28 December 1885

We have been working most of the day. I however skated for an hour while Berger was getting his dinner. The ponds are frozen again, and I hope the ice has come to stay. If so I shall skate a great deal after this week.

I am practicing on the typewriter as I have written 32 pages of my thesis off with it, and must finish the rest about as much more. It certainly makes a better manuscript than pen and ink, I cannot however as yet write quite as fast, nor could I compose very well on it, I imagine. If I had a private secretary I should translate something into English[1]—Wundt's psychology for example (which has been translated into French and Russian) but I do not feel equal to writing so much.

[1] JMC's earlier translation projects (for example, Lotze's *Gründzuge der Psychologie*) had come to naught; see 3.1.

5.130 To Parents, Leipzig, 4 January 1886

(This letter was misdated "Mon. Jan. 3rd." One paragraph of gossip has been omitted.)

I handed in the thesis today, "Psychometrische Untersuchungen". Did you ever expect to have a son who would write a paper with such a name?[1] I fear I shall still have trouble before the English version is in order—I must now finish copying it, and send it to Prof. Robertson. I suppose then I shall have to rewrite and shorten it, as Mind has never printed such a paper.

[1] See also 3.7: "I am thinking of writing a book on Psychometry. Did you think you would ever have a son, who would write a book on Psychometry?" The title of JMC's dissertation probably reflects his admiration of two scientists whose research was important to him. The first word, "*Psychometrische*," almost definitely indicates how highly he thought of Galton's work in their common area of interest (see 3.7). "*Untersu-* *chungen*" probably shows JMC's acquaintance with Exner's "*Experimentelle Untersuchungen*" using the reaction-time method (see 5.57 and 5.68) and his high opinion of them; Wundt and his other students did not use this word to describe their work. In the first five volumes of *Philosophische Studien*, only ten out of 95 articles included "Untersuchungen" in their titles; of these, four were by JMC and one was by Harry K. Wolfe.

5.131 To Parents, Leipzig, 9 January 1886

I am glad to say that I have finished an English version of my thesis, and shall send it away tomorrow. It makes 56 pages with the type-write and twenty pages of tables, so you see the mere copying was quite a task. Now I must take up some preparation for the examination—though I shall be interrupted by one thing and another. I am making experiments with Dr. Lehmann and shall have proofs to correct.[1] I do not propose troubling about the examination. There is the unpleasant possibility of being conditioned, but no amount of work during the next two months would especially lessen this. As to the grade, that on the thesis is always considered the most important, and if asked what my grade was I shall always give that, as it will in any case be higher than for the examination.

[1]On Alfred Lehmann see 5.114.

5.132 To Parents, Leipzig, 11 January 1886

I have begun today to read up for the examination. It seems absurd to try to learn all about physics, biology, and philosophy in two months. I know considerably less about physics than I did at the end of my junior year at college, nor shall I know any more than I did then by the examination. Biology (including zoology) is in a still worse way, and I'm sure I don't know where to begin. I cant find any suitable text book at all. In philosophy I am really worse off then in either physics or biology, because more will be expected from me. I know enough in some directions, and almost nothing in others.

I suppose the best way is to take things easily, as we do death and other unwelcome but inevitable matters.

5.133 To Parents, Leipzig, 18 January 1886

You speak, Papa, in your letter received yesterday with approval of the idea of my studying medicine. It would be curious if you have two sons studying medicine, neither being decided to practice it: yet I imagine we should thereby show ourselves standing "in the foremost files of time. The physician would seem to have the best role in this last act of the play—the man who studies the workings of health and disease in the world, and does what he can to improve a bit the state of things. Of course he need not be an M.D. still less need he visit people at $2 apiece, doing what plenty of others would like to do and can do just as well.

I have a curious feeling respecting those people who point out as an example to their children the successful and respected citizen who for money tries to deceive twelve fools and wrong an honest man, and look down with unbounded contempt on those who think and feel and dream, who try and fail. Those people are a hundred, I am one: why should they not be right? They are, perhaps, but their world is not mine.

Well, to come back to my studying medicine. I suppose I should have to study for about two years in order to get the degree. About one half of what I should have to do lies in the immediate line of my work, and I should want to do it sooner or later; the other half lies more remote, but time spent on it would be by no means wasted. It would certainly be a second string to the bow—and perhaps the stronger of the two. I'm sure that's worldly-wise enough to make up for ¶2.

At all events my plans for the immediate future are not concerned, and we can talk it all over when I come home.

5.134 To Parents, Leipzig, 21 January 1886

My thesis will, I think, be printed in full in the April and July numbers of "Mind." Mind is, as you know, not only the best philosophical magazine in the world, but stands higher than any other magazine devoted to a special department. My paper is perhaps the longest Mind has ever printed, as also probably the most unreadable. I am further paid a high compliment by the fact that my paper will first have appeared in German: an English editor would only reprint such a paper under very exceptional circumstances. As I have said already, however, this paper is not at all better than the others previously published.

I was at Prof. Wundt's this evening with several others. The ice still lasts.

5.135 To Parents, Leipzig, 22 January 1886

I think about the most important question which we may hope will be settled in the next 100 years is concerning the use of stimulant drugs.[1] I have thought a great deal on the subject, and, made a great many observations and experiments, but of course what one person can do in this direction is very little. Unfortunately before this question can be in even the least degree scientifically answered, we shall partly or entirely have lived our lives and for good or evil used or abstained from these drugs. There is a certain difference between the drugs, between say coffee and tobacco on the one hand, and alcoholic drinks and opium on the other, in that in the first case even constant use of the drug cannot intoxicate beyond a certain limit, whereas in the latter the intoxication can be carried further. This is however an artificial distinction, coffee can be used in a form as intoxicating as opium or champagne, and doubtless an alkoloid could be prepared from tobacco giving equally marked effects.

I have myself but little doubt that continued intoxications is very injurious—it only being a matter of degree between tobacco and brandy; occasional stimulation of the brain is however quite another matter, and a most perplexing question. When we use our arms the muscle is used up, but the substance is renewed from the blood, and is far better than the old. We get a better appetite, and practically use up, not

[1]It is not at all clear why JMC raised this topic with his parents at this time, more than two years after he started to use such drugs.

ourselves, but say 5 cts worth of beef. It is quite the same with the brain; when we think or study the brain substance is used up, but it is renewed from the blood, and those who think and study get a better brain than those who don't. Now suppose you drink a cup of coffee, the brain cells are excited. You feel better, you can and do work to better advantage. The brain stuff is used up, but if you give it food and rest (especially in sleep) it is restored, and there are the best reasons for supposing that the brain is healthier, than if the person had foregone the pleasure and profit due to the cup of coffee. If however the amount taken is so great as to use up the brain to an extent from which it does not readily recover, or if even a small amount is used so regularly that the brain is given no chance to recover, the stimulation becomes intoxication and is I doubt not injurious.

On the one hand if we do not use these drugs we lose a great deal, perhaps the very best of life, if we use them we run the risk of hurting ourselves from overuse. I for my part think it is best to accept them as part of God's world. If there is one chance in a thousand of a man's becoming a drunkard let him bravely take the chance, for the additional breadth and depth his life will be given in the other 999 chances. Nor am I by any reasons sure that, after a man has begotten his children, he may not to his best good use these drugs even to an extent injurious to his health. When I come to die I had rather used myself all up, having enjoyed and suffered the most I could.

5.136 To Parents, Leipzig, 29 January 1886

I received my thesis back from the university authorities this evening. I really ought not to have it until I pass the examination, but they have lent it to me, so that it can be put in type. Prof. Wundt has made a good lot of corrections—nearly all verbal however.[1] I wish I knew what grade they have given me. I don't much fear that it will not be "admodium laudable" (magna cum laude). This is a very high grade. Fine was very much pleased to get it and he is one of the ablest mathematicians in America, and two years older than I.[2] There is also a possibility of my having been given the egregia (summa cum laude) grade. Hart in "German Universities" says of this "There was still another, nominally the first . . . It was given, however, very seldom, and only to such candidates as displayed extraordinary knowledge. The last instance of its conferment had occurred eight or ten years before."[3] Hart makes a great fuss over the examination—says he worked on it ten hours a day for I don't know how long, broke down in health &c.[4] I am taking the matter very easily. Day

[1] This statement is not entirely true; see 5.142.
[2] On Fine see 5.111.
[3] Hart, *German Universities*, pp. 235–236. JMC's quotation is accurate. For similar comments on the rarity of this highest grade, see George M. Stratton to George H. Howison, 15 July 1896 (Howison papers).
[4] Hart, *German Universities*, chap. 13, "The Final Agony of Preparation," pp. 192–216.

before yesterday, most of the time I worked, I spent making experiments on the temperature sense. Last night I began to write a novel![5] As I did not go to bed until two, I got up at ten this morning. This afternoon I spent making experiments in the laboratory. This evening I have been fussing over a paper by Pierce and Jastrow of Johns Hopkins Univ.—they having given me an inviting chance to criticize them.[6]

[5]Notes on these experiments and evidence of this novel have not been found.

[6]Charles Sanders Peirce and Joseph Jastrow, "On Small Differences of Sensation," *Memoirs of the National Academy of Sciences* 3 (1884): 75–83. On the authors see introduction to section 2 and document 2.41. JMC later made

extensive use of Peirce's and Jastrow's ideas in a major paper he published with George S. Fullerton (*On the Perception of Small Differences* [Publications of the University of Pennsylvania, Philosophical Series, No. 2, 1892]; *Man of Science*, vol. 1, pp. 142–251). See postscript.

5.137 From William Cattell, Philadelphia, 29 January 1886

(Several paragraphs of gossip have been omitted.)

Do you see the New Princeton Review?[1] The first No. (Jan!) has two or three pages in the scientific movem't in German Philosophy[2]—has much to say about Wundt, & refers to "the younger men in the school at Leipsic from whom much is to be expected" naming "Kraepelin, Lorenz, Tischer & Mr. J. M. Cattell."[3] Dr. McCosh has an art. on "What an American Philosophy Shd Be".[4] If I thought yr Reading Room w^d not have a copy I w^d send you mine.—if you wish it let me know.

[1]In January 1886 the *Princeton Review* began its fifth series, changed its name to *The New Princeton Review*, and for the fifth time published an issue labeled "volume 1, number 1" (Frank Luther Mott, *A History of American Magazines*, vol. 1, *1741–1850* [Cambridge, Mass.: Harvard University Press, 1938], pp. 529–530).

[2]"The Scientific Movement in German Philosophy," pp. 148–152, an unsigned article in the "Criticisms, Notes and Reviews" section.

[3]Page 149. On Lorenz see 5.75. Emil Kraepelin (*World Who's Who in Science*, p. 965) was a psychiatrist studying with Wundt. While at Leipzig, he offered under the auspices of the medical faculty a series of demonstrations of psychophysical methods (*Verzeichniss der*

Winter-Halbjahre 1883/84 auf der Universität Leipzig zu Haltenden Vorlesungen, pp. 11, 17). Later, at the University of Munich, Kraepelin was primarily noted for his psychiatric nosology. Ernst Theodor Furchtegott Tischer (Poggendorff 6: 2668) was a student of Wundt's who earned his Ph.D. for a dissertation on the perception of sound ("Über die Unterscheidung von Schallstärken," *Philosophische Studien* 1 [1882]: 495–542).

[4]Pages 15–32. This article, a critical discussion of the current state of American philosophy, stated that "The American youth of the present day who wishes to carry on research goes for a year or more to a German university," and urged that American universities change so as to make such European study unnecessary. See 4.14.

5.138 From William Cattell, Philadelphia, 31 January 1886

(Several paragraphs of gossip have been omitted.)

I shall look for yr thesis with special interest:—that is, the English version.

I wish it could be got into, say, the Popular Science Monthly (Dr. Youman's).[1] They do publish articles from the European Reviews. But it is not likely they w^d print yours unless some "suggestion" was made; and some help in the matter, that is, finishing the translation. Perhaps you w^d not think the game worth the candle:— but if you thought it worth while to send me the translation of both the articles (you spoke of yr thesis being published in two Nos. of the Phil. Studien) when you send me the German, I will see Prof. Y. personally & find out what he will do—I mean of course only the translat. of the first art. at the time it appears—& the second when the next no. appears. It w^d appear as "translated from the Ph. Studien &"— And if this strikes you favorably you might let me know at once, so then I could see Prof. Y. in advance. If this can be accomplished I think it w^d bring you very favorably before the "scientific" public here.[2]

[1] Edward Livingston Youmans, a disciple of Herbert Spencer (see 2.3), had founded *Popular Science Monthly* in 1872 as "an American journal which should bring to thoughtful readers without scientific training some knowledge of scientific thought in Europe." From 1877 on, William Jay Youmans, an M.D. from the University of the City of New York, assisted his ailing elder brother in editing the journal. (See *DAB* 20: 615–617; Frank Luther Mott, *A History of American Magazines*, vol. 3, *1865–1885* [Cambridge, Mass.: Harvard University Press, 1938], pp. 495–499.)

[2] *Popular Science Monthly* later reprinted an article by JMC, "The Time it Takes to Think" (32 [1888]: 488–491) that had earlier been published in London (*Nineteenth Century* 22 [1887]: 827–830).

5.139 To Parents, Leipzig, 4 February 1886

I was at Prof. Wundt's this evening and afterwards took a walk and sat in a café until rather late with several of the fellows who were at Prof. W's. It is very good to know able men interested in one's own subject—I shall much feel the loss of this in America. We are so made that it is very difficult to do original work when quite alone. Well, perhaps it does not make much matter. During one year I shall have written twelve papers, and perhaps there is no great use in making it 12 × 30. What I have done this year could probably have been done by almost any one 10 or 20 years hence, and there is no good of racing with others anxious to do the same thing.

5.140 To Parents, Leipzig, 23 February 1886

I called on Prof. Hankel this morning (in evening dress!),[1] *he wishes me to come and be examined Friday at 5. I'm not sorry the day is different, if I must be examined in two editions; it would however have been better if the examination had been conducted in the usual way, as Prof. Wundt would scarcely have let me be conditioned, and would have recommended the higher grade if there was any question as to what it should be. The grade is not however of such great matter, as the really important grade (on the thesis) has already been settled.*

As to failing that is really very unlikely—though not more so than in the case of the fellowship at Baltimore—about as unlikely as that I shall die this year. Neither contingency is especially pleasant to think on, but I shall not in account of either of them sleep less soundly than usual tonight. Indeed on the contrary suposition (owing to four hours skating this afternoon) I shall arrange to sleep an extra hour.

[1] This was the traditional way to present oneself for final examinations (Hart, *German Universities*, pp. 224–225).

5.141 To Parents, Leipzig, 26 February 1886

1/3 of the examination is over at all events! Prof Hankel was really very nice, and I suppose there is no question about my having passed. The examination was such that it would not have been easy to have altogether failed, or to have passed an especially good examination—it covered the whole ground of elementary physics. If he had examined me more thoroughly on one subject, according to the subject, I might have done very well or very badly. He did however in some cases go into what seemed to me unnecessary minutiae, especially as to the people who found special laws and apparatus. I was considerably troubled by the foreign language, often not being able to express myself, or worse, saying the opposite of what I meant. I only spent about 30 hours in reading up, and am glad I did not trouble myself further. I skated this morning and afterwards read the papers until it was time to dress.

Your Loving Son
Jim (4/6 Ph.D.)

5.142 To Parents, Leipzig, 4 March 1886

(One paragraph of gossip has been omitted.)

I was at Prof. Wundt's this evening. They have been very kind to me this winter. He spoke very highly of my thesis—he said he had never yet given the extra grade, but that the substance of my thesis deserved it, the corrections he had to make in the form however made it impossible to give it.[1]

[1] See also 5.136. Wundt's changes were more substantial than formal. Many years later, in an undated and unpublished paper entitled "War and Education," JMC wrote, "I called my doctor's thesis 'The Time Taken Up by Cerebral Operations' and it was published in 'Mind' under this title, but Wundt made me change cerebral to psychic or mental wherever it occurred." JMC's physicalist perspective was derived primarily from his reading of Bastian's *The Brain as an Organ of Mind* (see 5.123); the work he did for the rest of his career was based on this perspective. He later even referred Wundt to Bastian's book, but the German professor was not convinced by its arguments (Wundt to JMC, 24 March 1887).

5.143 Journal, Davos, 22 March 1886

I am once more wandering over the face of the earth, and must record my whereabouts here. I left Leipzig Sat. night Mar. 6th and spent Sun & Mon at Munich; Tues I went to Landeck, Wed I walked & rode to Nauders, Thurs I walked to Schulz and Fri to Lavin, Sat over the Fluela here.[1] *I have spent a week here pleasantly enough, sledding, skating & doing nothing. Took dinner at the Symonds Monday.*[2] *As to the winter at Leipzig, what shall I say? I passed the Ph.D exam Mar. 1st on philosophy, zoology & physics, getting the grade cum laude; on the thesis admodium laudible. I wanted egregria on the latter.*[3] *Until Xmas I worked on the thesis & afterwards for the examination but at no time very hard. I did not give more than one good month's work to each. I was often in no mood for work, being swept hither and thither like a dead leaf in a storm. I used to go to the laboratory Wed. and Sat. afternoons, but did not myself do much work. The best part of the winter was the skating which lasted two months—I did not miss a single day. I fenced and swam with tolerable regularity. It does not seem to me that I went out much or read much: I scarcely know how to account for all the mornings & evenings. I was not at the theatre more than a dozen times. The fact is I often simply waited for the next day, using too much[4] chocolate and tobacco for pastimes. I kept house, having late dinner*

[1] Landeck, Nauders, Schuls, and Lavin were all small towns in southeastern Switzerland (Baedeker, *Switzerland*, pp. 404–421). The Flüela Pass connects southeastern Switzerland to Davos (ibid., pp. 353–354).
[2] John Addington Symonds (*Oxford Companion to English Literature*, pp. 795) was an English historian who, owing to poor health, resided in Italy and spent much time at spas such as Davos. JMC had met him in the summer of 1881 when he had first visited Davos, with his parents (see 1.12 and 1.13).
[3] See 5.136 and 5.142.
[4] JMC had first written "often" here, but crossed this word out and inserted "too much."

served in my rooms—I often omitted it however not having much appetite. I had intended to entertain, but had not the needed energy. Findlay Muirhead & I were right good friend, James M. also was there the last month.[5] I saw something of three men from the laboratory, Lehmann, Neiglick and Lange[6]—I gave them hashish and caffein. These drugs I also took once myself, and two or three times intoxicating doses of wine. Mr. & Mrs. Phillipps were good friends of mine & about the handsomest people I have met, giving one the idea that they had dropped out of fairy land. I was good friends with Miss Hopkins, Miss Bodington, Miss Brown, Miss Pease and others (Milles, Knight, Short, McDonald, Shepperd, Mills, Hauser, Dodds &c) but was not "in love" all winter.[7] I quite drifted apart from my German Frls of other years. I wonder what is to come next. Well I must be back at Leipzig the first of May and finish up some work, then I look to go to Norway returning home by way of England in the Fall. I look to study medicine during the winter at Phila. and should like to spend the next winter in Paris. So much for two of the 30 years I may expect.[8] I imagine I can do nothing but go on drifting, thinking, doubting—I may write on scientific & ethical questions, I cannot tell just how much ability I have. I do not put myself very high up, still I find few others who are much above me. I may practice medicine, or become connected with a college or journal. It doesn't make so very much difference after all—nothing does. I am at the foundation a sceptic—the whole of my philosophy is 0. In the first story I have some metaphysical ideas, but don't think them worth working out. The sum of them is God, the world and I are one and the same. I really live in the second story and call myself a scientist. Here I am an atheist, a materialist, a fatalist and a socialist, and make the world a logical whole albeit built on nothing.

[5]On the Muirheads see 5.24 and the introduction to section 1.

[6]On Lehmann see 5.114. Gustav Ludwig Lange (Poggendorff 7: 19–20) was a student at Leipzig who later became Wundt's second *Assistent*, and who developed the concept of sensory and muscular reactions (see 5.17 and postscript).

[7]These women were all students at the Leipzig Conservatory.

[8]Again, JMC's meaning is unclear. See 1.15.

From Davos, Cattell traveled to Italy, where he met G. Croom Robertson and toured with him and his wife through the ruins of Pompeii. There they were joined by Harry K. Wolfe, Cattell's friend from Leipzig, who had studied classical archaeology in addition to psychology (JMC to parents, 18 April 1886). From southern Italy Cattell headed north through Rome to Florence.

5.144 To Parents, Florence 25 April 1886

(Two paragraphs of gossip have been omitted.)

I sent from Rome the application I enclose.[1] You may not perhaps approve of my device to put Pres. Gilman and Prof. Hall between a Scylla and Charybidis, but I don't think you will mind the facts my letter contained being brought forcibly to their notice. Anyhow in the light of my success since, and in view of the fact that I am not in immediate need of a college position, I hope my Baltimore episode will not further trouble you. For my own part I can truthfully say that I have never regretted it, even taking into account the worry it has caused you. I trust you will let me live my life, only sorrowing in case I do what you think dishonorable or wrong. It is not needed that we should understand all the languages in which God can be praised.

[1] The enclosure was a letter dated 1 April 1886 addressed to Daniel Coit Gilman, applying for a fellowship at Johns Hopkins in psychology or biology. It was signed "James McKeen Cattell, Ph.D., Assistant in the Psychological Laboratory, Leipzig," and attached to it were excerpts of letters to JMC from Wundt and Francis Galton about his work. This letter may not have been sent to Gilman; no record of it has been found in the Gilman papers. A year earlier, on 18 April 1885, JMC had applied to Gilman for a fellowship and had asked Martin, and probably others of his professors from Johns Hopkins, for their support (Gilman papers; correspondence with JMC). This letter had contained a threat: "If I find . . . that personal considerations had influenced the withholding of the fellowship, I shall bring suit against the authorities of the University." (It had also included a letter of recommendation from Wundt; see 5.88). The reaction at Baltimore had been strong. Martin had passed his letter from JMC on to Gilman, and the faculty had passed a resolution that,

"owing to the unbecoming and insulting conduct of Mr. James M. Cattell toward authorities of the University. . . . The Academic Council refused to receive him here after as a member of the University in any capacity." JMC and Gilman had then exchanged several insulting letters on the matter, and Gilman had even checked with G. S. Morris his memory of one incident JMC had cited in his accusations (Gilman to JMC, 15 June 1885; JMC to Gilman, 10 August 1885; Gilman to Morris, 25 August 1885; Morris to Gilman 9 September 1885 [Gilman papers]). The exchange apparently ended at this point, but it had been so bitter that a historian who only saw the faculty resolution concluded correctly that "feelings on all sides must have run very high" (Mark Beach, "Professors, Presidents, and Trustees: A Study of University Governance, 1825–1918" [Ph.D. diss., Department of History, University of Wisconsin, 1966; University Microfilms no. 66-1262], pp. 239–240).

From Florence, Cattell went to Venice, Vienna, and Dresden. He arrived back in Leipzig on 4 May 1886.

5.145 To Parents, Leipzig, 5 May 1886

(Two paragraphs of gossip have been omitted.)

The laboratory opened this afternoon, eleven new men applying for admission.[1] At the opening of the semester I must go there some to get them started, but you know this is not like an ordinary laboratory, nothing but original investigation being undertaken. I myself shall do a little but not very much experimental work.

I have seen the Muirheads, but have no other special friends here. I don't somehow like Germans—I am not sure that they are yet to be reckoned as a civilized nation. Miss Hopkins whom I liked as well as any young lady is gone, and others are gone or going.[2]

I have not heard from Harry, nor could I find the arrival of his steamer announced in the papers.[3]

[1] JMC was again *Assistent* in the psychological laboratory (*Personal-Verzeichniss der Universität Leipzig*, no. 109, summer semester 1886, p. 32).
[2] On Myra Hopkins see 5.71.

[3] Henry Cattell had sailed from America a few weeks earlier, planning to spend some time at Leipzig with his brother; he arrived on 7 May.

5.146 To Parents, Leipzig, 8 May 1886

Harry does not give me a very encouraging account of the second year at the university: he thinks the third year will be better.[1] I could enter the third year and graduate with Harry, but should have to work very hard to pass all the examinations, and should have to begin at once. I am ready to work hard next winter, but it would be unfortunate should I leave the papers I am now working at and writing, and further it would not be wise to give up the vacation this summer.

When I said the examination here was not difficult, I did not mean that I could pass it in less than at Phila, only that more weight is laid on original work, and less on elementary subjects for examination such as chemistry, anatomy and materia medica. In either case, that is at Phila. or here, I should look to make the degree in two years from now or in 1½ year's work beginning from next winter. Medical students at Leipzig usually study five years.[2]

If you only come over this summer we can talk about what I had better do. I had about decided simply to take the last two years at the University, and pass all the examinations, if they would not excuse me from them. If however Harry does not

[1] Henry Cattell was a medical student at the University of Pennsylvania at this time (see 5.6 and 6.72).
[2] William Cattell had written earlier (5 April 1886) to ask whether a Leipzig degree in medicine would enable his son to practice in the United States. Foreign medical graduates

did have to "be passed upon for licensure" by the faculty of the Medical Department of the University of Pennsylvania after 1881 (Richard H. Shryock, *Medical Licensing in America, 1650–1965* [Baltimore: Johns Hopkins University Press, 1967]).

come to Europe the year after he graduates, then I might study at Phila. this winter, either as special student or in any class, not troubling myself over examinations, and come the next year to Germany and pass the examination. Or, I might stay here this winter (if you come over in the summer) and then enter the last year at the university. I should also have great advantages at London or Paris, as I can get into the great hospitals.

Fortunately there is no hurry about the matter, as in either case I shall not begin until the Autumn.

5.147 To Parents, Leipzig, 11 May 1886

(Several lines of gossip have been omitted.)

My trip to Italy cost me nearly $300—one cannot well travel constantly for less than $5 a day—I did once but scarcely ate properly and denied myself other things, which decreased not only the comfort, but also the profit of travelling. This time I nearly always stopped at 2nd class hotels, but found them quite as clean and comfortable as the Anglo-American institutions at about half the expense. Of course I don't mean my trip cost me $300 extra, but $100 extra; my expenses in Leipzig being $100 a month. I did not get an excursion ticket on account of the cholera fearing the R.R. might be quarantined. There was a little cholera in Venice all winter (news of which was hushed up) and I fear it is now becoming more serious.

The expenses connected with my work have been rather large—I enclose for example one bill—yet my thesis was printed (including costs) and 210 copies were given me quite free. 210 copies must be given to the university. This would otherwise have cost me $100 to $200 more. I also employed Heise nearly all winter and paid a draughtsman and others.[1] I gave Berger $25 for coming on and helping me translate the thesis, and must give him more for work he is now doing and is going to do for me. All this is in some sort an investment, which I am sure will pay me a very large per cent.

I estimated my expenses this year at $1150 (the same as Harry's counting $500 for board, rooms washing &c), and expected to spend the $300 of my own on extra travelling, apparatus, books &c. I fear the expense above referred to in connection with my work will not be covered by this, but my letter—$1750—would give me an abundance, if it were not for two items. The first of these is that I have quite supported Berger all year—indeed I did last year also—then however I gave him the money, this year I have lent it to him (except the amount above mentioned), I am really glad to do this, as it has put him in a position which will be of advantage to him all his life, and there is scarcely a chance that it will be lost. I have in this way lent him $250 and will probably lend him $50 more: from now on however he gets a salary—last year he did not receive a dollar though he was a regular teacher. This $300 I shall ask you to lend me, and will give it back to you when Berger pays

[1]On Heise see 5.36n.2.

me. A second item is things I shall buy to save next year, and which must be charged to next years account—I shall bring home enough clothes, for example, to last me a long time. I shall thus, I fear, need an extra draft, which however will not be used up, only advanced. If you come over, as I am hoping, you can of course give me what I may need.

You must tell me if my handwriting is illegible.[2] I can write better but have a great deal of writing to do, and am pressed for time.

[2]JMC's handwriting had become less distinct than it had been earlier in his stay in Europe, but it had yet to reach the illegibility of his later years.

5.148 From Elizabeth Cattell, Philadelphia, 11 May 1886

(One paragraph of gossip has been omitted.)

Your letter from Florence just received,[1] but I must say I cannot conceive how you could again apply to Dr Gilman for a Fellowship. I should have thought after the way in which they have treated you in Baltimore, you would never ask a favor, much less desire to go back, to be associated with those who I think refused to you the Fellowship in 1883, because they did not want you at the University. Not because you were not bright and talented, but because your manners, or mode of conducting yourself to them was unpleasant and disagreeable. I dread reading your letter to your Papa, for he is far from being well, he is working too hard, and I know he is going to be greived. O I do not wonder at it, but as you say we must let you live your life, but would you want to feel that we were not interested in your future welfare. This would be impossible. We enjoyed your letter, and I am delighted to think that you feel that we have done all we possibly could to give you a thorough education. I know we have tried to do all that laid in our power, and dear Jim you have always been such a great comfort to us both, and I know you always will be; you little know how we lean upon you. You will be surprised when I tell you, we have been sitting this evening, trying to make up our minds to cross the Atlantic. Dr Mitchell tells your Papa a sea voyage, would be the very best thing he could do for his nervousness.[2] If I could make up my mind such would be the case I would be willing to cross the Ocean, though we could only stay on the other side about six weeks.

[1]The letter referred to is 5.144.　　　　[2]On S. Weir Mitchell see 3.3.

5.149 To Parents, Leipzig, 19 May 1886

I am seriously thinking of going to England next month. I have you know all along intended to spend Sept. there, but that seems to be a bad month. I want chiefly to see people; and then the universities at Oxford and Cambridge are closed, and London too is deserted. But it has been the thought of meeting you, which is inclining me especially toward England now. I could meet you whether you come via Liverpool or direct to Antwerp—the fare from London to Liverpool and Antwerp respectively being the same.

I find my work here is in very good shape. There is one paper that I can about get ready in a couple of weeks[1]—but there is not the slightest hurry about it, as I should much rather print it in "Mind" than any place else, and my thesis will continue to drag its slow length along in "Mind" for the present. Then there is another paper, the work for which is only half done, so it cannot be written for the present.[2] I had thought of a third paper on the laboratory here, but it would require considerable preparation, as I should have to study up all the work that has been done there; my real reason for inclining to drop it however, is that I could not write it without criticising a good deal and that would offend Wundt.[3] It will not cost me more to go to England now than later, (the fare from here is $16) but it is a dear country to live in.

[1] Probably "Experiments on the Association of Ideas," *Mind* 12 (1887): 68–74. See 5.151.
[2] This paper has not been identified.
[3] JMC eventually did publish a not-very-critical paper on "The Psychological Laboratory of Leipsic," *Mind* 13 (1888): 37–51.

5.150 To Parents, Leipzig, 20 May 1886

We were both at Prof. Wundt's for supper this evening, and shall both be invited there every other week through the semester.[1] You must give me all due credit for getting on so well with Prof. Wundt. Personally he is a nice man, but it has needed careful steering through our psychological relations.

I send the second half of my thesis reprinted from the "Studien." It is to be hoped you admire the cover—which cost $10. Have you read it yet, Mama?

We have been thrown suddenly from early spring into the middle of summer. The thermometer was 78° this evening, which is very high for Leipzig.

[1] "We": JMC and Henry Cattell. Wundt later often invited his students' relatives to dine with his family (see "In Memory of Wilhelm Wundt," *Psychological Review* 28 [1921]: 171–172, 176).

5.151 Journal, Leipzig, 25 May 1886

(A short account of Cattell's trip through Switzerland and Italy has been omitted.)

Tues back at Leibigstr. 9. Laboratory opened Wed (5ᵗʰ) Harry came Fri evening, and we have since been living together. I have been preparing a paper on the Association of Ideas and continuing experimental work.[1] I have been feeling right well for the past 3 weeks, but have used drugs every day—on the plea it is time of experiment. I have been on the river a good deal—mostly alone but several times with young ladies—Miss Bodington, Miss Mills, Miss Knight[2]—last night with Miss Owen, whom I did not know before but whom I shall take care to see again.[3] So I was born 26 years ago today. When I come to die I think I shall only be able to say as I do now: who can tell the meaning of it all?

[1] The paper mentioned is "Experiments on the Association of Ideas," *Mind* 12 (1887): 68–74.

[2] See 5.143.

[3] On Josephine Owen see 6.87.

5.152 To Parents, Leipzig, 27 May 1886

We both received letters from you this morning, Mama. We are delighted to learn that you have engaged your passage, and I trust we shall spend a pleasant and profitable summer together. You sail not so very long after receiving this, but of course you will get more letters from me. I shall either meet you at Wildungen,[1] or if I go to England at Antwerp.

I am sorry you think I should not have applied for a fellowship at Baltimore. You naturally do not understand my motives very well, or you would not think I either expected or wanted the fellowship. I think it was not given me before from personal motives; there might however then have been a difference of opinion as to my merits; now there is scarcely any room for this, and I have those who refuse it on personal grounds in my power, if I wish to exert it.

I have never hid my belief that it is better to struggle through life than to slide through it. If at any time at your wish I refrain from what I think to be best, the sacrifice is made for your sake, not for mine.

[1] A German town with a mineral spa (see 4.2).

5.153 To Parents, Leipzig, 31 May 1886

(Brackets indicate wormholes.)

I am almost sorry to leave Leipzig, as my life moves along as quietly and pleasantly as it is likely to any place, and I find that I still have a good deal of work to do. Of course I can finish this in the Fall, but I should rather then devote myself altogether

to medicine. I paid $2 today for a course of lectures. [   ] I might
[   ]. Before I leave I shall [   ] laboratories and see what facilities I should have in dissecting and histology—I must devote myself largely to these
next winter. I suppose I shall be put to considerable expense for books, apparatus
&c. I shall perhaps buy a microscope with the money you gave me.

5.154 From Elizabeth Cattell, Easton, 3 June 1886

(Several paragraphs of gossip have been omitted.)

Mr. Fisler said yesterday some gentleman (he did not seem inclined to mention his
name) has been inquiering about you,[1] to know what kind of a Professor you would
make for the Ladies Quaker College at Bryn Mar,[2] I think you told me Miss Thomas
of Baltimore was there as Principal.[3] Harry makes a mistake, we did receive copies
of your Thesis that was published in "Mind" and "Brain".

[1] Samuel L. Fisler was secretary and treasurer of the board of trustees of Lafayette College (*Biographical Catalog of Lafayette College*, p. 129). The "gentleman" is probably James E. Rhoads (*DAB* 15: 530–531), a medical graduate of the University of Pennsylvania who devoted himself to philanthropic causes from 1862. He was the first president of Bryn Mawr College, and played a major role in selecting its faculty.

[2] Bryn Mawr was founded in 1880 by the Society of Friends as the coordinate college (for women) to Haverford College (Barbara Alyce Farrow, *The History of Bryn Mawr, 1683–1900* [Bryn Mawr, Pa.: Bryn Mawr College 1962], pp. 67–71).

[3] Martha Carey Thomas (*NAW* 3: 446–450), the sister of Henry M. Thomas (see 2.9), had studied unofficially at Johns Hopkins in the late 1870s, enrolled for a time at the University of Leipzig in the early 1880s, and earned a Ph.D. at Zurich in 1882. At Bryn Mawr, she was professor of English and dean of the faculty (1885–1894) and president (1894–1926). See Helen Thomas Flexner, *A Quaker Childhood* (New Haven, Conn: Yale University Press, 1940); Marjorie Housepian Dobkin, ed., *The Making of a Feminist: Early Journals and Letters of M. Carey Thomas* (Kent, Ohio: Kent State University Press, 1979).

5.155 To Parents, London, 8 June 1886

I got into London a couple of hours ago and am stopping at Bailey's, having written
ahead for a room.[1] It was rather a long trip straight through from Leipzig, but the
weather was favorable and I slept a good deal.

This is exactly the right time to be in London: the season is at its height: all the
picture galleries &c are open; there is opera in Italian and English: parliament
rejected last night the farmers bill &c.

I find notes from Mr Galton & Prof. Robertson: the former invites me to lunch
today: the latter to dinner on Sat. to meet Mr. & Mrs. Bain.[2]

[1] See 3.1.

[2] On Galton see introduction to section 6; on Robertson see 5.111; on the Bains see 5.89.

5.156 To Parents, London, 12 June 1886

(One paragraph of gossip has been omitted.)

I have been as you may suppose quite busy here, as besides my proper psychology—as a part of it if you please—I interest myself in things generally. I lunched with Mr. Francis Galton on Tuesday, and attended with him the meeting of the Anthropological Institute, of which he is president.[1] I spent most of Wednesday in the reading room of the British Museum—an excellent place for study. I took dinner with Mr & Mrs Alex Bain. Thursday I was at the art galleries—Grosvenor & Academy, and at the meeting of the royal society met Prof. Foster (Cambridge) and Prof. Burdon Sanderson (Oxford)—with Dr. Ferrier the greatest English physiologists.[2] Heard Gounod's Faust in the evening. Yesterday I was at the Kensington museum, heard Rubenstein's farewell piano recital, and the Lyceum Faust.[3]

You sail two weeks from today and ten days after we may hope to be together. I trust the summer will be all we could wish. We shall have much to tell and talk about. Until then goody-bye.

[1] On Galton see introduction to section 6. The Anthropological Institute was established in 1871 by the amalgamation of the Ethnological Society, which had been founded in 1843, and the Anthropological Society, founded in 1863. It combined what in twentieth-century terms would be called cultural and physical anthropology, though under Galton's presidency its journal stressed the latter. (See J. W. Burrow, "Evolution and Anthropology in the 1860's: The Anthropological Society of London," *Victorian Studies* 7 [1963]: 137–154.)

[2] Michael Foster (*DSB* 5: 78–83) was professor of physiology at Trinity College, Cambridge, and founder of the "Cambridge School of Physiology." On Ferrier see 5.110; on Burdon-Sanderson, 5.91.

[3] Anton Grigorevich Rubenstein, a Russian-born pianist of great reknown, toured Europe from 1885 through 1887 giving farewell recitals (*Grove's* 7: 295–296). The "Lyceum Faust" was a version of the Faust story by the English playwright William Gorman Wills (*DNB* 21: 518–520) that played with "conspicuous success" at the Lyceum Theater from late 1885.

William and Elizabeth Cattell joined their sons at Antwerp, and the family went to Britain together. Sometime during their stay, in late June or July 1886, James secured an appointment at Cambridge for the coming fall. It is not clear exactly how this appointment was arranged, but it seems to have been based on his extensive contacts with the British scientific community and to have been related to his interest in medicine (see 5.121). Curiously, James Cattell did not mention Donald MacAlister—the man most involved in arranging the appointment—in his journal or in his letters to his parents, and it is not clear how the two had met. MacAlister (*DNB*, 1931–1940: 557–558), in 1886 the medical tutor at St. John's College, Cambridge, had studied physiology in Leipzig in 1881, and may have met Cattell then. He obtained the M.D. degree from St. Bartholomew's Hospital in London in 1884, and returned to Cambridge that year to renew his contacts with St. John's. He later was to play an important role in Cattell's career at Cambridge (see 6.62, 6.133, and 6.140). In 1895 MacAlister married his distant cousin Edith

Florence Boyle Macalister, the daughter of Alexander Macalister, professor of anatomy at Cambridge—the first woman James Cattell had ever spent more than five minutes with alone (see 1.3). (Edith MacAlister, *Sir Donald MacAlister of Tarlbert* [London: Macmillan, 1935].)

In August 1886 James Cattell returned to Leipzig and prepared for his move to Britain (JMC to parents, 1 August 1886). From Germany he went to Norway for the walking tour that he had long planned.

5.157 To Parents, Fagerburg, Norway, 20 September 1886

(Four paragraphs describing Cattell's tour of Norway have been omitted.)

I can scarcely realize that I have been here nearly seven weeks. It seems like a long vacation, especially when I add to it the time I spent with you and the six weeks in Italy in the Spring. I had however been working for the three preceding years, and the work was of a peculiarly trying nature. It is considered especially exhausting to make experiments on ones self, and the preparation for the press was difficult for one so young, requiring a thorough knowledge not only of psychology but also of physics and physiology. Hereafter it will be easier for me to write. You know I printed ten papers and have three more partially ready. I am very glad that I have been able to do this work, but I sometimes wish I could have spent first a year out of doors—on a farm say. I was never in perfect health and could seldom work to the best advantage, often not at all. Brain-work accomplished is by no means proportional to the time spent on it—more can sometimes be done in a day than it others in a week or month. Darwin only worked three hours a day, and I suppose Shakespeare did not spend the average of an hour a day in writing. Even now I think it would be gain if it were possible for me to spend a whole year in the country—I could do 1/30 more work in each of the 30 years I am likely to live,[1] and should in all probability add a year to my life—not to old age, but to putting off old age. However it is much easier to follow the beaten track, than to make side excursions: and is possibly wiser when others, as well as one's self, are concerned.

[1] JMC had earlier referred to a life expectancy of 30 years from the date of his writing (see 1.15). Like that reference, this one is unclear.

After James Cattell had gone to Norway his parents returned to America, where William Cattell resumed his search for a position for his son.

5.158 From William Cattell, Philadelphia, 27 September 1886

(This letter was dated "Mon., 27th.")

In my letter yesterday I referred to the desirability (as it seems to me) of having some of yr articles in "Mind" or "Brain" republished here—say in The Popular Science Monthly—It w^d call the attention of College Men to yr work & incidentally pave the way for the offer of some place.[1]

I shd like especially to see reprinted the art. from Mind *Vol. XI. No 42 with its continuation.*[2] *Of course the illustrations are a necessity; & I would gladly pay for having them made here, if you cannot send electrotypes.*

Perhaps this may not strike you as it does me;—but please write at once & let me have yr views.

At all events send me at least thirty copies of the "off-print" of your next article[3]—*in the same style as the former. I think it is best without cover (just as the last) only you will of course state that is "continued from &c"—or "2^d Part". And of course if you think favorably of the reprint & have the electrotypes send them at once. I am by no means sure that Prof. Youmans will find these articles to his mind: but it is worth while to make the effort.*

From what I have learned since I left home all *the Colleges are poor! and all have departments* now, *that need enlarging. They will not entertain new projects.*

I am still inclined to think yr future life w^d be more independent in a Profession—than as a Professor. You will be more yr own master & have really more time for original researches (if that shd continue yr bent) if you were a specialist say in Neurology: —*and if you c^d only secure a Lectureship, in some standing in the Faculty here, you c^d pursue yr Medical studies with less expense & with more dignity—than if you entered the university as a student. But write to me at once—and let me know where to direct yr letters in Cambridge.*

[1] This paragraph summarizes the 26 September letter.

[2] These two issues contained the first two installments of the English version of JMC's dissertation, "The Time Taken Up by Cerebral Operations," *Mind* 11 (1886): (no. 42) 220–242, (no. 43) 377–392. Issue no. 42 was published in April 1886, no. 43 in July.

[3] The third installment of JMC's dissertation (*Mind* 11 [1886]: 524–538), which was to be published the next October in issue no. 44.

6

The Psychologist as Scientific Celebrity: Cambridge, England, September 1886–December 1888

When James McKeen Cattell arrived, in autumn 1886, Britain's intellectual community was ready to welcome a scholar with his background. Many who had had their beliefs shaken in the aftermath of the Darwinian revolution were looking for a way to reconcile the traditional concerns of religion—especially the ethical teachings of Christianity—with the recent scientific developments and theories.[1] Such an interest had attracted many British intellectuals (including Alfred Russel Wallace, the co-discoverer of natural selection) to spiritualism, and psychical research had become a major intellectual movement.[2] Similar concerns led others to follow the development of the new physiological psychology in Germany.[3]

Before coming to Cambridge, Cattell had met Alexander Bain (see 5.89) and G. Croom Robertson (see 5.111), both of whom had visited the Continent in part to observe the work of Wundt and his students. Others with similar interests were to welcome Cattell, who knew more than anyone outside of Germany about the new psychology. Cattell's visits in England in August 1885 (see 5.111) had made him aware of their interest in his work, and during his last year at Leipzig he had exchanged letters with several English scientists.[4]

[1] See Frank M. Turner, *Between Science and Religion: The Reaction to Scientific Naturalism in Late Victorian England* (New Haven, Conn.: Yale University Press, 1974).

[2] See Alan Gauld, *The Founders of Psychical Research* (New York: Schocken, 1968); Malcolm Jay Kottler, "Alfred Russel Wallace, the Origin of Man, and Spiritualism," *Isis* 65 (1974): 145–192.

[3] See Lorraine J. Daston, "British Responses to Psychophysiology: 1860–1900," *Isis* 69 (1978): 192–208.

[4] See Francis Galton, "On Recent Designs for Anthropometric Instuments," *Journal of the Anthropological Institute* 16 (1887): 2–11. See also, for example, Galton to JMC, 24 February 1886 (Cattell papers).

To be sure, Cattell's experimentalist orientation did not correspond directly to the philosophical concerns of those who welcomed him, and his ultimate influence on turn-of-the-century British philosophical thought was negligible. Nevertheless, in September 1886 Cattell arrived in England as something of a celebrity.

The English scientific and scholarly institutions were also ready to support Cattell's scientific interests. Cambridge had undergone a period of reform; earlier in the 1800s it had been in an intellectual slump, out of which it had gradually emerged in the middle of the century.[5] Political preference as a factor in appointments had declined in importance, sinecures had been abolished, the colleges had been opened to "dissenters" against the Church of England, and women's colleges had been established. The reforms most important from Cattell's point of view were the broadening of the curriculum and the recognition of the importance of research and scholarship. Cattell never held an official teaching function at Cambridge (though he did teach; see 6.165) and was never a candidate for a Cambridge degree (though he sometimes considered pursuing one; see 6.49). Still, the changes of the previous fifty or so years had done much to make his stay at Cambridge valuable to Cattell.

Until the mid-nineteenth century, Cambridge had always required the mathematical tripos, a qualifying examination, of all candidates for the honors Bachelor of Arts degree.[6] Other honors examinations had been introduced such as one in classics (1851); but to be eligible to take these examinations a candidate had first to obtain honors in mathematics. The abolition[7] of such requirements between 1850 and 1860 made it possible for a Cambridge student to major in the moral sciences (moral philosophy, political economy, modern history, general jurisprudence, and the laws of England). By the 1870s it was possible to concentrate within moral sciences on moral and political philosophy, mental philosophy, logic, and political economy[8]—just the subjects in which Cattell was most interested.

The reform of the Cambridge triposes could not have taken place without an accompanying change in the curriculum, and this change required a

[5] On the emergence of modern Cambridge see Denys Arthur Winstanley's trilogy: *The University of Cambridge in the Eighteenth Century* (Cambridge University Press 1922), *Early Victorian Cambridge* (1940), and *Later Victorian Cambridge* (1947). The influence of the emerging German university system (see introduction) on the reforming of Cambridge has not been well studied, despite George Haines IV, *German Influence upon English Education and Science, 1800–1866* (New London, Conn.: Connecticut College [Monograph no. 6], 1957). But see also Gerald L. Geison, *Michael Foster and the Cambridge School of Phys-iology: The Scientific Enterprise in Late Victorian Society* (Princeton, N.J.: Princeton University Press, 1978), chap. 4, pp. 81–115.

[6] The name "tripos" was taken from the three-legged stool on which a participant in disputation sat during the degree ceremony. Even the standards of mathematical tripos had been lowered by 1800, and that test underwent its own reform (see *Early Victorian Cambridge*, pp. 157–160).

[7] Tanner, *Historical Register*, pp. 602, 703, 737.

[8] *Later Victorian Cambridge*, pp. 185–188.

significant shift of political power from the colleges, where tutors had traditionally provided most instruction, to the university, which was emerging as much more than a collection of colleges. Between 1866 and 1896, at least ten new professorships were established within the university rather than within the colleges.[9] The best known of these is probably the Cavendish professorship of experimental physics, founded in 1871 and held successively by James Clerk Maxwell, John William Strutt (Lord Rayleigh) (see 6.31), and Joseph John Thomson (see 6.116). University professorships were also established in zoology and comparative anatomy (1866), in animal morphology (1882), and in physiology, in surgery, and in pathology (all 1883). As Cattell planned to study medicine in Cambridge, the existence of these chairs was important to him. Equally important was the establishment in 1883 of at least twenty university lectureships, the holders of which taught students from all colleges.[10] These new positions not only expanded the range of what was taught at Cambridge but increased the degree of specialization. There were five new lecturers each in mathematics and history, and three in physiology. Most of these were younger men, and many of them were to become good friends of Cattell.

In the 1880s, then, Cambridge was an intellectually stimulating place. Many student societies met regularly to discuss the topics of the day, and even Baedeker (*Great Britain*, p. 225) noted their importance. For Cattell the most important of these was the Moral Sciences Club,[11] which centered around Henry Sidgwick, fellow of Trinity College and Knightsbridge Professor of moral philosophy in the university. Sidgwick (1838–1900) is in many ways typical of the men interested in Cattell and his work. He had followed the development of the new psychology in Germany, had played a major role in the introduction of the moral sciences at Cambridge, and was an active investigator of psychical phenomena.[12] Sidgwick was a member of what has been called the "intellectual aristocracy," a large group of intelligent, middle-class men and women whose families had begun to intermarry by the end of the 1700s.[13] An acquaintance with one member of this group typically brought introductions to others. In Cattell's case, Sidgwick was important in providing entrée into the British intellectual community.

[9]Tanner, *Historical Register*, pp. 99–107. See also C. P. Snow, *The Masters* (New York: Scribner's, 1951), pp. 359–374.

[10]Tanner, *Historical Register*, pp. 117–130.

[11]Little has been written about this club or about most of the others that were active during the period; perhaps the documents here will stimulate interest.

[12]*EoP* 7: 434–436; Arthur Sidgwick and Eleanor Mildred Sidgwick, *Henry Sidgwick: A Memoir* (London: Macmillan, 1904); Shel-

don Rothblatt, *The Revolution of the Dons: Cambridge and Society in Victorian England* (New York: Basic Books, 1968), pp. 133–154; Gauld, *The Founders of Psychical Research*, pp. 88–136; J. B. Schneedwind, *Sidgwick's Ethics and Victorian Moral Philosophy* (Oxford: Clarendon, 1978).

[13]Noel G. Annan, "The Intellectual Aristocracy," in J. H. Plumb, ed., *Studies in Social History: A Tribute to G. M. Trevelyan* (London: Longmans, Green, 1955), pp. 241–287.

London was as much a center for the intellectual community as Cambridge. The capital, which was only about seventy-five minutes away by rail, regularly attracted Cattell and other Cantabrigians. Almost all felt the draw of London's art and music, but for many the major attraction was University College. This school, which had been established in the 1820s to provide education to religious dissenters, had by the 1850s grown into an intellectual center rivaling Cambridge in richness.[14] Cattell had already met Robertson, professor of mental philosophy at the college, and his contacts with *Brain* had led to at least an acquaintance with some of the physicians at the University Hospital (see, for example, 5.110).

Also important to the British intellectual climate were independent scholars not connected with any formal institution. London during the 1880s was the home of many such men working in areas of interest to Cattell. A prime example is Shadworth Hodgson, an Oxford-trained metaphysician of independent means who devoted his life in London to the study of philosophy.[15] The coalescence of informal and formal groups around the independent scholars enabled them and university people to get together and learn from each other. For example, the Aristotelian Society for the Systematic Study of Philosophy gradually formed around Hodgson. Organized informally in 1880, within a year or so the society presented records of its meetings in the "Notes and Correspondence" section of *Mind*; by 1887 it was publishing formal proceedings and had membership lists and by-laws.[16] Another such group was the Society for Psychical Research, which included many members Cattell admired.[17] Cattell did not join this group. He would later accuse most psychic researchers with fraud,[18] and he criticized some of the psychic experiments he saw in England (see 6.32). However, he did take part in the activities of such formal associations as the Neurological Society and such informal groups as the Psychological Club, which sprang up around Robertson.

In many ways, the person who most influenced James Cattell during his stay in England was Francis Galton, an independent scholar located in London and a member of the "intellectual aristocracy" (he was a first cousin of Charles Darwin). Galton had a wide range of interests and made contributions to many fields.[19] An early interest in geography had led Galton into anthropology, and, under the influence of his cousin's ideas, he had begun by the 1860s to study individual and racial differences. This work led him

[14]Hugh Hale Bellot, *University College, London, 1826–1926* (University of London Press, 1929).

[15]*EoP* 4: 48–49.

[16]See *Mind* 6 (1881); 151, 603; *Proceedings of the Aristotelian Society* 1 (1887–1891). See also Hodgson to William Torrey Harris, 15 July 1881 (William Torrey Harris papers, Hoose Library, University of Southern California).

[17]Gauld, *The Founders of Psychical Research*, passim.

[18]JMC, "Esoteric Psychology," *The Independent*, 45 (1893): 316–317.

[19]See *DSB* 5: 265–267 (an extremely disappointing sketch); Karl Pearson, *The Life, Letters, and Labours of Francis Galton* (Cambridge University Press 1914–1930), 3 vols.

to investigate the heredity of human traits, and stimulated him to develop the concept (and ideology) of eugenics, the systematic biological improvement of the human race.[20] He had established his Anthropometric Laboratory in 1884 at the International Health Exhibition, and after the Exhibition closed, he transferred the Laboratory to the South Kensington Museum, from which evolved later the Victoria and Albert Museum and the Science Museum. At this laboratory, for a fee of three pence, individuals could have a full range of bodily measurements taken and have their abilities to perform certain physical tasks measured (see postscript and figure 16.) Galton hoped to use these "anthropometric" data in his investigations of human differences.[21]

But Galton had concentrated on physical characteristics, and he soon looked beyond these to search for a way to investigate psychological differences. For a number of years he had considered the various forms of reaction-time experiments to represent such a technique,[22] and his interest in how these experiments had been performed led him to Cattell. Perhaps Galton found Cattell's name in the literature, or perhaps Bain had mentioned the young American to Galton. It is also possible that Cattell had been brought to Galton's attention by Galton's friend John Shaw Billings, who probably saw Galton in London in August or September 1884, soon after he had met Cattell (see 4.10). In any event, by February 1886 Galton was quoting Cattell's letters to him at the Anthropological Institute.[23] Cattell would later call Galton "the greatest man I have ever known,"[24] and throughout his stay in England he often looked to Galton for inspiration.

James Cattell is best known in the history of psychology as the first person to use the experimental techniques developed in Germany to investigate the (normal healthy adult male) mind to study the differences between people. It often is assumed that Cattell went to Wundt in 1883 with this interest in what became known as differential psychology fully formed.[25] He himself wrote much later that his early interest in athletics had led him to a concern with individual differences in ability.[26] But as the documents dating from

[20]Pearson, *Galton*, vol. 1, pp. 70–215; Galton, *Hereditary Genius* (London: Macmillan, 1869); Galton, *English Men of Science: Their Nature and Nurture* (London: Macmillan, 1874).

[21]Pearson, *Galton*, vol. 2, pp. 357–386; Galton, "The Anthropometric Laboratory," *Fortnightly Review*, new series, 31 (1882): 332–338; Galton, "On the Anthropometric Laboratory at the Late International Health Exhibition," *Journal of the Anthropological Institute* 14 (1885): 205–221; Galton, *Inquiries Into Human Faculty and Its Development* (London: Macmillan, 1883).

[22]Galton, "Notes and Calculations about Re-

action Time, 1878–83," unpublished notes (Galton papers).

[23]Galton, "On Recent Designs," pp. 7–8. JMC had called on Galton during one of his earlier visits to London (see 5.156).

[24]JMC, "Psychology in America," *Science* 70 (1929): 335–347.

[25]Edwin G. Boring, *A History of Experimental Psychology*, 2nd edition (New York: Appleton-Century-Crofts, 1950), pp. 532–540; Boring, "The Beginning and Growth of Measurement in Psychology," *Isis* 52 (1971): 626–635.

[26]Cattell "Autobiography," p. 634.

his stay in Leipzig make clear, though Cattell *may* have had a small interest in the differences between experimental subjects (see 5.17), his major concerns had been experimental procedure and precision. His dissertation had mentioned the major differences between his and Berger's reactions, but (as he noted much later) these were cited merely as epiphenomena.[27] Within a few years after meeting Galton, Cattell was doing differential psychology and had coined the term "mental test" for his measures of individual differences.[28] It can be argued that Cattell had probably had some slight interest in the differences between people while at Leipzig, but that he had found no outlet for it in Wundt's laboratory. It is also possible that Wundt had actively discouraged such interest. In any case, what Galton did was to give Cattell the framework of physical and physiological anthropometry in which to develop his interest in individual differences.

In Cattell's day it was unusual for an American scholar to study at an English university. Americans interested in "liberal culture" and the arts had always followed developments in England closely,[29] and many American writers and painters spent time in England,[30] but most scientists and scholars did not feel the same attraction. The English universities did not generally offer the opportunities for research and scholarship available in Germany, and besides they were expensive. Because Cattell's experience as an American scientist in England was relatively rare, his papers from this phase provide a valuable glimpse into the English university and intellectual life of the time.

[27] JMC "Cerebral Operations," in *Man of Science*, vol. 1, p. 58: "It was my object in the experiments here under consideration rather to eliminate these sources of variation than to investigate them." See also Cattell "Autobiography," pp. 629–630, and 3.12.

[28] JMC, "Mental Tests and Measurements," *Mind* 15 (1890): 373–381.

[29] Laurence R. Veysey, *The Emergence of the American University* (University of Chicago Press, 1965), p. 196.

[30] Richard Kenin, *Return to Albion: Americans in England, 1760–1940* (New York: Holt, Rinehart and Winston, 1979).

6.1 To Parents, Cambridge, 27 September 1886

(One paragraph about Cattell's Norwegian tour has been omitted.)

Here I am where I shall be for a good while and whence I shall write many letters to you. However I am as yet only in Cambridge, not in quarters of my own.

Mr. Heitland, who is going to arrange about my rooms is not here, which is a pity, but he is expected back tomorrow.[1] I have however seen several of the people I am interested in, and am going to work in anatomy at once.

I took dinner with Mr. Langley in Trinity College[2]—I shall describe the Hall dinner to you sometime. It is pleasant and sociable but takes lots of time—tonight most of them stayed together until 10.30.

Everything seems to be made very easy and pleasant for me—I trust it will be a good year.

[1] William Emerton Heitland, who lectured at Cambridge in Greek and Roman history, was to be JMC's tutor at St. John's College (see 6.2; Heitland, *After Many Years: A Tale of Experiences and Impressions Gathered in the Course of an Obscure Life* [Cambridge University Press, 1926]; Venn, *Alumni Cantabrigiensis*, vol. 3, pp. 322–323).

[2] John Newport Langley was a fellow of Trinity and University Lecturer in Physiology from 1883 through 1903. After 1903 he was University Professor of Physiology, succeeding Michael Foster. (Venn, *Alumni Cantabrigiensis*, vol. 4, p. 95; see also 5.156).

6.2 To Parents, Cambridge, 28 September 1886

I will write to you now every evening, but you understand it is only a way of saying good-night, and of mentioning perhaps a fact or two as to the day which may seem to bring us a little nearer together. I should neither be able to find things of interest with which to fill up a letter, nor have time to write it.

Just now there is however a good deal I might tell you. I shall only say however that I have seen Mr. Ward and Mr. Heitland—the men (with perhaps Prof. Foster) of most importance to me.[1] Mr. Ward is considered the foremost psychologist in England. He was offered the professorship at Baltimore afterwards given to Dr. Hall.[2] He wishes to work together with me in experimental psychology.

[1] James Ward (*DNB*, 1922–1930: 884–887), educated at London and Cambridge, was a fellow of Trinity College who was to become professor of moral philosophy in the university in 1897. In 1870 he had studied at Göttingen with Lotze. On Foster see 5.156.
[2] Ward had not in fact been offered the position, but had been interviewed for it by James Carey Thomas and found (despite Ward's strong religious beliefs) insufficiently orthodox (Hugh Hawkins, *Pioneer: A History of the Johns Hopkins University, 1874–1889* [Ithaca, N.Y.: Cornell University Press, 1960], p. 189).

Mr. Heitland thinks I will be made Fellow-commoner at St. Johns.[3] *This is an honour equivalent to a fellowship—there is at present no other at Johns nor as far as I know in Cambridge, most of the colleges not having the appointment. The only one there was at Johns last year has been made professor at Manchester. A fellow-commoner used to be a nobleman or such like who was allowed to sit at the fellows table &c—the last one at Trinity was the eldest son of the Prince of Wales, but the new statutes, which made great changes in the university, have done away with this, and several of the colleges have made it a fellow by courtesy.*[4] *Mr. Heitland has also rooms I can have, but I have not seen them.*[5]

I began work at once this morning in anatomy, studying bones.

[3] The College of Saint John the Evangelist, usually referred to as St. John's or even as Johns, was by 1886 Cambridge's second-largest college, after Trinity. With Trinity and King's, it was among the richest. St. John's was starred by Baedeker (*Great Britain*, pp. 444–445), and its buildings and library were extensive. (Edward Miller, *Portrait of a College: A History of the College of St. John the Evangelist, Cambridge* [Cambridge University Press, 1961]).

[4] Through the 1870s Fellow-Commoners had not been popular. Many were wealthy young men who "did not take degrees, wasted their time and money, and set a bad example to the other undergraduates" (Denys Arthur Winstanley, *Early Victorian Cambridge* [Cambridge University Press,

1940], pp. 414–415). By the 1880s, the title was conferred on men older than undergraduates, and often well-educated, with whom the colleges did not know what to do. (See Heitland, *After Many Years*, pp. 106, 110.)

[5] JMC lived in Room E7 in the Fourth (or New) Court, built of stone in 1831 across the River Cam from the older buildings of the college and linked to them by the Bridge of Sighs (Miller, *Portrait of a College*, p. 76; Baedeker, *Great Britain*, p. 445; G. C. Moore Smith, *List of Past Occupants of Rooms in St. John's College* [Cambridge: The Eagle, 1895], p. 58). Heitland apparently had charge of the rooms off Staircase E, as he lived in Room E5 (Moore Smith, *Past Occupants*, p. 45).

6.3 To Parents, Cambridge, 30 September 1886

I am tonight in college though not in my own rooms. The rooms reserved for me are very good except in that they face the north. I have a good sized study and bed room, a little kitchen room, and a tiny room in a tower. They will cost with service about $150—less than I paid at Leipzig; but I am put to expense for furniture. I buy at an estimation what I wish of the furniture of the previous occupant, and in the same way sell it when I leave. I shall have to pay about $150 and even then the rooms will not be comfortably furnished.[1] *I could not take the carpets as they were worn in holes. I wish I had my rugs.*

I have been fussing all day over the rooms and my boxes, which are here, 1000 lbs in weight and costing $25 carriage.

[1] Expenses at Cambridge were always high, and many observers decried this fact (see Hart, *German Universities*, pp. 330–332; Eliz-

abeth A. Osbourne, ed. *From the Letter-Files of S. W. Johnson* [New Haven, Conn.: Yale University Press, 1913], p. 76).

6.4 To Parents, Cambridge, 1 October 1886

(A brief postscript about JMC's mailing address has been omitted.)

I am still busy with affairs, rather than with work proper. I am having the floor of my rooms stained and book-shelves put in—it will cost respectively $15 and $20, but could not well be helped. I hope of course to get most of the money back from the man who follows me. I have had to pay for paint, paper &c.

I lunched with Mr. Ward and we talked about work; we shall probably continue with my apparatus work such as I did in Leipzig. I think they will give me a room in the physiological laboratory.[1] Both Mr. Ward and Prof. Foster think I ought not to give up my work, but of course the more I work in psychology, the less time will I have for medical studies—I shall not however neglect these.

[1]JMC's idea that he could get a room in this laboratory was unrealistic. In 1886 the Cambridge Physiological Laboratories occupied a building that had opened in 1879 after several years' debate as to what facilities it should house. At Michael Foster's insistence, most of the fittings and furniture, and their arrangement, were modeled after the laboratories of Carl Ludwig, professor of physiology at Leipzig. By the mid-1880s, these laboratories had become so crowded that Foster was unable to teach comfortably and "by October 1886, Foster was practically frantic." (Gerald L. Geison, *Michael Foster and the Cambridge School of Physiology: The Scientific Enterprise in Late Victorian Society* [Princeton, N.J.: Princeton University Press, 1978], pp. 162–173, 306–309.)

6.5 To Parents, Cambridge, 3 October 1886

This letter was misdated "Sun. Oct. 2nd.")

I seem now a part of the university and went to church today in a night-shirt. I am practically a fellow-commoner, though I do not seem to have been elected. I wear a gown different apparently from any one else in the university.[1] It seems curious to wear a cap and gown. They must be worn to dinner and to all university exercises, and all day Sunday though "taking a walk in the country" buys a dispensation. The gown costs $10, the cap $2.50 and the night shirt $5. I must pay $125 as caution money, which seems a good deal, but of course does not matter much.

See figure 13; Tanner, *Historical Register*, p. 196.

6.6 From Elizabeth Cattell, Philadelphia, 5 October 1886

(Several paragraphs of gossip have been omitted.)

The Trustees of the University met last Tuesday, and unanimously elected you Lecturer in Psycho-Physics. M[r] Fraley the President of the board of trustees said, he was delighted that you were elected to fill this position, that he had had you on his mind for three years, proving that he had remembered the talk your Papa had with him so long ago.[1] I enclose a slip from the Press you will find you are to be associated with scholarly men.

[1] See 2.17.

6.7 To Parents, Cambridge, 7 October 1886

I get most excellent food here and the meals are very conveniently arranged. Breakfast and lunch I have in my room. The people who look after me ("Gyp" and "bedmaker") set the table and wash the dishes.[1] I have certain things, bread, butter, jam, fruit &c always in the room and can order what I want in addition from the college kitchen this is brought punctually and without any trouble to me. I have thus oatmeal for breakfast; but nothing else unless I want meat for lunch. It is thus very convenient to entertain.

Dinner is at 7.15 in a magnificent hall. The undergraduates sit at long tables and the fellows at a raised table at the end of a room. The dinner is almost too good— soup, fish or entreé, most with vegetable, dessert, and cheese. It is pleasant socially the men sit in the order in which they come in, so you have various company or can manage it to sit by whom you please. After dinner the fellows go up to the "combination room" another magnificent room and have coffee staying until about nine. Dinner cost two and six (60 cts.) I don't know what the kitchen prices are—I suppose about the same as at a restaurant, but things are much better.

[1] "Gyp" was Cambridge slang for a college servant who regularly attended more than one student (*OED*).

6.8 From Elizabeth Cattell, Philadelphia, 8 October 1886

(This letter was misdated "Sept. 8th 1886." Several paragraphs of gossip have been omitted.)

The next letter you receive after this, will probably be an important one, from Dr Pepper,[1] in which he will offer you a Lectureship in the University here. You will lecture only to post Graduates, the salary will be small $250, for the lectures, for the first year, but you will have most of your time for study and you can still go on with your investigations. Your Papa has worked hard for this, as we have felt anxious to have you live at home with us, and this may be a fine opening for the future, they are perfectly willing you should remain abroad another year, and we are glad of it, as we know you would rather go on with your studies.

[1] William Pepper, provost of the University of Pennsylvania (see 2.17).

6.9 From William Cattell, Philadelphia, 9 October 1886

(One paragraph of gossip has been omitted. Except for the postscript, this letter was typewritten.)

Your Mother in her letter yesterday gave you an indication of the enclosed letter.[1] It seems to offer just the career I had hoped for you. Indeed, I felt like placing your acceptance in Doctor Pepper's hands at once: but at your age the decision must be made by yourself.

I can conceive of only one ground for your hesitation: and that is, the salary. You would undoubtedly feel freer and more independent in a position where you could support yourself: but our position is, as a family, somewhat exceptional: your dear Mother and myself want you with us: we need you. It may not be for many years, for we are growing old (I am growing old very fast) and it will be a solace in our declining years to have again at home our dear, kind, thoughtful son. And the expenses of living being thus provided for, the $300 salary, with your own income,[2] will make you independent in all your personal expenditures. Besides, I think with Doctor Pepper the salary will soon be raised. I am sure you and the authorities will always be in accord and that you will make the department a success; so that it will not be long before a full Professorship will be established.

Meanwhile, you have just what I have longed to secure for you—a recognized and honorable position among scholars and teachers: no class-room drill; plenty of time, with fine laboratory opportunities for study and research. And of course you can attend all the lectures in the Medical Department that you may desire—not as one of the "students," but as one of the Staff. I would hardly know how to choose for you a position that promised more advantages—or so many—to say nothing of our joy in having you at home.

Doctor Pepper explained to me in our interview the necessity for the prompt reply to which he refers in his letter. They want to get our Circulars, as well as to announce the Post-Graduate Course in the forthcoming Catalogue. Several Professors are already engaged—but not one of them with a salary. They accept positions in the Post-Graduate Course for the honor. So please send me a Cablegram as soon as you receive this; —one word will do: simply, Yes. And to provide against errors of transmission, you had better give my full name, together with our address, 222 So. 39th street. I will add no more, but shall write to you soon again.

My dear Son
I hope you have rec[d] the letters—of which these are Copies[3]—but I send the Copies as letters So Miscarry & this is important.

---

[1] The "letter yesterday" is 6.8; the "enclosed letter" has not been found.
[2] From the Warren Foundry (see introduction).

[3] William Cattell had sent two copies of this letter (one to Brown, Shipley & Co. [see 4.2] and one to Cambridge; see 6.10), and added this handwritten note to one of them.

6.10 From William Cattell, Philadelphia, 11 October 1886

(This letter was dated "Monday, 11th." One paragraph of gossip has been omitted.)

On Saturday I mailed you an important letter from Dr Pepper: —sent it as I shall this to Brown Shipley & Co.—as letters do miscarry I mail you another copy direct to Cambridge—You will let us know yr address there as soon as you are located: —meanwhile of course you will inquire for letters at the P.O. or leave with them yr Cambridge address. I send you also (to Cambridge) two numbers of "Science"—(not Youman's Popular Sc. Monthly) which contain articles on Psycho-physics that may interest you.[1]

The more we think of Dr Pepper's offer the more we are pleased: —I hope it will strike you as favorably.

[1] The articles were Joseph Jastrow, "An Easy Method of Measuring the Time of Mental Processes," *Science* 8 (10 September 1886): 237–241; James Hervey Hyslop, "Psycho-physics," *Science* 8 (17 September 1886): 259–262. Jastrow was JMC's friend from Johns Hopkins (see 2.41). Hyslop (*Who Was Who in America*, 1897–1942, p. 616) had studied philosophy in Leipzig (1882–1884) and had taught at Lake Forest College and Smith College before going to Johns Hopkins (fall 1886), where he earned a Ph.D. From 1889 he taught at Columbia University, where he was to be a colleague of JMC.

6.11 Journal, Cambridge, 12 October 1886

(An account of a walking tour through Norway has been omitted.)

Here I am: sometimes I think for the 30 odd years left me.[1] I am a fellow-commoner of St. Johns College. Life is here made as easy as may be, people mostly seem able to forget what an uncomfortable world it is. I am quite at a loss what to do. My first plan was to study medicine—now I sometimes think I should act chiefly with reference to a fellowship, but it is a long way and slippery.

[1] Again, this point is unclear. See also 1.15.

6.12 To Parents, Cambridge, 13 October 1886

I am now going to try to live quite regularly. I get up at 7:30 and go swimming; eat breakfast at 8.15; from nine on to one or two hear lectures and work in laboratory. From 2 or 2.30 as long as possible exercise out of doors. Afterwards till dinner read papers novels or such like and see people; dinner at 7.15 combination room till 9; 9 till 12 read and write.

There is just so much time in each day and I shall try not to crowd things by agreeing to write a paper by a certain day etc. I am quite determined now and as long as I am young to take plenty of exercise. I am, I confess, at a loss how to divide time between philosophy and medicine.

6.13 To Parents, Cambridge, 14 October 1886

I said in my letter last night that I should not engage to write a paper at a given date, and curiously enough this morning I received a letter from the Sec'y of the Aristotelian Society asking me to read a paper on psycho-physic research.[1] *I shall accept as I have been intending to write a paper on this subject, and it will in many ways be an advantage to read a paper before this society, which is the best philosophical society in England. Bain, Romanes, Hodgson, Ritchie, and Alexander, almost the first psychologists in England were five of the seven who read papers last year.*[2]

I played tennis this afternoon with Forsyth a fellow of Trinity.[3] *He plays about like Harry.*

[1] On the society see introduction to section 6. In 1886, the "honorary secretary" was Edward Hawksley Rhodes (see *Mind* 9 [1886]: 599).

[2] George John Romanes (*DSB* 11: 516–520) was a Cambridge-educated scientist who studied physiology with Michael Foster and later wrote extensively on comparative psychology and the concept of evolution. On Hodgson see introduction to section 6. David George Ritchie (*DNB*, 1901–1911: 208–209) was an Edinburgh-educated philosopher who in 1886 was a fellow at Jesus College, Oxford. From 1894 he was professor of logic and metaphysics at St. Andrew's University. Samuel Alexander (*DNB*, 1931–1940: 3–5) was an Australian-born philosopher, educated at Balliol College, Oxford, and from 1882 a fellow of Lincoln College, Oxford. Later, from 1888 through 1891, he tried to work in experimental psychology and studied the subject in Germany. However, in 1893, he became professor of philosophy at Manchester; he held this chair until 1924. Among the others who had spoken before the Aristotelian Society the previous year were Herbert Wildon Carr, later professor of philosophy at King's College, London (*Who Was Who* [Br.], 1929–1940, p. 224), and Aubrey Lackington Moore, a well-known author of books on theological topics (*DNB* 13: 788–789).

[3] Andrew Russell Forsyth (1858–1942) was a university lecturer in mathematics (1883–1895) who would succeed to the Sadleirian professorship in 1895 (Venn, *Alumni Cantabrigiensis*, vol. 2, p. 541; Tanner, *Historical Register*, p. 118).

6.14 To Parents, Cambridge, 19 October 1886

Your letter, Mama, received today changes and settles everything.[1] *I am glad that I can soon come home to stay—this is the great reason why a position at the university is so welcome; then there are other advantages over most colleges. Though in no hurry to take it, I imagine I am better fitted for a university position than for anything else. I could become a physician but that would require years of preparation and waiting, and I could scarcely hope to be successful unless I devoted myself wholly to it, and that I should rather not do. I could support myself by writing, but that is proverbially uncertain and uncomfortable. Whereas with a university position either of the last mentioned can be taken up as far as seems desirable.*

You will perhaps be still more pleased at the present offer, when I say that if I had not received a satisfactory position, I should with your consent have proposed staying on here three years in all. Indeed it is rather hard for me to give this up. I

[1] The letter referred to is 6.6.

am not eligible for a fellowship here unless an A.B. of the university, and cannot take this degree until I have been in residence here three years. But I think it is about as certain as most things that in three years I should be given a fellowship.[2] I should then receive $1000 to $1250 a year for six years, and could live where I please and do what I please. $7000 or $8000 seems a great deal to earn by two years spent in the pleasantest way possible. Nor is this all. If I then stayed on here the fellowship would be renewed for 20 (14 years more) years, and after that doubled for life.

You see I give up much to be with you. If it were not for you and if I were not married I should be better off here than anywhere else in the world.[3]

[2] See 6.11.   [3] JMC was, of course, not married at this time.

6.15 To Parents, Cambridge, 20 October 1886

I have not heard from Dr. Pepper, but unless I misunderstand your letter, Mama the matter is definitely decided. My immediate as well as future course is very considerably changed. If your letter had been 24 hours later I should have paid $50 to be matriculated, it takes place tomorrow, but is not necessary for me if I do not want to pass any examination.

I shall under the circumstances study less and probably spend more money, which sounds rather bad; but if the position is given me, I shall have time enough to read, whereas while here it will be a great advantage to see as much as possible of things and people—going for example oftener to London than I should otherwise have done. It will also be advisable to be especially careful of my health, going to bed for example at eleven instead of twelve.

6.16 To Parents, Cambridge, 24 October 1886

I was invited to lunch today at Mr. Ward's to meet Mr. Sully, who is lecturing for Prof. Robertson at London this winter.[1] Prof. Robertson has been very ill, and it is doubtful when he can resume work.[2] Mr. Ward is not at all strong, being often interrupted in his work by bad health. After Prof. Bain and Mr. Spencer these are

[1] James Sully was another close friend of Bain's (like Robertson) who had studied at Berlin and Göttingen. In 1892, upon Robertson's death, he succeeded to Robertson's chair. (Sully, *My Life and Friends: A Psychologist's Memories* [London: T. Fisher Unwin, 1918].)

[2] Robertson had always been weak, and had required leaves of absence earlier. In 1888 he offered his resignation to University College, which did not accept it. (*DNB* 16: 1294–1295.)

the two most eminent psychologists in England.[3] *If such eminance is only to be purchased by loss of health from an egoistic point of view one is better off without it.*

I naturally think much of the University position, and have to rearrange many plans for the immediate as well as for the more distant future.[4]

[3] On Herbert Spencer see 2.3.
[4] In answer to JMC's queries on this point Robertson noted that "the time which is left to you before entering upon your duties ought to be of special value to you, and will be so, I fancy, the more according as you employ it in widening your psychophysical and philosophical interests. In this view, your attendance at the courses of [Moral Science] lectures you mention is much to be approved of." (Robertson to JMC, 6 November 1886). See also 6.35 and 6.50.

6.17 To Parents, Cambridge, 26 October 1886

(This letter was typewritten. One paragraph of gossip has been omitted.)

You see my writing-machine arrived safely. I have written out on it the paper for the January "Mind",[1] *and tonight wrote to Prof. Wundt and Berger. Prof. Wundt is always very thankful to me for having told him about the machine; he uses it nearly always, and writes about as much as any one living;*[2] *he has just published an ethics.*

[1] "Experiments on the Association of Ideas," *Mind* 12 (1887): 68–74.
[2] On JMC, Wundt, and the typewriter see 5.1. Others complained about Wundt's extensive writing, claiming that he used verbiage to hide shallowness (for example, see Ralph Barton Perry, *The Thought and Character of William James* [Cambridge, Mass.: Harvard University Press, 1935], vol. 2, p. 68).

6.18 To Parents, Cambridge, 28 October 1886

One way and another things are expensive here.[1] *Lectures cost 50 cts each which seems a good deal especially after Germany where they cost 7 cts. Fortunately I don't have to pay for dinner if I am absent and give notice three hours in advance. The dinner is given (free) to fellows in residence, but if they are absent without giving notice they must pay for it—at first sight a curious, but in reality a sensible arrangement. The fees for lectures etc. will I imagine be over $200, and the sessions together do not make up over six months. The subscription for all sorts of athletics together seems to be $5 a term, which is less than I had expected.*

[1] See also 6.3, 6.5, and 6.42. At Leipzig, $500 was said to be enough for an American student to live well for a year (Hart, *German Universities*, pp. 393–394).

6.19 To Parents, Cambridge, 30 October 1886

(Brackets indicate wormholes.)

I dined this evening at Mr. Ward's. Mrs. Ward is, as I think I have mentioned, sister of Prof. Martin at Baltimore.[1] She was lecturer at Girton College and has had [   ] in "Mind".[2] She is [   ] dresses and [   ] but neither she nor [   ] all strong.

Will you tell [   ] or if you do not know find out, whether the authorities (1) will allow me to sign myself Univ. of Penna and (2) whether they would like to have me do so.

[1]Mary Martin Ward, daughter of a Congregationalist minister, had married James Ward in 1884 (*DNB*, 1922–1930, p. 887).
[2]Girton was the first Cambridge college established for women (1869) (R. Brimley Johnson, *The Cambridge Colleges* [London: T. Werner Laurie, 1909], pp. cxiii–cxiv). JMC was incorrect, as Mary Martin Ward had been connected with Newnham College (the second Cambridge women's college, founded in 1870) since 1879, when her performance on the moral sciences tripos was rated equal to first class (Mary Agnes Hamilton, *Newnham: An Informal Biography* [London: Faber and Faber, 1936], p. 110). Before her marriage, Mary Martin had published at least two papers in *Mind*: "Mr. Gurney on the Utilitarian 'Ought'," 7 (1882): 554–558, and "On Some Fundamental Problems in Logic," 8 (1883): 183–197.

6.20 From William Cattell, Philadelphia, 31 October 1886

(Several paragraphs of gossip have been omitted.)

And what family talks we do have about this opening for you at the university! I wonder if it strikes you as it does us! We shall probably hear from you about this by tommorow's mail. To us it seems just the very thing:—except of course the salary. But so long as the W. F.[1] Goose continues to lay the golden eggs this will make no difference; and then to earn a larger or a living salary you w^d of course have more teaching—irksome duties &c. &c. It is just the place for you to study with every opportunity and inducement for original research. Young Fullerton (who has the Psychology-adjunct) attends the medical lectures with a view of taking an M.D. here.[2] You c^d do this—and with no tuition to pay this is equivalant to an addition of $125 or more to yr salary. I understand the Board meets next Tuesday when the appointment will be made. Dr. Pepper & I have had frequent conferences as you may suppose:—but this is a subject I could not exhaust in a score of letters!—God bless you my dear son.

[1]Warren Foundry (see introduction).   [2]On George S. Fullerton see 2.17.

6.21 To Parents, Cambridge, 1 November 1886

I was this evening at "An Ethical and Political Conference" at Prof. Sidgwick's;[1] he is a very entertaining and able man, indeed I think his "Ethics" is the most valuable philosophical book written by a man now living.[2] The meeting was right interesting this evening. I am however in doubt as to going regularly, as if I do I must myself write a paper, which I don't much fancy.

I was in the library today for the first time. It is I believe one of four libraries which receive all books printed in England. The arrangements seem convenient, except that I cannot take books out of the building.

[1] Henry Sidgwick (see introduction to section 6).
[2] *The Methods of Ethics*, 3rd edition (London: Macmillan, 1884). Others shared JMC's opinion of this book (see *EoP* 7: 434–436; Noel

G. Annan, *Leslie Stephen: His Thought and Character in Relation to His Time* [Cambridge, Mass.: Harvard University Press, 1952], p. 206).

6.22 To Parents, Cambridge, 2 November 1886

There is nothing in particular to write. I corrected today a proof for "Mind".[1] It is short and tolerably readable, more so than anything I have printed. I received a pleasant note from Mr. Galton, asking me about my plans; also program of the Aristotelian Society announcing the papers to be read by Bain, Romanes, Cattell and other eminent philosophers.[2]

People keep calling on me. I know as many people here as I did in Leipzig and the society is in every way pleasanter.[3]

[1] See 6.17.
[2] See 6.13.

[3] See 6.27.

6.23 To Parents, Cambridge, 10 November 1886

It has several times occurred to me that I should like to go to Germany this Xmas. There are different reasons, none conclusive, but considerable if taken together. I received yesterday a letter from Prof. Wundt which again turned my thoughts that way.[1] If I am to start a laboratory at Phila. I should like to talk the matter over with him. I should also like to see what work is now being done in the laboratory both for the above reason, and because I am about to write a paper on "Recent psychophysical research"[2] Besides Prof. Wundt I should like to see the other men in the laboratory especially Dr. Lange my successor,[3] whom Harry knows. I should like also

[1] Wundt to JMC, 7 November 1886. In this letter Wundt congratulated JMC on his appointment in Philadelphia, discussed the possible publication of an article by JMC in *Philosophische Studien*, passed on the greetings

of Wundt's wife and children, and asked JMC to offer his regards to Professors Ward, Sidgwick, and Waldstein (see 5.116).
[2] See 6.13.
[3] Gustav Ludwig Lange (see 5.143).

to see Krille the man who made my apparatus, as I may want to order apparatus from him both for America and for Cambridge.[4] As to further business I should like to see Berger; he is making some experiments for me, and I may want to translate something for the "Studien".[5] Beside the business I have a longing to be once more in Leipzig before returning home. It is partly the wish to be where I have spent half my life for the past six years, partly because I have friends there, partly on account of the music and (if in Dresden or Berlin) pictures etc. The only reason for not going was the expense, about $50 more than stopping here, or $25 more than if I should spend a week in London. However it was a mere possibility until today. But today I asked Mr. Ward, if he did not think he ought to take a rest this Xmas, going for example to Germany. As I have written he has not been strong. He said he would like to go, if I should go with him and his physician recommended it. He studied under Lotze and Wundt ten years ago, and has not been in Germany since. So if his physician thinks it will be good for him we shall probably go. I think Mr. Ward is the ablest psychologist living, though he has not of course done as much as Bain and Wundt.

[4] JMC had earlier written to his father, urging him to speak with Pepper about the possibility of having the University of Pennsylvania pay for European-made instruments to be used at the university. This letter has not been found but its content is implied in William Cattell to JMC, 11 November 1886. See 6.97, 6.112, and 6.129.

[5] Berger later helped JMC translate his last *Mind* article (see 6.17 and 6.41) into German for *Philosophische Studien* ("Die Association unter wilkürlich begrentzten Bedingungen," 4 [1888]; 241–250), as Wundt had suggested in his letter cited above.

6.24 To Parents, Cambridge, 15 November 1886

I received this morning the copy of the "Times" announcing my election at the university. My life seems in some sort arranged now, and perhaps it is best so. Can you not also find a wife for me? If that were fixed I should have nothing further to do, but to wait until I die. Will you please send me a catalogue of the university? When does the Autumn term open and when shall I have to deliver the lectures? I should like to know in case I go to Germany at Xmas whether you think I had better secure some apparatus. I suppose I shall be given a room. Two small rooms would be better than one large one. I'm sure I don't see where any students are to come from. At Baltimore they offer the bribe of two fellowships, and I should recommend a friend of mine to go to Germany.[1]

[1] In fact, JMC did so later, when he recommended that one of his students at the University of Pennsylvania (Lightner Witmer, later best known for his work in clinical psychology) go to Leipzig for his Ph.D. rather than stay in Philadelphia and earn the degree under JMC's direction (Robert I. Watson, "Lightner Witmer: 1867–1956," *American Journal of Psychology* 69 [1959], 680–682).

6.25 To Parents, Cambridge, 18 November 1886

I dined this evening with Forsyth in Trinity.[1] He is one of the ablest mathematicians here, and one of the youngest fellows of the royal society.

I am going up to London tomorrow with Mr. Ward to attend a meeting of the psychological club at Prof. Robertson's.[2] It is a great pity there is no train from London later than ten—it is only 1¼ hours away yet one must stay overnight if he wishes to spend the evening in town.

[1] On Forsyth see 6.13.
[2] See introduction to section 6. Robertson had earlier (on 6 November 1886) made sure to invite JMC to this meeting.

6.26 To Parents, London, 19 November 1886

I am down in London this evening. I came on with Mr. Ward, getting here at six. We drove to Prof. Robertson's and dined there, the meeting being held afterwards. The meetings this winter are to consider original psychophysical research and to discuss how psychological terms are used and should be used. The best psychologists belong to the society and what they decide as to terminology will be carried out, so such a meeting was very interesting. Besides others whom I already knew I met Prof. Adamson, Mr. Hodgson (Pres. of the Aristotlian Society) and Mrs Bryant.[1] The latter is one of the ablest women in England and very interesting.

[1] Robert Adamson (*DNB*, 1901–1911: 15–17), educated at Edinburgh and Heidelberg, was from 1876 through 1893 professor of philosophy and political economy at Owens College in Manchester. On Hodgson see introduction to section 6. Sophie Willock Bryant, educated at Bedford College, London (B.Sc., 1881), had been the first woman to earn a D.Sc. from the University of London (1884). She had been married and widowed in about a year's time (1869–1870) and was Mathematical Mistress (later Headmistress) at the North London Collegiate Institute for Girls. (John S. Crone, *A Concise Dictionary of Irish Biography* [London: Longmans Green, 1928], p. 21; *Who's Who* [Br.], 1901: 208–209).

6.27 To Parents, Cambridge, 21 November 1886

I dined this evening with Mr. Venn in Caius (pronounced Keys) College.[1] It is the third in size of the colleges here. Mr Venn is the author of a celebrated book "The Logic of Chance"[2] and is considered a very able man. It is very pleasant and profita-

[1] John Venn, (*DSB* 13: 611–612; *DNB*, 1922–1930: 869–870) had been educated at Gonville and Caius College. From 1857, when he graduated, until his death he was a fellow at Caius. From 1862 he had lectured on moral science at Cambridge, where he had played a major role in the introduction of the moral science tripos. He is best known for his work in logic and the theory of probability, which led to his development of what are today known as Venn diagrams.
[2] London: Macmillan, 1866.

ble to be acquainted with all the men most eminent in the subjects in which I am interested.[3] I think this is likely to be the pleasantest & most useful year in my life so far.

[3] See also 6.22. In many ways this was to be the main benefit to JMC of his stay in Cambridge.

6.28 To Parents, London, 22 November 1886

I am, as you see, in London again. I came down this morning. First I spent an hour with Mr. Galton, discussing things psychological; I then met Muirhead and we lunched together—Muirhead was about the best friend I had at Leipzig, he is now completing a Baedeker for England.[1] In the afternoon I was at the South Kensington. In the evening I was at the Aristotelian Society; for which of course I came. Ritchie an Oxford friend read the paper.[2] Afterwards I was at the rooms of the president.[3]

[1] On James F. Muirhead see introduction to section 1, 5.24, and 5.143.
[2] On David George Ritchie see 6.13.

[3] On Shadworth Hodgson see introduction to section 6.

6.29 To Parents, Cambridge, 24 November 1886

(Two paragraphs of gossip have been omitted.)

I have been writing today an essay on "The Measurement of Pleasure" which I must read at Prof. Sidgwick's next Monday. It is an interesting and difficult question. What do you think? Can we ever say that one pleasure is twice as great as another?[1]

[1] This essay has not been found. However, JMC was later to express other thoughts on this question: "Most men will think that a just king is happier than a tyrant, but few will agree with Plato in considering him 729 times as happy." ("On the Perception of Small Differences," in *Men of Science*, vol. 1, p. 152).

6.30 To Parents, Cambridge, 27 November 1886

There was rather an interesting meeting in our college this evening anent a "settlement" the two universities have made in the East End of London. They have built a hall, and about twenty university men (ie graduates) live there, and while carrying on their ordinary work, try to see as much as possible of and do as much as possible for the people about—perhaps the least happy folk in the world.[1] *Since, we have been in one of the tutors rooms till late or rather early, discussing this and other things that are amiss.*

[1] This settlement—known as Toynbee Hall, in honor of Arnold Toynbee (1852–1883), the Oxford economic historian who had been involved in what is now called social work in the East End of London before his early death—had been founded in 1884, and by 1886 was regularly sponsoring meetings and lectures of all sorts (Arthur C. Holden, *The Settlement Idea: A Vision of Social Justice* [New York: Macmillan, 1922], pp. 11–14; John Alfred Pimlott, *Toynbee Hall: Fifty Years of Social Progress, 1884–1934* [London: J. M. Dent, 1935]). JMC's interest in the work of Toynbee Hall was derived from his acquaintance with Donald MacAlister, who was extremely active in the settlement's affairs (Edith MacAlister, *Sir Donald MacAlister of Tarlbert* [London: Macmillan, 1935], p. 103), and might be considered in relation to his allusions to socialism during his days in Germany (see 5.36 and 6.108).

6.31 To Parents, Cambridge, 29 November 1886

The college rowing is about over.[1] *I rowed in a "funny" this afternoon and nearly upset.*[2] *I had to get out and empty out the water. Such a boat is only a few inches wide and is only kept right side up by the oars.*

I dined at the Sidgwick's this evening. I think I have said Prof. S. is one of the foremost philosophers living, and the ablest writer on ethics. Mrs Sidgwick is also distinguished.[3] *She has published papers with her brother-in-law Lord Raleigh, who is probably the ablest English physicist, she is also secretary of the Society for Psychical Research, treasurer (practically head) of Newnham College etc.*[4] *She is a niece of the Prime Minister's.*[5] *They are further both very clever talkers and interesting personally.*

[1] The Lady Margaret Boat Club, in which JMC was active, had always played an important role in the social life of St. John's (Heitland, *After Many Years*, p. 110).

[2] "Funny" was Cambridge slang for a narrow, light boat (*OED*).

[3] Eleanor Mildred Balfour Sidgwick (*DNB*, 1931–1940: 811–812) was active in several learned societies and involved with women's education.

[4] John William Strutt, Lord Rayleigh (*DNB*, 1912–1921: 514–517; *DSB* 13: 100–107) had been educated at Trinity College, Cambridge, where he was later a fellow. From 1879 he held the Cavendish Chair of Experimental Physics, which he resigned in 1884 to become secretary of the Royal Society.

His wife Evelyn was a sister of Eleanor Sidgwick, with whom Lord Rayleigh published, among other articles, "On the Specific Resistance of Mercury," *Philosophical Transactions* 174 (1882): 173–185. On the Society for Psychical Research see the introduction to section 6. On Newnham College see 6.19, note 2. Eleanor Sidgwick had been treasurer of Newnham from 1880 and vice-principal from 1880 through 1882. Later, from 1892 through 1910, she served as principal. (Johnson, *The Cambridge Colleges*, pp. cxiv–cxv.)

[5] Eleanor Sidgwick's mother was the sister of Prime Minister Robert Arthur Talbot Gascoyne-Cecil (*DNB*, 1931–1940: 811–812).

6.32 To Parents, Cambridge, 1 December 1886

I was out at Girton College this afternoon, Mrs Robertson, who is Bursar, having written asking them to show & tell me things.[1]

This evening I was at a mesmeric proformance. It was only tolerable the really remarkable phenomena of mesmerism being mixed up with trickery. I caused considerable excitement by announcing that the boy in "thought transference" was not blindfolded. When I blind-folded him the bank-note trick could not be done.[2]

[1] Newnham and Girton Colleges arranged for Cambridge lecturers (both official and unofficial) to present their work to the women students (Mary Agnes Hamilton, *Newnham: An Informal Biography* [London: Faber and Faber, 1936], pp. 109, 124).
[2] See also introduction to section 6.

6.33 From William Cattell, Philadelphia, 1 December 1886

(Several paragraphs of gossip have been omitted.)

I send you by this mail a copy of Science *containing a paper that will be of some interest to you.[1]*

I have not been able yet to see Dr. Pepper anent the apparatus;[2] but I spoke to one of the Trustees my neighbor Mr. Potts,[3] who seemed to favor the proposition. Suppose you wrote me a letter about it that I can show the Dr?—and mention the amount: —was it $500 or $1000.

I also enclose Jim the official notice of yr appointment which the clerk has just sent me. You ought to reply to Dr. Pepper himself—briefly—stating that you have rec[d] the notice of yr appointment and also his letter to me—& that you accept &c—to enter upon yr duties in the Fall.

Harry will send you the new Catalogue as soon as it is issued.

[1] Joseph Jastrow, "Experimental Psychology at Leipzig," *Science* 8 (19 November 1886): 459–462.
[2] See 6.23.
[3] William Francis Potts, a Philadelphia iron merchant (Charles Morris, ed., *Makers of Philadelphia: An Historical Work* [Philadelphia: L. R. Hamersly, 1894], p. 274).

6.34 To Parents, London, 5 December 1886

I have spent the day here not doing very much.[1] It is an excellent idea for university men to build a Hall and live here in the East End, and try to get acquainted with the poor people about. I cannot tell exactly how much they are doing, apparently less

[1] In addition to visiting Toynbee Hall (see 6.30), JMC was in London to hear George Romanes (see 6.13) read a paper at the Aristotelian Society on "Neo-Kantianism in Relation to Science" the next evening (JMC to parents, 6 December 1886; *Mind* 12 [1887]: 319).

than their reports lead one to think. I imagine all benevolent schemes must take into account the thorough selfishness of people, both of them that give and of them that get.

The poor in London are not criminal, nor starving. The great pity would seem to be the dull, dreary lives of constant labor, with little chance and less capability for enjoyment.

6.35 To Parents, Cambridge, 8 December 1886

Have I told you that I have stopped dissecting?[1] After receiving the appointment I felt things to be considerably changed. On the one hand it seemed desirable to devote all my time to psychology, on the other I shall have good opportunity to dissect & study medicine at Phila. Further I feel if I get along well as a lecturer it may not be worth the while to prepare for an M.D. examination. It would, as I have said, take about a year, beyond what physiology, anatomy and pathology I need for psychology.

[1]JMC especially disliked dissection, which he called "tedious" and "trivial" (JMC to parents, 12 October 1886).

6.36 To Parents, Cambridge, 12 December 1886

I walked out to Prof. Foster's this afternoon; he lives about four miles from Cambridge.[1] A good many university men go out to call Sunday afternoon. One really does not see so very much of men belonging to other colleges, on the other hand men of the same college are constantly thrown together. This evening I was in Hogg's rooms with three other of the younger fellows, Stout, Love and Mathews.[2] It would be hard to find as good men as these at Phila. and even if one knew them, one would see them but occasionally. I have not played billiards or cards, nor indeed any game, except twice foot-ball and twice tennis. There is very little drinking here. Wine and beer are on the table, but except at "feasts" the majority to which I of course belong, do not take them. I however usually drink coffee after dinner, I learned to do this in Norway, where it was very good, more especially when half

[1]On Michael Foster see 5.156.

[2]Robert Wallace Hogg (Venn, *Alumni Cantabrigiensis*, vol. 3, p. 406; see also 6.52) was a fellow at St. John's from 1885 and, from 1887, mathematics master at Christ's Hospital in London. George Frederick Stout (*DNB*, 1941–1950: 844–845; *EoP*, 8: 22–23), JMC's exact contemporary, was a fellow of St. John's who lived on the same staircase as JMC (see 6.2). He had earlier studied philosophy with Ward, and remained at Cambridge through 1896 as a lecturer. He then taught at Oxford and Aberdeen before becoming (in 1903) professor of logic and metaphysics at St. Andrew's University. Augustus Edward Hough Love (*DSB* 8: 516–517) was a fellow at St. John's from 1886 through 1899, a distinguished mathematician and, after 1898, Sedleian Professor of Natural Philosophy at Oxford. George Ballard Mathews (Venn, *Alumni Cantabrigiensis*, vol. 4, p. 359) was a fellow at St. John's from 1884 while holding the chair of mathematics at Bangor, Wales.

cream. While I am about it, I suppose I had better plead guilty to smoking once in a while. I have done this ever since the year at Baltimore, but as it has only been on rare occasions, I have not felt it necessary to confess.

The question as to the use of these stimulants is most perplexing.[3] They undoubtedly give much pleasure, and may at times be of great use. I am inclined to think occasional use, more especially as one grows older, is profitable.

[3] See 5.135.

6.37 To William Cattell, Cambridge, 13 December 1886

In your letter received this morning, you say you will speak to Dr. Pepper anent the advisability of my securing some apparatus while over here for the proposed psychophysical laboratory.[1] Of course instruments are necessary for experimental research, but much can be done with inexpensive and self-made apparatus, more especially if certain pieces, a Kymograph or recording turning-fork for example, can be borrowed, when needed, from the physical or physiological department.

But if it could be afforded a small appropriation for apparatus would be welcome. I believe $500 a year is given at Baltimore, and they propose giving the same amount here. Prof. Wundt has $250 but adds to this from his private income. Unless we have more students at Phila. than we can expect, I should think an appropriation of $500 would last for three years. Some instruments are useful for demonstration and general use, other apparatus must be devised for special lines of research. I got a set of my own apparatus for the Army Medical Museum at a cost of about $175.[2]

If it is thought the university would later grant an appropriation, I could in the meanwhile buy what apparatus could be secured to best advantage while I am over here.

You can of course show this to Dr. Pepper, if you think it worth the while.

[1] See 6.23 and 6.33.   [2] See 5.64.

6.38 To Parents, Cambridge, 13 December 1886

I have decided, after much undeciding and redeciding, to go to Germany. I am not yet sure whether I ought to spend the money or not, but so much has been invested in me already, that a little more may help the rest. Traveling expenses will be about $50, $30 railway, and $1 a night at hotels, food costing about the same as here. Besides this there will be of course other expenses—for which however I should look to get more than an equivalent return—I should be very glad to pay for an opera seat here in Cambridge if it were only possible.

Your letters received this morning and the fact that most of my friends are leaving Cambridge, finally decided me to go too.[1]

I have written to the Secy. of the Univ. and enclose a letter anent apparatus.

I have received the copy of "Science" which gives a good summary of my paper.[2] *The writer, now I think fellow or assistant at Baltimore, wanted my position at Phila; fortunately his brother was elected, at the same time I was, lecturer in Assyrian.*[3] *You say something, Mama, about Pres. Gilman and this review; did you know he is president of the company publishing "Science"?*[4]

I trust, Papa, your trip West has been successful and not trying to your health. I'm glad you are taking German lessons, Mama. We must read together when I get home.

[1] The letters referred to are 6.33 and Elizabeth Cattell to JMC, 3 December 1886.
[2] See 6.33.
[3] Joseph Jastrow's brother, Morris Jastrow, Jr., was from 1885 professor of Semitic languages at the University of Pennsylvania

(*Who Was Who in America*, 1897–1942, p. 630).
[4] Elizabeth Cattell had written, "I do wonder how President Gilman felt when he picked up The Science Monthly, and saw how many columns Mr. Jastrow had given you." (3 December 1886).

6.39 To Parents, Göttingen, 17 December 1886

How strange and how natural it seems to be in Göttingen once more! Göttingen is the first place I lived in away from home, and that winter somehow stands out more prominently in my memory than any other. The winters before that are quite blurred, so that I must count up in order to remember in which one of them any given event occurred. I have the winter at Baltimore well in mind, but the winters at Leipzig are somewhat confused and monotonous.

I left Hanover at noon getting here in two hours and a half. Everything seems so familiar, strangely, sadly familiar—the walk around the wall, which I used to take every day was full of interest. I heard G. E. Müller lecture[1]*—he is Lotze's successor and a very able man. After the lecture I went home with, and he took me to see another professor. Then I went to Reinsgruben B1, and stayed to supper.*[2] *Things are much now as they were then, but two of the sisters are away. Did you know one of them had married a son of Dr. John Hall?*[3]

I look to go on to Mühlhausen tomorrow.[4]

[1] George Elias Müller, educated at Berlin, Leipzig, and Göttingen, was best known for his psychological work on vision and memory (*IESS* 10: 523–525).
[2] JMC visited the Bartlings, with whom he had lived while studying at Göttingen (see 1.4).

[3] Jenny L. E. Bartling, in 1884, married Thomas Cuming Hall (*Who Was Who in America*, 1897–1942, p. 507), the son of John Hall, then president of the University of the City of New York (see 4.10).
[4] To the home of Gustav Berger.

Cattell then spent almost a week at Mühlhausen, working with Berger and visiting with Berger's family and fiancée. An unexpected blizzard that lasted more than three days forced Cattell to stay with his friend longer than he had planned, but he eventually set off for Leipzig.

6.40 To Parents, Leipzig, 24 December 1886

A Merry Xmas and a happy New Year! I must send over the ocean the old greeting I am not at home to speak. I trust next year there will be no need to write what can so much better be said. Less you think I am confusing our ways with German, I must add that it is really Xmas day, I being up late.

I called on the Wundts and Heinzes this morning, and was invited to spend the evening with both families.[1] I accepted both in some sort. I lit the tree &c and was at supper at Prof. Wundt's and after the children had gone to bed went to Prof. Heinze's. I gave the children Max and Lolo Wundt and Margarethe Heinze English illustrated books and bonboneries and was in turn given presents—which you can see when I come home.

I dined with Muirhead[2]—I have not seen any one else whom I know. The English keep coming and going. I have outlived several generations.

[1] See 5.106 and 5.125.   [2] Findlay Muirhead (see 5.24 and 6.62).

6.41 To Parents, Leipzig, 26 December 1886

It is tolerably cold today. The ponds are frozen and there seems to be good chance that we may skate on them tomorrow. I did not come to Germany for the skating, but I imagine if I had been sure there would be none, I should not have come.

I dined at Prof. Wundts again today, and have since seen some of my English acquaintances.

I shall mail you with this two proofs.[1] I scarcely think it worth the while to try to have either of them reprinted in America, but neither would it do any harm, if one or the other of them was wanted anywhere. The "Reprinted from" must not however be omitted. I have with Berger translated (while at Mühlhausen) the paper on the Association of Ideas into German for Prof. Wundt.[2] I did not care especially to do this, but he really seemed to want it.

[1] Of "Experiments on the Association of Ideas," *Mind* 12 (1887): 68–74, and "Abstracts of British and Foreign Journals: Wundt's Philosophische Studien," *Brain* 9 (1887): 553–556.
[2] See 6.17 and 6.39.

6.42 To Parents, Leipzig, 2 January 1887

(One paragraph of gossip has been omitted at the beginning of this letter.)

I have moreover received my Cambridge bills. The college bill is £39.15—$200, and this does not include furnishing my rooms, breakfast & lunch, nor many other expenses to be reckoned as university—anatomy, reading room ($10), subscriptions almost necessary &c.

The college expenses are thus not far from $1000 a year—that is six months! I pay in some ways slightly more than the undergraduates—12 cts more for dinner, but they nearly all take lessons at $2.50 each, many drink wine etc.

You see it is a very expensive matter which explains why there are only 6000 students in England and five times that many in Germany & America. Of course the large number of scholarships &c must not be forgotten. My college gives $30 000 a year to undergraduates.

6.43 To Parents, Leipzig, 4 January 1887

(Two paragraphs of gossip have been omitted at the beginning of this letter.)

I was however at the laboratory this afternoon, one party having begun work. The assistant, Dr. Lange, whom Harry knows, is not yet back.[1] He gets M900 this year, whereas I who started the post had nothing! So gehts im Welt!

[1]On Gustav Ludwig Lange see 5.143.

6.44 To Parents, Cambridge, 13 January 1887

(Two paragraphs of gossip have been omitted.)

Here I am once more. I was away a day over four weeks.

I am on the whole glad I went to Germany. I saw a great deal of Wundt and something of other philosophical people,[1] looked at what was going on in the laboratory etc. Then I had the music and pictures. The skating was also welcome. And I was glad to be at Leipzig once more before the "Wanderjahre" are gone.

I spent (not including a few purchases) about $125. It would not have cost much less than $75 if I had stopped here, spending in few days in London. My finances are in a bad way. I owe $350 here—$150, the cost of the furniture of the room, I get back mostly, but not until next winter.

[1]In addition to meeting Georg Elias Müller at Göttingen (see 6.39), JMC had visited Berlin and met Friedrich Paulsen (*EoP* 6: 60–61), a distinguished philosopher and theorist of pedagogy (JMC to parents, 12 January 1887).

6.45 From Elizabeth Cattell, Philadelphia, 14 January 1887

(Several paragraphs of gossip have been omitted.)

We received eight letters from you last Monday, and your proof on "The Association of Ideas",[1] I read it with a great deal of interest, and I hope you will send us a few copies of "Mind". Your Papa left us to go to Baltimore this morning, he will remain there until next Monday working up the cause of Releif, he has not been quite so well this week, he has been working too hard, I want him upon his return to run off and take a little rest, he has been steadily at work ever since our return from Europe. By legacies in the last month from different sources the Board of Releif has been left $12000. It pleases your Papa very much.

Your Papa has been working on our accounts for 1886. To think we had an income of $12000.[2] From January 1885 to Jan. 1886 you drew $2100, but I remember you said you had loaned Berger some, and that you were obliged to place at the University $250. Our trip (Harrys included) across to the other side, came to about $2500, ($1500 more than we would have spent if we had remained on this side of the Atlantic), but I think it was well spent, as it did us a great deal of good. I think we have saved up about $2500, something new for us, though in 1885 we put a great deal in furniture. Mr Drake & Mr Fisler is to give us $1000, for the ground in front of our old home.[3] I fear this year our income will not be so large, as I think we will not receive our January divedend from Warren Foundry. This year our notes will be finished up, and I fear they do not intend giving us any more. It always seemed so nice to have them laid aside for emergencies, but I feel with you very thankful, that while you boys needed it for your education, we had full, and plenty.

[1] See 6.41.
[2] In 1886 the average annual earnings of an American nonfarm worker was $453.00. The Consumer Price Index (as calculated by the U.S. Bureau of Labor Statistics, with 1967 taken as 100) was 27 in 1886 and 225.4 in October 1979. Hence, the Cattells' $12,000 income in 1886 had the purchasing power equal to that of about $44,000 in 1967 and over $100,000 in 1979. (U.S. Bureau of the Census, *Historical Statistics of the United States, Colonial Times to 1970* [Washington,D.C.: Government Printing Office, 1975], vol. 1, pp. 165, 211; *Monthly Labor Review* 103, no. 1, [January 1980]: 85).
[3] In February 1887 Howard Drake donated this plot of 0.195 acres to Lafayette College (Skillman, *Biography of a College*, vol. 1, insert map). On Samuel Fisler see 5.154.

6.46 To Parents, Cambridge, 18 January 1887

I paid today my college bill it was:

Steward* 15 0 7

(*Includes besides dinner, "establishment charges", "univ. capitation tax", "gas", "water", "shoe cleaning", ect. the separate items are not on the bill.[1]

Kitchen Act.	———		
Milkman	1	4	2
Bedmaker	2	13	4
Laundress	1	16	
Rent	7		
Coals	1	4	1
Tuition	1	10	
Private Tutor	———		
Public Lectures	6	11	
College Adm[n].	2	3	
Surgeon	———		
Upholsterer	10		
Valuation of Rooms	———		
Repairs of Rooms	———		
	———		
	39 – 12 – 2		

That would be at the rate of $600 for college expenses—not including them all. Oxford and Cambridge are as I have often said expensive places—intended chiefly for rich men's sons. They have recently instituted non-collegiate students, as such one can get along cheaper, but not very pleasantly. £250 is considered a fair sum for a student, but by no means a liberal allowance. You see how this is for me for example

College Bills	*$600*
Lunch, Breakfast etc.	*100*
Other Univ. Expenses	*100*
Cloathing	*200*
Furnature & Repairs	*50*
Personal	*200*
	1250

[1]This note was written along the side of the page.

The above estimates are not especially large. Under "other univ expenses" are such items as reading-room & club $20 subscription to athletic clubs $21 fees to servants $25 ect. Food besides dinner & milk must cost about $100 (perhaps somewhat less) although I eat no meat, and do not even have cream with my oat-meal. Cloathing costs $200 even over here Three suits 75 overcoat 25 boating tennis etc things $35 cap, gown & surplice $20, 4 pairs shoes 20 tennis & boating shoes 10. Flannel shirts 10 Hat 10 Gloves & cravets 10 undercloathing 15 = $235.[2] I had to pay $150 for furniture (which is reestimated when I leave) also to have the floor stained, put in bookshelves etc. also linen, dishes etc. Add to this deterioration of new furniture I have bought and I dare say it will make $50 & more. $200 is not extravagant for personal expenses. I go up and down to town, must pay for swimming, boating, tennis, fencing etc, go to concerts etc. lamps postage, etc. Men who come to see me must be given cigars costing 12 cts each and tea (fortunately not wine), some little must be given away etc. So the $200 would only with strict economy include such items as dentist $60 boxes from Leipzig here $25 etc. The college bill may be slightly lower next time the £2.3 admission I suppose is once for all: and I shall probably hear fewer lectures. Still I fear my expenses will be about $1250—which does not include the vacation. This would cost perhaps $50 in addition stopping here or $100 travelling, but then the estimate for personal expenses $16 a month, becomes too small. Thus if I go to the continent at Easter & to Scotland in July my expenses will be about $1550 to which must be added $100 to get home with. You accused me in a recent letter (wrongly) of not looking matters financial fairly in the face, but I always do this however unpleasant.

I am sorry I should cost you so much this year. It troubles me however less in so far as I hope soon to earn my own living. I could have gotten along at Leipzig on $500 less (and should have received $275); still I am glad I came here. If I do not marry, I dare say, $500 will someday not be to me such a very important matter, and I shall thank you for having given it to me, when it was of the most use.

[2]The total is actually $230.00.

6.47 To Parents, Cambridge, 19 January 1887

I must put an adenda to my financial letter of yesterday—to the effect that the estimate there given was for current expenses, not including anything I have bought or may buy. I have spent a good deal on furniture (sofa, carpeting, chest of drawers etc) I needed the things here, but of course bought them for home, being glad to be able to get them in free of duty. I have also bought a few things in the way of art or quasi-art, and should like to buy more. Suppose I must give two wedding presents a year—I can get for $5 or 10 here what would cost $10 or 20 at home. Then I have bought books, and should buy more. I shall have to pay besides postage etc 25% duty on all books I get from Europe. I have had to buy a number of expensive textbooks, Ferrein's anatomy, Heath's id, Foster's Physiology, Ferrier's Functions of the Brain, Volkmann's Psychologie, Sidgwick's Ethics, Gurney's Sound etc.[1] Then I have bought some editions of the poets—these latter I shall take as a Xmas present from you, or with more I should like to have, with the money from Grandfather's estate.

I may also want to buy some apparatus. Then I have advanced $150 for the furniture & fixtures of my room & shall not get this back until next winter. I have also deposited $125 caution money, but suppose I get this back when I leave.

It got suddenly warm yesterday. I skated on water, but today even that is not possible.

[1] "Ferrein's anatomy": *Cours de medecine pratique redige d'apres les principes de M. Ferrein par M. Arnault de Nobleville*, based upon notes taken by students attending the anatomy lectures given by Antoine Ferrein at the Paris Faculty of Medicine and first published in 1769 by Louis Daniel Arnault de Nobleville (*DSB* 4: 589–590). The transcription of "Ferrein's" is uncertain. "Heath's id": Christopher Heath, *Practical Anatomy: A Manual of Dissections* (London: J. Churchill, 1864; 6th edition, revised by Rickman J. Godlee, 1885). "Foster's Physiology": Michael Foster, *A Textbook of Physiology* (London: Macmillan, 1877; 4th edition, 1884). On Foster see 5.156. "Ferrier's Functions of the Brain": David Ferrier, *The Functions of the Brain* (London: Smith, Elder, 1876; 2nd edition, 1886). On Ferrier see 5.110. "Volkmann's Psychologie": Wilhelm Fridolin Volkmann, *Lehrbuch der Psychologie von Standpunkte des philosophischen Realismus und nach gene-tischer Methode* (Halle: 1856; 3rd edition, Cöthen: O. Schulze, 1884–1885), the standard German text in psychology before the publication of Wundt's books (*Pre-1956 Imprints* 641: 487). "Sidgwick's Ethics": Henry Sidgwick, *The Methods of Ethics* (London: Macmillan, 1874; 3rd edition, 1884). On Sidgwick see introduction to section 6 and 6.21. "Gurney's Sound": Edmund Gurney, *The Power of Sound* (London: Smith, Elder, 1880). Gurney was an independent scholar best known for his work in psychical research (*DNB* 8: 799–801).

6.48 To Parents, Cambridge, 24 January 1887

You told me last summer they might possibly make me lecturer at Bryn Mawr.[1] I should, I suppose, have accepted such a position, albeit with no great satisfaction. Now however things are different and if I had the option I should be willing to lecture or teach there say twice a week. I think it would be good training for me as I have so little to do at the university, and I should like to earn more money. I cannot support myself comfortably I mean buy books, apparatus, travels etc. on what the university gives me, even though I get $300 from Bailey,[2] and live at home at your expense. But I really should prefer to teach or lecture more than I shall have to at the university.

[1] See 5.154.

[2] JMC had apparently invested some money with his classmate Morton S. Bailey in Colorado (see 2.51).

6.49 To Parents, Cambridge, 25 January 1887

(One paragraph of gossip has been omitted.)

It has often occurred to me since receiving the appointment at Phila. that the more time I spend in doing over here what I could not at home, the better. If it had not been for this I should have stayed here Xmas, and should only have proposed taking as much vacation at Easter as seemed desirable for health. But now much is changed. There is no more reason—on the whole less—for working hard now than at any other time in my life. If I did not have a position I should feel it desirable to work as hard and publish as much as possible, in order to attract attention to myself and get either a place in America or a fellowship here.[1] Now I feel that if I have any ideas they can as well be worked out and published next winter as now. From a worldly point of view I suppose this would be the course most likely to promote my personal interests at the university.

So I have proposed all along to go to the Continent at Easter—from the languages, art, institutions etc there is so much to learn and enjoy—all beyond my reach when I return to America. But it has recently occurred to me that it might be wise to leave Cambridge altogether at Easter. My first term was worth by far the most to me, but after this they become of less and less value. I do not mean that it becomes a less comfortable place to live in, or a less satisfactory place for work, but these are about what I have at home. One gets pretty much what there is in a man in six months—I do not think I have much more to learn from Ward, Sidgwick, Foster etc. I suppose I need not explain what I mean by this, though it might be misunderstood.

[1] JMC would have loved being appointed a fellow at St. John's, and regularly hinted as much to his parents. For example, a few days earlier, he had written to his parents about William Ritchie Sorley, a Trinity fellow then lecturing in philosophy at Cambridge who was later to succeed Sidgwick as Knightbridge Professor of Moral Philosophy (*DNB*, 1931–1940: 827–828; *EoP* 7: 499–500): "He did something that I had thought of. After having completed his studies at Edinburgh, he came here and enterred as a Freshman and so got a Fellowship" (JMC to parents, 23 January 1887). See also 6.11, 6.14, and 6.73.

6.50 To Parents, Cambridge, 26 January 1887

I am only hearing seven lectures a week this term. There are not many one cares to attend. Out of the seven three are on political economy, which does not concern me very nearly.[1] Two by Foxwell who is lecturer and fellow of this college and professor of political economy in London,[2] and one by Sidgwick. Then I hear a lecture on

[1] Political economy was an important part of the moral sciences curriculum (see introduction to section 6).

[2] Herbert Somerton Foxwell (*DNB*, 1931–1940: 293–295).

Ethics by Sidgwick and three a week on psychology by Ward; these last are the only ones I really care about.

Forsyth a fellow in mathematics at Trinity dined with me this evening.[3] He is I imagine the youngest F.R.S.

[3]On Andrew Russell Forsyth see 6.13.

6.51 To Parents, Cambridge, 27 January 1887

I am sorry I drew so much money last year—it is more than I had thought—though I knew it had been an expensive year. Besides Berger, whom I have paid a great deal during the past two years, and in addition lent $295, the Ph.D with printing cost about $200. The trip to Italy in the Spring, to England in June, and to Norway were all expensive. I have further spent about $100 in books, $200 in furniture, and $100 in other things of permanent value.

As I have said several times I do not consider the money I have spent over here wasted or even quite gone, but rather as invested in myself. From a purely financial point of view I imagine that if I were going to practice law or medicine what I have learned over here (beyond what I should have learned at home) would in the course of my life be worth to me $100 000. And the gain is none the less real because I do not propose taking up a money-making profession. Even as it is I shall probably earn in any given decade—say from forty to fifty—as much owing to my experience as I have spent in gaining it.

If needs be, I think I can economize without suffering, but I appreciate the advantages from money. Living, even on its economical side, is not such a simple matter. If I proposed trying to get a fellowship here, I am sure the money I have spent in furnishing my room prettily, would be a sound, indeed a brilliant investment.

If I only have myself to look after, I think I shall never need to worry about money matters.

6.52 To Parents, London, 29 January 1887

I came here this morning.[1] I lunched with Hogg, whom I recently mentioned as being a master at Christ's Hospital. Hodgson who lives with him is one of the chief workers in the Society for Psychical Research[2]—you will probably have a chance to

[1]JMC was in London to attend a meeting of the Psychological Club (see also 6.25 and JMC to parents, 28 January 1887).
[2]On Robert Wallace Hogg see 6.36. Christ's Hospital was a boy's school in London founded by Edward VI for orphans and the sons of poor families (Baedeker, *London*, pp. 91–92). Richard Hodgson was an Australian-born lawyer who had studied at St. John's, where, under Sidgwick's influence, he had become interested in psychical research (Alan Gauld, *The Founders of Psychical Research* [New York: Schocken, 1968], pp. 202–203). The Society for Psychical Research was one of the more formal such groups to form in London in the late nineteenth century (see introduction to section 6).

see him, as he is going to America to look after affairs psychical.[3] *I called on Dr. de Watteville editor of Brain,*[4] *and went to Prof. Robertson's where I dined with Prof. Bain. The best psychological people here attend the meeting, so it is pleasant for me to be present and see them.*

[3] In 1887 the financial and other affairs of the American Society for Psychical Research were in chaos. Hodgson went to America to straighten out this situation, and stayed until his death to manage what became the American branch of the Society for Psychical Research. (R. Laurence Moore, *In Search of White Crows: Spiritualism, Parapsychology, and American Culture* [New York: Oxford University Press, 1977], pp. 144, 147, 175.)
[4] On Armand de Watteville see 5.110.

6.53 To Parents, Cambridge, 31 January 1887

(Several paragraphs of gossip have been omitted.)

I did not write last night—I could not conveniently—in fact forgot about it while it was possible. I went to see Muirhead late in the evening, and stopped the night with him.[1] *I had intended to return to Cambridge yesterday, but finally decided to stay over. I spent a good part of the day at Bethlehem and St. Thomas hospitals.*[2] *The former is Bedlam, the famous lunatic asylum. The director was very kind and offered to let me study there, which would be very useful to me. I think I should gain more by living in London next term than by stopping here. It is only six weeks long—all interest being given to the boat races. But on the whole I think the continent would be the most profitable and less expensive.*

[1] On James F. Muirhead see introduction to section 1.
[2] Edward Geoffrey O'Donoghue, *The Story of Bethlehem Hospital from its Founding in 1247* (London: T. Fisher Unwin, 1914); Pliny Earle, *A Visit to Thirteen Asulums for the Insane in Europe* (Philadelphia: J. Dobson, 1841); Eildh M. McInnes, *St. Thomas' Hospital* (London: George Allen and Unwin, 1963); Baedeker, *London*, pp. 286–288.

6.54 To Parents, Cambridge, 6 February 1887

Would it not be worth the while to bring to the notice of Dr. Pepper and possibly others of the powers that be at the university, the facts I mentioned in a recent letter. Namely that I have been asked to stay here on the grounds that I should probably be made fellow of this college and university lecturer on psychology.[1] *Also that if I had stayed at Leipzig I should receive $250, private room and appropriations for apparatus, with no other duty then to carry on psychophysic research.*[2] *I*

[1] JMC might reasonably have hoped for an eventual appointment as a Cambridge fellow (see 6.49), but there is little evidence that he had been promised one or that it had been suggested that he stay at Cambridge to earn one (however, see 6.79).
[2] See 6.43. JMC started this sentence with "Perhaps," but changed it to "Also that."

naturally want a salary from the university large enough to support me (I am willing to do more lecturing, teaching or even other university work) and should also like the titel of "Professor," and of philosophy or psychology.

I have been elected a member of the Neurological Society which is considered an honor.[3] They have asked me to take part in a discussion (afterwards to be printed) on the "Sense of Effort," but I shall not be able to do this.[4]

[3] The Neurological Society was organized in 1885 to support research. A few months after JMC wrote this letter the society assumed publication of *Brain*, which had been published independently for about ten years. (Armand de Watteville, "Editorial Note," *Brain* 10 [1888]: 1–4; *Yearbook of Scientific and Learned Societies—1890* [London: C. Griffin, 1890], p. 204).

[4] This discussion was to be a continuation of one on "The Muscular Sense," held 16 December 1886 and opened with a talk by H. Charlton Bastian, "The 'Muscular Sense;' Its Nature and Cortical Localisation" (*Brain* 10 [1887]: 1–89). (On Bastian see 5.123 and 5.142.) This talk had led to much disputa-

tion; the remarks of Ferrier (see 5.110) and Jackson (5.116) were especially critical (ibid., 89–118). Bastian's reply (ibid., 119–137) was equally disputatious, and a follow-up discussion could not be held until 21 May 1891, when Augustus Desire Waller, a physiologist at the University of London, presented "The Sense of Effort: An Objective Study" (*Brain* 14 [1891]: 179–249). This controversy is reviewed by E. G. Jones, "The Development of the 'Muscular Sense' Concept during the Nineteenth Century and the Work of H. Charlton Bastian," *Journal of the History of Medicine and Allied Sciences* 27 (1972): 298–311.

6.55 To Parents, Cambridge, 7 February 1887

(This letter was misdated "Mon. Feb. 6.")

Mr. Ward and I are thinking of preparing together a course of practical or experimental psychology to correspond with Huxley & Martin's Biology and Foster and Langley's Physiology[1]—short laboratory text books with which Harry will be acquainted. If we carry out our plan Ward will write chapters on Touch and Sight, and I on Hearing and the more purely psychological matters, Time of mental processes, Association of Ideas etc.[2]

Such a book would be useful in teaching and learning psychology, and would be of advantage to me from a personal point of view.

[1] Thomas Henry Huxley and H. Newell Martin, *A Course of Elementary Instruction in Practical Biology* (London: Macmillan, 1875; new edition, 1883); Michael Foster and John Newport Langley, *A Course of Elementary Instruction in Practical Physiology and Histology* (London: Macmillan, 1876; 5th edition, 1884). These books were part of the series "Macmillan's Manuals for Students" (*Pre-*

1956 Imprints 262: 290). Each was written by a junior scholar who later earned a major reputation (on Martin see introduction to section 2; on Langley see 6.1), in collaboration with a distinguished scientist (on Foster see 5.156).

[2] This project was never completed. Drafts of several chapters are among the Cattell papers. See also 6.156.

6.56 To Parents, Cambridge, 17 February 1887

I was at a party this evening at Mr. Horace Darwin's; there are three sons of Charles Darwin here, and they are about the nicest people in Cambridge; they all three have remarkably pretty and clever wives.[1] This Mr. Darwin (and I dare say the others also) has a house beautifully built and furnished. The leading university people were there tonight; there was dancing, but it was not taken part in by many. They were mostly leaving when I came away at 11.30.

It continues cold—last night the water in my bed room froze nearly an inch thick.

[1] Horace Darwin (*DNB*, 1922–1930: 238) was Charles Darwin's fifth son. Educated at Trinity College, Cambridge, and later apprenticed to an engineer, he was in 1887 president of the Cambridge Scientific Instrument Company. As early as 1886, this company was making sets of anthropometric apparatus, some of which were exhibited to the Anthropological Institute at the same meeting at which Galton discussed some of the instruments JMC was using at the time (February 1886) in Germany. (H. Darwin, "Exhibition of Anthropometric Instruments," *Journal of the Anthropological Institute* 16 [1887]: 9–11; Galton, "On Recent Designs for Anthropometric Instruments," ibid., 2–9.) The other sons of Charles Darwin at Cambridge were Francis, a fellow at Christ's College, and George Howard, Plumian Professor of astronomy and experimental philosophy (*DSB* 3: 581–584; *DNB*, 1912–1921: 144–146; *DNB*, 1922–1930: 237–238). Horace was married to Emma Cecilia Farrar, daughter of the first Lord Farrar; Francis to Ellen Crofts, who had placed in the second class in the 1879 moral sciences tripos and who was then lecturer in English literature at Newnham College; and George Howard to Maud DuPuy of Philadelphia (*DNB*, 1912–1921: 144–146; *DNB*, 1922–1930: 237–238; Mary Agnes Hamilton, *Newnham: An Informal Biography* [London: Faber and Faber, 1936], p. 111).

6.57 To Parents, Cambridge, 18 February 1887

I read tonight at the meeting of the Moral Sciences Club a paper on "The Time it Takes to Think." The same paper metamorphosed to "Recent Psychophysical Researches" I read at the Aristotelian Society next Monday. I got it written very comfortably, it being no great matter and I having plenty of time. I suppose I can write ten lectures next winter easily enough. I should like to know when I am to give them—once a week after the Xmas vacation.

The cold weather we have been having for the last two weeks has gone. I am sorry to say, now I fear we may look for rain and fog.

6.58 To Parents, London, 20 February 1887

I had to be here tomorrow, and so came on this afternoon in order to hear Mrs Bryant make an address before the Ethical Society at Toynbee Hall. I think I have mentioned Mrs Bryant the Ethical Society and Toynbee Hall, all of which are more or less interesting.[1] I feel that this last year I ought to see as much as possible of

[1] On Sophie Bryant see 6.26. The Ethical Society was another of the informal London intellectual groups; see introduction to section 6. On Toynbee Hall see 6.30.

things which will be out of my reach at home, sometimes I have energy to do this, sometimes not.

As I have said a night at this hotel is comfortable, costing with breakfast $1.25.[2]

[2] The Midland Grand Hotel, located at St. Pancras Station, was recommended by Baedeker (*London*, p. 6).

6.59 To Parents, London, 21 February 1887

(Two paragraphs of gossip have been omitted.)

I lunched with Hogg and Hodgson[1]*—two right good men—you will see the latter this winter.*

Then I was at the National Hospital for Paralitics and Epileptics for a clinic.[2] *I saw Drs Horseley and Beavor two of the leading men here—that is of the younger men.*[3]

Afterwards I was at a queer place. The Headquarters of the Woman's Suffrage Association.[4] *The Sec'y is very nice—I went, however, to meet a Mrs Costello, a number of whose friends I know. She is a Philadelphia Quakeress, married to a man rather prominent in English politics and social movements.*[5] *She is considered very clever and beautiful. I dine with them on Thursday.*

I had intended to come to town again on Wednesday to attend a meeting of the Neurological Society—I find now that the Messiah is to be given Wednesday night. I dont know which to attend. There is so much at London.

I dined with Muirhead, and then read my paper at the Aristotelian Society.[6] *They*

[1] Robert Wallace Hogg (see 6.36) and Richard Hodgson (see 6.52).

[2] This hospital, known as the National Hospital, Queens Square, since 1947, was established in 1860 by a group of London neurologists (Gordon Holmes, *The National Hospital, Queens Square, 1860–1948* [Edinburgh: Livingstone, 1954]; MacDonald Critchley, "The Beginnings of the National Hospital, Queens Square (1859–1860)," in *The Black Hole, and Other Essays* [London: Pitman, 1964], pp. 155–173).

[3] Victor Alexander Haden Horsley (*DNB*, 1912–1921: 270–271) was from 1886 a surgeon at the National Hospital for the Paralyzed and Epileptic. Charles Edward Beevor (*DNB*, 1901–1911: 126–127) practiced both at University College Hospital and at the National Hospital as a consultant neurologist.

[4] The London Society for Women's Suffrage, established in 1867 (Ray Strachey, "*The Cause:*" *A Short History of the Women's Movement in Great Britain* [1st edition, 1928; reprint edition, Port Washington, N.Y.: Kennikut, 1969], pp. 110–113, 267–269).

[5] Mary Logan Whitall Smith Costelloe, the niece of Mary Whitall Thomas, whom JMC had known in Baltimore (see 2.25), had studied at Smith and Cambridge and married Benjamin Francis Conn Costelloe, a lawyer who also wrote extensively about Roman Catholicism. In 1900, after Benjamin Costelloe's death, Mrs. Costelloe married the art critic Bernard Berenson. (Sylvia Sprigge, *Berenson: A Biography* [Boston: Houghton Mifflin, 1960], pp. 14, 19, 67, 104; *NAW* 3: 313–316.)

[6] See 6.57.

seemed to like it better than I had expected, and the discussion was kept up until nearly eleven. I was then with Prof. Dunstan in the President's rooms for a while,[7] and here I am.

[7]Wyndham Rowland Dunstan (*DNB*, 1941–1950: 227–228), a chemist lecturing at Oxford in the 1880s, was one of the founders of the Aristotelian Society.

6.60 To Parents, London, 7 March 1887

You see I am again writing from London. I heard lectures at Cambridge this morning and came on here this afternoon. The ride 1¼–1½ hrs is not especially irksome. I of course travel 3rd class. If I had plenty of money I should travel 1st; I do not like 2nd.

There was quite an interesting meeting at the Aristotelian. The paper was on Lotze;[1] Bosanquet who translated Lotze into English was present.[2] I myself took considerable part in the discussion.

[1]The paper was by Andrew Muter John Ogilvie, a graduate of University College, London, and an electrical engineer (*Who Was Who* [Br.], 1916–1928: 791).

[2]Bernard Bosanquet (*DNB*, 1922–1930: 91–93) was a scholar of independent means who in 1884 had published translations of the second editions of Lotze's *Logik* and *Metaphysics*.

6.61 To Parents, Cambridge, 12 March 1887

I lunched today with Miss Scott and dined at Prof. Sidgwick's. Miss Scott is very clever and nice, she is keeping house, Sir Rowland and Lady Wilson being away.[1] Prof. and Mrs Sidgwick are said to be the most interesting people in England.[2] It would take many words to describe them, and only lead to failure. Mrs. Sidgwick's brother has just been appointed Sec'y for Ireland the most important post in the cabinet.[3] Another brother, who died a little time since very young, was considered the ablest man of science in England.[4] There were other people there tonight, among them the Master of Emanuel.[5]

[1]Miss Scott, otherwise unidentified, but probably a student at Girton or Newnham, was living with the family of Rowland Knyvet Wilson (*Who Was Who* [Br.], 1916–1928: 1137), a fellow of King's College and university reader in Indian law (see JMC to parents, 9 December 1886).

[2]See also 6.21.

[3]Arthur James Balfour, an M.P. from 1874 and later (1902–1905) Prime Minister (*DNB*, 1922–1930: 41–56).

[4]Francis Maitland Balfour, who had published extensively in embryology (*DNB* 1: 970–972; Henry F. Osborn, "Francis Maitland Balfour," *Science* 2 [1883]: 299–301).

[5]The Reverend Samuel G. Phear was Master of Emmanuel College from 1871 through 1895 (*Who Was Who* [Br.], 1916–1928: 834).

6.62 To Parents, Cambridge, 13 March 1887

Muirhead has been spending the afternoon and evening with me.[1] He is showing me about Cambridge. He is going to Leipzig soon, but will return with his brother in about two months and they will then take a house in London and have several of their sisters staying with them. Two of their sisters are to be sent to an endowed school in which Mrs. Bryant (whom I have several times mentioned) is the chief teacher.[2] Muirhead will continue to edit Baedeker from London. A number of men, Macalister, Sorley, Foxwell etc were in my room this evening and stayed, till one.[3]

[1]James Fullerton Muirhead. The brother mentioned in the third sentence is Findlay Muirhead.
[2]See introduction to section 1.
[3]On Donald MacAlister see p. 215; on W. R. Sorley see 6.49; on Herbert S. Foxwell see 6.50.

6.63 To Parents, Cambridge, 14 March 1887

This will be my last letter from Cambridge for some time. I look to go to Chester tomorrow. I should rather go to Spain or Paris but think my present plan is on the whole the wisest. I intend to walk along the coast all the way from near Liverpool to Land's End.[1]

Muirhead and I prepared the Cambridge Baedeker this morning and he went back to London at two.[2] I hope you will see him in America someday.

I have been paying bills and making other arrangements to leave. I shall take no luggage except a knapsack. I wear knickerbockers and a flannel shirt. I am of course going alone, but may meet several friends who are to be in the same region.

[1]JMC accomplished this plan, walking through Wales and arriving at Land's End on 13 April 1887 (JMC to parents, 13 April 1887). He returned to Cambridge soon afterwards.
[2]During his walk JMC noted that "Muirhead gave me the proof sheets, and also maps, of the Baedeker he is preparing and this has been a great convenience" (JMC to parents, 9 April 1887). See also Muirhead to JMC, 19 April 1887 and 3 May 1887.

6.64 From William Cattell, Baltimore, 13 April 1887

(One paragraph of gossip has been omitted.)

The enclosed letter from Dr. Twining, one of the editors of the Independent, explains itself.[1] He thought you were at home: and I wrote to him you were in England: but he replied that they w^d wait for the "review" and asked me to forward the letter to you. I don't suppose the compensation will be much & as your name will not appear of course you wont get much credit for the work. However I shd really like to have you do this and make a hit—for Dr. Ward (the Ed^r in Chief) has expressed to me a very great interest in you and it is well for you to have such a friend.[2] In such reviews, three or four columns of the Independent, you of course "pad" the the articles by something of yr own in the subject as well as a ref. to what Prof. Ladd says or dont say.

But let me advise you as to one thing—be sure to emphasize what is good:— and let the "padding" be scholarly.

At any rate I suppose you will want to see the book—as it is the latest English publication in yr line, so I will have it sent to you. Please reply to Dr. T. at once.

[1] Kingsley Twining (*Who Was Who in America*, 1897–1942, p. 1261), a Congregational minister, was literary editor of *The Independent*, a weekly journal of liberal Congregationalism (Frank Luther Mott, *A History of American Magazines*, vol. 2 [Cambridge, Mass.: Harvard University Press, 1938], pp. 367–369). In the letter mentioned Twining asked JMC to review the first American volume on experimental psychology—George Trumbull Ladd, *Elements of Physiological Psychology: A Treatise of the Activities and Nature of the Mind from the Physical and Experimental Point of View* (New York: Scribner, 1887). JMC's review, "Ladd's Physiological Psychology," eventually appeared in the literature section of *The Independent* (39 [30 June 1887]: 9–10). Ladd, a graduate of Western Reserve College and Andover Theological Seminary, was in 1887 a professor of philosophy at Yale (see Eugene S. Mills, *George Trumbull Ladd: Pioneer American Psychologist* [Cleveland: Case Western Reserve University Press, 1969]).

[2] William Hayes Ward (*DAB* 19: 442–443) was in 1887 superintending editor of *The Independent*.

6.65 To Parents, Cambridge, 19 April 1887

I have been looking over the catalogue of the University with interest.[1] I am glad to have my name in it; except financially it is a satisfactory position and good opening. I think on the whole I am best off at home, and imagine I shall be able to do what I undertake. As I have written, both at Leipzig and here, I have had good openings.[2] At Leipzig I should have had a salary this year and excellent laboratory facilities. Here I could probably get a scholarship worth $500 a year, and before it expired a fellowship worth $1500 and at the same time a university lectureship with salary and

[1] The University of Pennsylvania. [2] See 6.54.

making the fellowship for life. But as I say, I shall be better off at home on your account and perhaps also for other reasons. I shall not look back with regret on these openings, but with satisfaction.

I have been chiefly interested in the course of the medical school. If I can get time I shall perhaps try to follow this, taking the degree at the usual time—3 years.[3] I know about half of what is required in each year, and can work faster than the average—so could perhaps keep up working 10 hours a week. This would leave me 10 hours for psycho-physical experiments and 10 hours (scarcely enough) for reading and writing psychology.

I hear different accounts of "my collegue" Fullerton.[4] He was studying at a Lutheran seminary when called to Phila, and has recently taken orders in the Episcopal Church, —although his views would scarcely seem to allow this—at least from my standpoint. He lectures at Bryn Mawr in the way I once suggested that I might do.[5]

[3] See 5.121.
[4] George S. Fullerton (see 2.17).
[5] See 5.154 and 6.48.

6.66 To Parents, Cambridge, 22 April 1887

I helped Mr. Ward lay out his tennis court this afternoon—he, like other fellows, does not play well—about like you, Papa. Mrs. Ward however plays right well—she is, you know, Prof. Martin's (Baltimore) sister.[1] Afterwards I called on Mrs Sidgwick who is very clever and interesting.[2]

Perhaps you had better send me a draft for $250, so that I get it by the end of this term, as my bill for this term together with things I look to buy will about use up my letter of credit. As I always say when writing about money matters, I am sorry to have spent so much during the past three years. It must however be remembered that I bring home with me about $2000 worth of things—they having cost about that, and being worth more than that to me, and of about that market value in America. If this total seems large to you I might analyse it by saying I have over 300 books worth about $500, pictures and works of art worth about $500 (the most valuable the drawing of the Sistine Madonna) apparatus worth $300, Furniture worth $400 (in America more) Clothing $500 (I have just counted that I have 15 suits and 20 pairs of shoes & slippers). That would be a total of $2000.[3]

[1] See 6.19.
[2] See 6.31.
[3] The actual total is $2,200.

6.67 To Parents, Cambridge, 24 April 1887

The lectures begin tomorrow—I look to hear more than last term. Two on psychology and two on Kant by Mr. Ward. Three by Prof. Sidgwick (Ethics). Two by Mr. Foxwell (Political Economy.[1] Two by Mr Archer-Hind (Plato) and three by Mr. Hart (Physics).[2] The latter is a review for the M.B. = American M.D examination, and being in this college will not cost me any thing. Indeed most of the lectures are not in the line of my present work, but I think I shall be repaid for attending them. I may not take the Plato. I shall only have ten hours a week for reading and writing on psychology. I crossed out experimenting[3] because I am doing something of the sort at one daily.

[1] On Herbert S. Foxwell see 6.50.
[2] Richard Dacre Archer-Hind, born R. D. Hodgson, was a fellow at Trinity College and a lecturer in Greek (*DNB*, 1901–1911: 49–50). Hart has not been identified. Where JMC wrote "physics" he meant "physic," the archaic word for medicine.
[3] In the preceding sentence JMC had first written "reading or experiment. . . ."

6.68 To Parents, Cambridge, 25 April 1887

(One paragraph of gossip has been omitted.)

I have written saying that I will write the review for the Independent.[1] Indeed I am glad to do so as it is in the direct line of my work, and I am glad to earn money. I don't however imagine it is a good thing to write too much anonomously for money. Some of the London papers pay $25 a column. This seems a good deal, but in the end one simply earns a comfortable living and does nothing else.

[1] See 6.64.

6.69 To Parents, Cambridge, 29 April 1887

A friend of mine Hogg is up and he and another man lunched with me.[1] As I had a good lunch I did not go in to dinner, thus saving the cost and time. My time seems so fully taken up that there is no chance for extra work. I have just received a letter from Prof. Robertson, asking me for a paper for "Mind" but I shall not be able to write it, at least not till next winter. I was also asked to take part in a discussion being carried on in "Brain" but had to decline this also. I have of course written nothing for next winter, nor do I look to, until I come home. I am taking it for granted, as I wrote you, that I shall not lecture until after Xmas.

[1] On Robert W. Hogg see 6.36.

6.70 To Parents, London, 2 May 1887

After hearing lectures at Cambridge this morning I came on here. I looked up Muirhead and we went together to the Academy, this being the opening day.[1] *The exhibition is considered better than usual—the most interesting picture is by an American, Sargeant.*[2] *Muirhead is a good man to see pictures with as he knows a good deal about them and has a correct taste—i.e. a taste agreeing with mine.*

Afterwards I dined with Mr. and Mrs. Costelloe, and she went with me to the meeting of the Aristotelian Society; a friend of her's, and mine, Alexander, reading a paper on Hegel's Rechts-philosophie.[3]

[1] The Royal Academy of Arts (see Baedeker, *London*, pp. 208—209).
[2] John Singer Sargent was an American por-

traitist (*Oxford Companion to Art*, p. 1041).
[3] On Mary Costelloe see 6.59; on Samuel Alexander see 6.13.

6.71 To Parents, Cambridge, 8 May 1887

It has been a glorious day—such as one scarcely finds elsewhere—an Italian day having a somewhat different charm. Spring has been coming on rapidly during the last few days; the trees are getting green, the fields are bright with flowers and the birds sing jubilantly. But spring is lovely everywhere, even in that combination of palace, poor-house, mad-house and hospital, which we call a "city".

The Fellows garden is wonderfully beautiful—the finest in Cambridge.[1] *It is called "the wilderness". It is quite a fit name, there is one fine piece of lawn; the rest is almost a bit of original forest, with immense trees, wild flowers etc. The garden is private; only fellows (including me) having keys.*

[1] Baedeker (*Great Britain*, p. 445) starred the grounds of St. John's and particularly noted

the Fellows' Garden—not surprising, as JMC helped prepare this description (see 6.63).

6.72 To Parents, Cambridge, 13 May 1887

(One paragraph of gossip has been omitted.)

I congratulate Harry heartily on his M.D. and more particularly on his prize in surgery.[1] *I should myself much rather be a surgeon than a general practitioner. The responsibility is perhaps not less for the former, but his business is much more evident and definite.*

[1] The nature of the prize is not known. Henry Ware Cattell was to have a distin-

guished career as a pathologist and a medical editor (*Science* 83 [13 March 1936]: 252).

I have a friend here, Strong fellow of Harvard, and holding (appointed to) a position in philosophy at Cornell,[2] his father is president of Syracuse Univ.[3] you may know him.

[2]Charles Augustus Strong (*DAB* Supplement 2: 638–640 was a graduate of Harvard who in 1886–1887 shared a Harvard Travelling Fellowship with George Santayana (*DAB* Supplement 5: 601–603), the Spanish-born philosopher who would later teach at Harvard. Earlier in 1887, Strong had written to JMC from Berlin for information as to whether he and Santayana might study at Cambridge later that year (Strong to JMC, 31 January 1887 and 16 February 1887). Eventually Strong and JMC would be colleagues at Columbia.

[3]Augustus Hopkins Strong (*DAB* 18: 142–143) was a Baptist minister who had studied at Yale, Rochester Theological Seminary, and Berlin. From 1872 through 1912 he was president of the Rochester Theological Seminary (not Syracuse University).

6.73 To Parents, Cambridge, 14 May 1887

I have been entertaining Strong most of the day. He is one of the best men I know. I invited a man Mackenzie to lunch with us,[1] and a number of men were in my room both this evening and last evening. I sometimes get tired of talking so much, but I learn a great deal both by hearing and in talking myself, as we talk much on philosophic and social questions. One's own thoughts become clearer through expressing them. I am sorry that I had so few friends at Leipzig (especially last winter) with similar tastes and knowledge.[2]

[1]John Stuart Mackenzie (*DNB*, 1931–1940: 578–579) was a student at Trinity College. In 1889 he was to win first class honors in the moral sciences tripos, and later he would be professor of logic and philosophy at University College, Cardiff. By enrolling at Trinity after studying at Edinburgh and Glasgow, Mackenzie followed a course JMC had mentioned as desirable (see 6.49).

[2]See 6.22 and 6.27. Mackenzie, too, later spoke highly of what he learned from his friends and acquaintances at Cambridge (*DNB*, 1931–1940: 578–579).

6.74 To Parents, Midland Grand Hotel, London, 16 May 1887

After hearing lectures at Cambridge, I came up to town at noon with Strong. I went to see Muirhead and we together visited the "American Exhibition" and "Wild West Show."[1] The latter at least is interesting; I believe it has been travelling in "the States".

[1]Annual exhibits of paintings by artists of various nationalities were held in London (Baedeker, *London*, p. 43). The "Wild West, Rocky Mountain and Prairie Exhibition," organized in 1883 by "Buffalo Bill" Cody (*DAB* 4: 260–261), toured Europe in 1887. A few days after seeing the show JMC wrote to Galton to ask, "could any [anthropometric] measurements be made on the North American Indians at the 'Wild West Show?'" (JMC to Galton, 25 May 1887, Galton papers).

This evening I was at the last Aristotelian Society meeting of the year—for which I have come up. Prof. Bain read the paper.[2] Then Strong, Sorley and I came to this Hotel. Strong goes to Paris tomorrow. Sorley is fellow of Trinity College and is at present doing Prof. Robertson's work here.[3]

[2] Bain's paper was entitled "The Ultimate Questions of Philosophy" (see "Notes and Correspondence," *Mind* 12 [1887]: 486).
[3] W. R. Sorley see 6.49.

6.75 To Parents, Cambridge, 17 May 1887

Back Again! This morning I went with Muirhead to Dulwich and we looked over the picture gallery for the new Baedeker.[1] This afternoon I was at Prof. Robertson's, who has just returned to the city. You know he has been very ill. He and Mrs Robertson are good friends to me.[2] Afterwards I dined with Mrs. Bryant and we went to the Psychological Club.[3] We are doing together an elaborate piece of work on the association of ideas.[4] When I come by the late train, I get back to my room by midnight. I seldom get to bed before this, but ought to, as I get up every morning at 7.30 exactly.

[1] On Dulwich see 1.3.
[2] See 5.111 and 6.16.
[3] See 6.26.

[4] This work was later published as "Mental Association Investigated by Experiment," *Mind* 14 (1889): 230–250.

6.76 To Parents, Cambridge, 24 May 1887

(One paragraph of gossip has been omitted.)

I am usually quite busy all day long. Getting up at 7.30 I must hurry to be ready for lecture at nine. Then I hear two or three lectures and go to the physical laboratory three days a week. The other days I have a couple of hours for reading (psychology) nearly the only time in the whole week. At about 1.30 I run three miles, and at lunch read the paper. This lasts till nearly three. Afterwards I can read a little but should not work hard. Then I am usually out for a while and dinner at seven fifteen comes before I know it. Dinner lasts till eight and combination room till nine. I then have three hours, but this is taken up about half the time by company. The other evenings I should like to read poetry and such like, but it is about the only time I have for extra work, such as a paper I have just finished and this book of Prof. Ladd's.[1] I am glad to find that the book is right good—better than I had expected. Did I tell you that I shall also review it for "Mind"?[2]

[1] The paper JMC had just written was "The Way We Read," a revision of a talk he had given at the Aristotelian Society on 21 February 1887 (see 6.77).

[2] See *Mind* 12 (1887): 583–589.

6.77 To Parents, Cambridge, 28 May 1887

You will be glad to hear that the "Nineteenth Century" will print a paper of mine. Mr. Knowles, the editor writes "the M.S. on "The Way We Read," which you so very kindly offer me for my Review, I gladly accept it."[1] And signs himself "Yours obliged and faithfully." He also makes the unusual request "I should feel most grateful to you if you could and would kindly expand it a little"![2] The Nineteenth Century is, as you know the best review in the world. The last number contained papers by Matthew Arnold, Gladstone etc. I suppose I shall be paid $5 a page, but my paper is very short.

The weather has not been so pleasant recently. It rains a good deal, and is very cold for this season. My fire is never out.

[1] James Thomas Knowles (*DNB*, 1901–1911: 407–409), an architect interested in philosophical questions, edited *The Nineteenth Century* as a journal focusing on topics of current interest.

[2] JMC had earlier presented this talk to the Aristotelian Society (see 6.76). On the fate of this paper in *The Nineteenth Century* see 6.131.

6.78 To Parents, Cambridge, 30 May 1887

(One paragraph of gossip has been omitted.)

I fear this letter will trouble you. I wish it had been possible for me to consult you on the matter, but I did not want to trouble you untill I had decided that it would be best for me to stay here until Xmas, and after that it was too late. The advantages of staying I give in the enclosed copy of a letter to Dr. Pepper. The only reason for not staying is the grief of being separated from you for four months longer. I shall come home if you want me or if the university authorities insist on it, but otherwise shall look forward to much gain from spending another term here.

I should look forward to working hard at Leipzig and here, with perhaps a month's walk in Tyrol. It would not be easy to overestimate what I have gained this year and I feel that even a few months more will be most useful to me.

The expense will not be a very serious matter, about $300 more than if I were at home. I can afford to pay this myself and know you would give it to me if it seemed for my good.

[The text of James Cattell's letter to Dr. Pepper follows.]

I write to ask a favor. I should be glad if you will grant me permission to stay in Cambridge until Christmas.

I understand that I am not expected to give the course of lectures until the winter term, and feel that I could prepare them better here than elsewhere. I have found this year to be the most profitable of my life, and think it would be well if I could stay here a little longer. I hear lectures in moral and physical sciences and work in the laboratories, and feel that I am learning the best methods of teaching and research. Personal intercourse with the men here and in London is also of great advantage to me. They want me to start a psycho-physical laboratory here in the Autumn and the experience I should gain from this would be of much service to me at Philadelphia. Mr. James Ward, probably the ablest living psychologist and I are planning the preparation together of a text book in psycho-physics. I have been asked to read papers before the Moral Sciences Club here, and the Aristotelian, Psychological, and Neurological Societies in London, and have promised articles for 'Mind' and 'The Nineteenth Century.' For these reasons I feel that one more term here would be very useful to me in my future works and think that the interests of the authorities of the University of Pennsylvania and my own are the same, namely that I should be as fit as possible for the post you have given me.

6.79 To Parents, Cambridge 31 May 1887

(Three paragraphs of gossip have been omitted.)

I trust you do not disapprove of my having decided to stay here another term. Yet I suppose you think it better that I should at last settle down. It is perhaps not to be avoided that we should look at things in a slightly different light, but we should probably agree that it is best for me to do as much good as possible and be as happy as possible in the world. I have become a different man altogether after having spent these years in Europe, from what I should have been if I had stayed at Easton like Russel Stewart or Dave Wagener.[1] Now wherever I am or whatever I do I am likely to lead—and I hope for good.

As I have written several times this has been the best year of my life, and I feel almost certain that it would add to my happiness and usefullness if I could stay here longer.

You perhaps fear that if I do not come home until Xmas, I shall not come then. I do not deny that I should like to spend another year here, but I shall not give up the position at Phila. and even if they were to give me leave of absence, I fear I could not afford it financially. Neither do I deny but that I sometimes think of the great advantages of having a fellowship here. It would be of the utmost advantage to me to take the Moral Science course here, and everyone seems to think there is but little doubt but that I should after this be given a fellowship.[2] That would be $500 a year for seven years, besides the prestige.

[1] Russell Stewart was an 1878 graduate of Lafayette who attended Columbia Law School and then settled in Easton to practice law (*Who Was Who in America*, 1897–1942, p. 1186.) On Wagener see 5.22.

[2] JMC apparently thought that completing the course and obtaining the degree would guarantee the fellowship. See 6.54.

6.80 To Parents, Cambridge, 6 June 1887

We had a garden party this afternoon—that is to say the fellows of the college gave such a party. Our "wilderness" is the best thing in Cambridge;[1] we also have the best court and the best combination room of all the colleges.

I have just finished the review for the Independent. It is to be hoped that they will like it. It took me about 20 hours to read the book and 10 hours to write the notice—but of course the time was not misspent. I received this evening a note from the Editor of Brain asking for a paper.[2]

I go to London tomorrow and to Germany next week.[3]

[1] See also 6.71.
[2] This was probably the request that led JMC to write the review "Recent Books on Physical Psychology," *Brain* 11 (1889): 263–266.

[3] In London JMC met with Galton and Robertson (JMC to parents, 8 June 1887; Galton to JMC, 3 June 1887).

6.81 To Parents, Cambridge, 9 June 1887

Work is now over here, and a week or ten days is spent in amusement. Afterwards people mostly go away, but return, many of them, for July or August—which I suppose is not a bad time to work.

I am rather troubled as to my own plans, —and I fear you are too. I am quite willing to do whatever you think best. On the one hand it seems best that I should return home to stay in August; on the other there are ardent advantages in staying here until Xmas. It would have been an immense loss to me, if I had not spent this year in England.

James Cattell stayed at Cambridge for a few days, and then on 13 June left for Leipzig with James Muirhead and George Stout. He immediately visited the Wundts; then, after a night at a hotel, he went looking for quarters (JMC to parents, 11 June 1887 and 15 June 1887).

6.82 To Parents, Leipzig, 16 June 1887

(Two paragraphs of gossip have been omitted.)

We have now lodgings "bei" Frau v. Janecka;[1] Harry will remember it as the familily with whom the Hanes lived. They have however now moved and are within a stones throw of the university, museum and theatre. I pay M 70 a month for two rooms—for the present Stout pays half of the study.

We have been at the laboratory today where much is going on to interest me. It is important that I should know the latest work—which may not be published for a year or two. It is very discouraging to do work which someone else has already done.

[1] Possibly Martha Jaenecka, who in 1901 lived at Elisenstrasse 28 (*Leipziger Adress-Buch für 1901* [Leipzig: Edelmann, 1901], p. 437).

6.83 To Parents, Leipzig, 19 June 1887

I have not called on any German acquaintances except the Wundts and Heinzes. As I wrote Prof. Wundt's little boy is very ill; the Heinzes were not at home.[1] There are still a great many American and English people here whom I used to know—not however many whom I knew well. I called on the Donaldsons this evening, and asked them to dine with me tomorrow.[2]

The weather has been wonderfully fine. Last night I walked around by the Grosse Eiche, once a favorite and almost daily walk, and this afternoon to Connewitz.[3]

[1] On JMC's relationships with the Wundts and the Heinzes see 5.106, 5.125, and 6.91.
[2] Henry Herbert Donaldson (*DAB* Supplement 2: 156–157; *DSB* 4: 160–161) was spending two years studying with European neurologists. His wife was Julia Desborough Vaux Donaldson.
[3] On the *Grosse Eiche* see 5.15n.1.

6.84 From William Cattell, Philadelphia, 19 June 1887

(Several paragraphs of gossip have been omitted.)

The postponement of yr return home was a great surprise to us. We had been eagerly looking forward to meeting our dear boy again in a few weeks and altho' the announcement of yr purpose is a great disappointm't to us we are both agreed that your decision is wise. I called on Dr Pepper and he is of the same opinion—that it is important for you & will be of advantage to the university for you to finish the work you have in hand at Cambridge, especially for you to be identified with the organization of the new Laboratory. So we have cheerfully decided that our "feelings" must not be allowed to interfere with what is for your best interests.

6.85 To Parents, Leipzig, 20 June 1887

(One paragraph of gossip has been omitted.)

We were in a boat and to Connewitz this evening with the Muirheads and four young ladies, Miss Owen, Miss Stewart and two Miss Whitefields.[1]

[1] All four young women were students at the Leipzig Conservatory: Agnes Stewart from Connecticut; Mabel and Margaret Whitfield from Ealing, England, and Josephine Owen from London (*Königliche Conservatorium*, pp. 53, 64). See also 6.87.

6.86 To Parents, Leipzig, 22 June 1887

(Two paragraphs of gossip have been omitted.)

There is not an immense deal to tell you. I try to read in the morning. I have read Ward's article "Psychology" in the Encyclopaedia Britannica[1] and am now reading a "Psychology" by Dewey.[2]

[1]Ward's long article in the ninth edition of The *Encyclopaedia Brittannica* ([Edinburgh: Adam and Charles Black, 1886], 20: 37–85) was his most systematic treatment of psychology before his last years, and was long

regarded as a major statement of his views (*EoP* 8: 277–278).
[2]New York: Harper, 1887. On JMC's relations with John Dewey see introduction to section 2 and postscript.

6.87 To Parents, Leipzig, 24 June 1887

I write to you as usual today, but shall not mail the letter for a couple of days, so that you may be sure of getting the other first.[1]

I scarcely know what more to tell you. It is not easy to describe any one, least of all the woman I love.[2] If there were things in which I do not think her perfect I should not tell you of them, and all the praise I could write, you would take as a matter of course. She looks both younger and smaller than she is—she must be 5ft 3in tall, perhaps a little more, and weighs 115 lbs—which is easy enough to write, but does not convey much. She was ill some when she was younger, having had scarlet fever three times, but is now in good health and very strong, at least has a immense deal of energy. We rowed down from Connewitz the other evening without stopping in 25 minutes—which Harry will tell you is very fast. Today we walked six or seven miles before dinner, she had a music lesson in the afternoon and we went to the theatre in the evening—it was Rienzi, the beginning of the Wagner cyclus—and she does not seem to be at all tired.

Now, what more shall I say? I wish most of all you could see her. Perhaps it will decide you to come over this summer—that would be altogether pleasant. You will be quite delighted with your daughter. I imagine you thought it possible that I might marry an actress, socialist, widow or such like—but every one who knows "Jo" is

[1]The other letter has not been found.
[2]Josephine Owen was a student at the Leipzig Conservatory when JMC met her on his birthday in 1886 (see 5.151). She had been born on 8 October 1865 in Sheffield, England, to Samuel Owen and Sarah Elizabeth Lofthouse Owen. The Owens were a middle-class English family of Welsh origin whose comfortable life was possible by investments in coal mining. They had little known appreciation for learning and art. However, Sarah Owen's brother, Samuel Hill Smith Lofthouse, had been educated at Oxford and trained as a lawyer, and his wife, Mary Forster Lofthouse, had been a

noted water colorist. The Lofthouses probably influenced Samuel Owen's decisions to allow his daughter to study music in Germany and to marry an American. By 1887 Josephine Owen's mother had been dead for some time, her elder sister Edith was involved in missionary work in the East End of London with the Salvation Army, her younger sister Grace lived with their father in Blackheath (a suburb of London), and her brother Ernest, the youngest child, attended school in Leipzig. (*DNB* 12: 72; Ernest Owen to Quinta Cattell Kessel, 11 August 1932.)

fond of her, and I am sure you will come to love her as much as if she were your own child. All her girl friends and the people she lives with are extremely fond of her. She lives in a nice family—the widow of a professor—Harry knows the son Dr. Bruns. The little girl keeps following her and looking at her all the time. She and her brother are very fond of each other. I imagine half the men I have seen with her have been in love with her.

You see I am trying to give objective rather than subjective opinion—the latter being under the circumstances, if not less true, at all events less likely to be accepted as accurate. By way of further objective fact I might say that before leaving England she passed a "Local Examination" and had studied Latin, Geometry etc. For the past two years she has been studying music here—and in America, at least, would be considered a very accomplished musician. She will give up continuous piano practicing—as we agree in thinking that it costs too much time—but will go on with her singing. I have not heard her sing—except in the choir at church—I imagine she has a good but not remarkable contralto voice. She has read a good deal—she understands and likes Shakespere, Goethe, Browning etc. She knows French as well as German. She likes to cook and sew. She designs her own frocks and dresses very prettily. She understands at once whatever I explain to her—she is anxious to help me in my work and will be quite able to. But it is not for these reasons that I love her—you will partly understand them when you see her, partly they will perhaps remain my secret.

6.88 To Parents, Leipzig, 25 June 1887

It is your birthday, Mama. I trust you are strong and happy, and are looking forward hopefully to the years to come. I think I can do more to add to your happiness now, than if I had come home to live unmarried. There is said to be a pleasure in looking back on past dangers, so I might tell you, that if I were to become an old bachelor, I should think it wisest to stay in Cambridge. In any case I should probably have killed myself sooner or later.[1]

I have loved Jo ever since I first met her—fourteen months ago.[2] I went away from Leipzig two weeks afterwards. It was after I knew her, that I planned to stay in Leipzig for the winter, instead of returning home as we had proposed. But I did not then think it possible to marry anyone, so decided it would be better to go to Cambridge. I came here Xmas to see her and since then we have written to each other. It was however only at about Easter that I made up my mind, if possible, to win her love and marry her. Since then I have been happier than ever before.

You may of course, and had better tell people of my engagement. I shall write to no one. I hope Cousin Jo[3] will be pleased with her namesake.

[1]See 1.16 and 2.12.
[2]See 5.151.

[3]Josephine Cattell Fithian Hitchcock (see 1.1).

I have given her a ring costing $100. Having had no experience, I do not know whether or not this is extravagent. However as she does not wear other rings, this one should be good.

I dined with her today and we went on the river.

6.89 To Parents, Leipzig, 26 June 1887

As I wrote you my plans are not changed by my engagement, as they were made with reference to it. If our marriage were put off for three years, and things could be arranged at Phila. it might be well to try and get the fellowship at Cambridge. In this case Jo would probably also study at Cambridge. I do not however know that there is sufficient reason for putting off our marriage. My salary at Phila. ought gradually to be increased—if not I must go elsewhere. You see I am fit to teach most things: there is no philosopher or psychologist of my age, with as good a reputation as mine, nor with as influencial friends as I have—so I have no fear of the future. Jo tells me she had a little money of her own—five or six thousand dollars. I should not in the least mind spending this and what I have at the rate of $1000 a year. It would last 10–12 years, and after that I should doubtless be having a good salary. There is also it seems a little money entailed to Jo at her father's death. I do not know whether or not he has money.

So it would be possible and perhaps wise for us to marry next year. In this case, though I want now more than ever to see you—I should not think it best to return home in August. Jo expects her father to take her to Tyrol. I should go with them, and we should then return to England, and I should work at Cambridge until Xmas. I should then come home and after doing the work at Phila., return to England and we should be married. I should of course be back for the Autumn term. You are hoping to come to Europe next summer after the Assembly, are you not?

But there is time enough to decide all this—of course Jo's wishes must be considered first of all.

6.90 To Parents, Leipzig, 28 June 1887

It seems natural to write to you mostly about my "betrothed," as she is mostly in my thoughts: still it is no new thing, as this has been pretty much the case for a year and more. But I think I have told you every thing I can well write. Jo's family is Wesleyian, but she does not like certain features in the Methodist church to which I also object. Here she attends the English service.[1] If we were living near you we should doubtless go to church with you. I, however, think the Presbyterian church lays

[1] See 3.11.

too much stress on theological dogmas, for which there are not grounds in the gospels.[2] *I also prefer the Church of England service, unless the clergyman is very different from the average.*

We were caught in the rain on the river today.

[2] This is JMC's first criticism of the Presbyterian Church in his letters to his parents.

6.91 To Parents, Leipzig, 30 June 1887

I called on Mrs. Wundt this morning. As I told you the little boy, Max, has had scarlet fever and she has been nursing him continually for six week. He is now getting well, but is still very weak. They have invited us—that is Jo, Stout and I to dinner on Sunday—this is very kind, as it is the first company they have had.

I had cabinet photographs taken of Jo yesterday.[1] *I will send one to you if they turn out well. Won't you please send me yours. I have none at all of you, Mama, and yours, Papa, I left at Cambridge. Has Harry a recent photograph? I should like a couple of copies of the "Independent" if they print my review.*[2] *I am now trying to review the book again for "Mind."*[3]

[1] See Figure 14. A cabinet photograph was the term for a smaller photograph, readily locked in a cabinet for safe keeping (*OED*).

[2] See 6.64.
[3] See 6.76.

6.92 To Parents, Leipzig, 2 July 1887

Mr. Owen is here, having crossed to see his daughter and, I suppose, more especially me. He is naturally somewhat anxious about his daughter marrying a man he had not seen—more especially owing to my being an American. He is, however, very nice and very fond of her. He does not quite like it that I did not ask his consent first, but that was out of the question, just as it was impossible to ask yours. He says he does not want to give his final consent until you have given yours, so that he may know that you will receive his daughter kindly.

He will stay here for a couple of days, and then return to England and in two weeks come back to Germany with his younger daughter and we shall all go to Tyrol. I shall return home on Aug. 20th or not as you think best.

6.93 To Parents, Leipzig, 3 July 1887

We were at church this morning. Tell Harry, Tuttiet has gone to Stutgart. Then we dined at Prof. Wundt's. There were there Dr. Lange, my successor, and Dr. & Mrs. Martius, he is privat-docent at Bonn and has come here for the semester to study under Prof. Wundt.[1]

Prof. Wundt is preparing a third edition of his psychology[2]—I should like to translate it into English[3]—but that is out of the question. Prof. Ladd's Physiological Psychology is scarcely as good as I might write.

[1]On Gustav Ludwig Lange see 5.143. Götz Martius did some work with Wundt on reaction times and was later appointed professor at Bonn (in 1896) and at Kiel (in 1905). See Martius, "Über die muskuläre Reaction und die Aufmerksamkeit," *Philosophische Studien* 6 (1888): 167–216; *Pre-1956 Imprints* 365: 577.
[2]The third edition of Wundt's *Grundzüge der physiologischen Psychologie* was published later in 1887, with a foreword dated September 1887.
[3]On JMC as a translator see 5.129. Wundt's book was not translated into English until 1904, when an English version of one volume of the fifth German edition was published by Edward Bradford Titchener. See postscript and Eleanore Wundt, *Wilhelm Wundts Werke* (Munich: C. H. Beck, 1927), p. 60.

6.94 To Parents, Leipzig, 4 July 1887

I am very glad, Papa, to receive your long and affectionate letter.[1] I shall write and have my passage transferred to the latter part of December; then I shall return to live at Phila., at least if things can be arranged so that I can be married in the Summer. I think I ought to live where I can do the most good—that is give as much happiness as possible to the world—including of course myself. At Cambridge I should have better opportunities and more encouragement for scientific work; in America I should have more influence if I saw fit to interest myself in education or politics. If I were unmarried, I think it would be wiser for me to stay in Cambridge. I could contribute as much to your happiness, probably, by coming to see you each year, as by living altogether with you. I should myself be more comfortable and in better condition to do scientific work.

As it is however, it seems wiser for me to return home—partly to be near you— partly because I should sooner have a professorship and proper salary. It is not unlikely that I should have to give up advanced scientific work: but I should of course teach and write.

You must not confuse my stay in Europe with that of floating Americans. I have been quite settled down—except with you and in Italy and on walking tours—I have travelled but little, and have never lived doing nothing. I am no more a "traveller" at Cambridge, England, than I should be at Cambridge, Mass.

Trusting that all things are working for good.

[1]The letter mentioned is 6.84.

6.95 To Parents, Leipzig, 7 July 1887

I worked hard again today, not getting dinner until 5, and finished my review for "Mind."[1] This evening we were at the Walküre the eighth of the Wagner Cyclus. It is very grand, with "Tristan und Isolde" and the "Meistersinger" reaching not only the highest place in opera, but perhaps the best that any art has attained.[2]

I have now no special work before me until I return to Cambridge—though while here I want to make some notes on the laboratory. When I go away I shall take some important book on psychology, probably Kant, and study it thoroughly. When I get back to Cambridge I shall have an immense deal to do.

[1] See 6.91.   [2] See introduction to section 3.

6.96 To Parents, Leipzig, 10 July 1887

(One paragraph of gossip has been omitted.)

We were at the English Church and afterwards called at Prof. Wundt's. The little boy still seems weak; they will soon go to the mountains. I dined with the rest at the Frau Professorin's; there was a big family—17. The Grossvater is over 80, and is still strong and talks well. Stout & I called on another old man yesterday, Prof. Fechner, 85 years old—but still able to think clearly—he had a paper in the last number of Wundt's "Studien."[1]

[1] On Fechner see 5.117. The paper mentioned is "Über die psychischen Mass principien und das Weberische Gesetz," *Philosophische Studien* 4 (1887): 161–230.

6.97 To Parents, Leipzig, 12 July 1887

(This letter was misdated "Wed. July 11th.")

I have received my bill from Cambridge—it is $180. I feel that under present circumstances I ought to try to economize, but I scarcely know where to begin. I am in no direction extravagant, but have become accustomed to live comfortably without thinking much about money.[1]

Just now at the end I have a lot to do. Today I have been mostly at the laboratory and at the machinists. I have made a list of all the apparatus in the laboratory, and gotten the prices. I have ordered two pieces of apparatus, one to cost about $75, the other $25. I am also having some charts of the brain etc. drawn for me. All this adds to my expenses—I hope however the university will ultimately buy my apparatus.[2]

[1] See also 6.3 and 6.45.
[2] See also 6.112. Many other Americans who studied with Wundt made sure to purchase apparatus for use in America before they left Germany (see George M. Stratton to George H. Howison, 23 June 1895, Howison papers).

On 14 July 1887, James Cattell, his fiancée, her father, and her brother left Leipzig for a trip through southern Germany and the Tyrol. Cattell anxiously awaited his parents' responses to the news of his engagement. One finally arrived from his mother a week later, and two weeks after that he heard from his father. Neither of these letters has been found, but Cattell's replies to them follow.

6.98 To Parents, Berchtesgaden, 21 July 1887

(One paragraph of gossip has been omitted.)

I am made very happy today by receiving your letter, Mama, anent my bethrothal. I knew you would be delighted when you see Jo, but could have easily understood, if before that you had been inclined to be cautious in expressing yourself. In fact, next to me, it is you who have best cause to be pleased. Jo seems to be a perfect daughter and sister—for the past week I have seen them together continually, and her father is nervous and not always reasonable.[1] I think you will come to love her, as you would have loved a daughter of your own.

[1] As Jo's sister later remembered, "our father and Jim never got on well together" (Grace Owen, "Memories of My Sister," copy owned by Marcel Kessel and Elizabeth Cattell Kessel, Storrs, Connecticut).

6.99 To Parents, Innsbruck, 3 August 1887

Your letter, Papa, gives both Jo and me great happiness. Now, as always, you and Mama have been most thoughtful and loving. I really think I have added very, very much to your future happiness; I myself shall probably be more what you would wish, than otherwise; Jo will, I am sure, become a real daughter to you; if there are little children it will be a great joy. Yet but few parents—not knowing her—would have written so kindly and affectionately.

I do not think I shall make any decision about the time of our marriage until I see you. Jo has $6000 of her own and her father tells me he will leave her about $20000.[1] I do not know what sort of an opening there will be for me at the university. I should want if possible to stay at Philadelphia, but if I cannot earn my living there I suppose I must go elsewhere. I do not think I need really worry about the future, as there seems to be no psychologist in America better than I.

Jo wishes me to send you her love.

[1] For a note on converting dollar value see 6.45.

6.100 To Parents, Partenkirchen, 6 August 1887

We drove hither this evening—about ten miles from Mittenwald, where we spent last night. This place is one of the grandest I know—it is on a broad and fertile plain, about which barren and very high mountains rise.

Our trip is coming to an end: Earnest's school reopens a week from Monday and Mr. Owen must be back in England by then—so we leave the mountains the middle of next week. I, too, shall go to Leipzig. As far as work is concerned I can do that to better advantage there than elsewhere; as I could make experiments in the laboratory, and the writing I must do—a paper on the Leipzig laboratory[1] and lectures on psycho-physics—can be done to best advantages at Leipzig. On the whole I should prefer a longer vacation, but if I spend all the afternoon out of doors, both at Leipzig and Cambridge, I may be as well off as if I take more holidays now. And I have plenty to do. The changed circumstances make it more important than it was before that I become known and earn money.

I am also glad to return to Leipzig on account of apparatus and drawings being made for me.

[1] The paper mentioned was presented to the Aristotelian Society on 21 November 1887 (*Proceedings of the Aristotelian Society for the Systematic Study of Philosophy* 1 [1891]: 75) and later published in *Mind* (13 [1888]: 37–51).

6.101 To Parents, Leipzig, 13 August 1887

I have made arrangements to take dinner and supper with Jo. Breakfast I shall have in my room. I drink coffee now for breakfast. I like both the taste and the effects, and suppose the harm is less than the pleasure and advantage.

I shall try to work as hard as I can in the morning. I must begin to prepare lectures for next winter; I should like also to write a paper for "Mind" and continue experimental research. I can perhaps find books in the libraries here, not to be had at Phila, or even at Cambridge.

6.102 Journal, Leipzig, 13 August 1887

(This is the last entry in the journal.)

It is ten months since I last wrote here. Much has happened. My seven "Wanderjahre" are now gone—the years of "Sturm und Drang". My life is set straight. There will still be much faltering and falling, but my way has become clear and is likely to be followed. It is true my passions and counter-passions are by no means quieted. I exult and despair, and cannot foretell the end. But forces are at work which will have their way. I scarcely know what to write on this last page which closes the account of these seven years. I was at Cambridge during session, with a number of excursions to London. Made many acquaintances & friends—some of them, I hope, for life. I was often very, very, miserable, but look on it as the best *year of my life. I rowed the first term & swam before breakfast all through the winter. I ran three miles a day during the third term. I worked on psychology. I heard lectures by Ward, Sidgwick MacAlister, Foster, Hill, Foxwell and Hart.[1] Wrote papers for Mind & Nineteenth Century.[2] My ideas became considerably cleared up. Xmas I came to Germany, was here at Leipzig mostly, but also at Dresden and Berlin.[3] Easter I took a long walk by the ocean washing the west of England—from Rhyl to Lands End, only descending the Rye in a boat and sailing from Chepstow to Minehead.[4] But the two great events of the year are yet to tell. I was appointed lecturer (Oct.) at the University of Pennsylvania[5] and have won Josephine Owen (Thurs. June 23rd) to be my wife.[6] It is thus that I have put away much and accepted much. This book is now to be closed and with it much of what it has told. I must still say: I know nothing, it is a child's dream, a madman's fancy, but I have accepted real life and its duty. I shall try to learn and teach, to make myself and others some little wiser and better. I want to know what is right and do it. I want to love all things, as I love her.*

[1] See 6.50 and 6.67.
[2] See 6.17 and 6.77.
[3] See 6.39–6.43.

[4] See 6.63.
[5] See 6.6 and 6.8–6.10.
[6] See 6.87 and 6.88.

6.103 To Parents, Leipzig, 14 August 1887

(One paragraph of gossip has been omitted.)

I have received the proof of my review for 'Mind' of Prof. Ladd's book.[1] It is probably better than the one for the Independent—Prof. Robertson praises it. Will you please send me three copies of the 'Independent' in question. They wrote saying a check would be sent me when the review was printed, but I have not received it. I shall be given $15 by "Mind"—as I think I have told you.

[1] See 6.76 and 6.64n.1.

6.104 To Parents, Leipzig, 15 August 1887

I have done today what I look to do usually during the rest of my stay here—indeed during the rest of my life. I got up at 7.30 and worked until about 11.30. Then I gave Jo a short fencing-lesson, and we worked together arranging experiments I have made on the association of ideas.[1] We dine at 1.30. Afterwards we talked a bit and she read for an hour to me. Then we went out to swim and row, spending about 4 hours out of doors. We have supper at 7.30, and afterwards read etc. We shall go to the theatre as often as good operas are given. I suppose on the whole it is easiest and wisest to live quite regularly.

[1] Josephine Owen Cattell would work with her husband on his experiments throughout their life together. The results of the work mentioned here were later published in JMC's paper with Sophie Bryant, "Mental Association Investigated By Experiment," *Mind* 14 (1889): 230–250.

6.105 To Parents, Leipzig, 26 August 1887

I received today some books which Stout sent me from Cambridge, so I can now go on comfortably with my work. I must try to get the article I have spoken of ready by the middle of Oct. so that it can appear in the Jan. "Mind." [1] *I must spend quite a good deal of time at the machinist's, as the two pieces of apparatus he is making for me are both somewhat complicated and expensive. The one I have invented, and on the other I am making certain improvements.* [2]

It has recently been clear and warm—I fear it may yet become unpleasantly hot. I stay here, you know, about a month longer—or rather less.

[1] The article on the Leipzig laboratory (see 6.100).

[2] The instruments mentioned are the gravity chronometer, which JMC had invented in fall 1883 (see 3.4) and later modified (see 3.10), and an improved version of the Hipp chronoscope (see figures 8, 10, and 15). JMC had Carl Krille (the Leipzig mechanic who made instruments for Wundt [see 5.8]) rewind the electromagnets of the chronoscope with wire coarser than that provided by the manufacturer (Peyer and Favarger of Neuchatel, Switzerland) to reduce its latent times of magnetization and demagnetization. These times had caused JMC trouble in the past (see 3.12). For a detailed discussion of these modifications that illustrates how JMC later used these two instruments at the University of Pennsylvania, see JMC and Charles S. Dolley, "On Reaction-times and the Velocity of the Nervous Impulse," *Proceedings of the National Academy of Sciences* 7 (1896): 393–415. See also Beatrice Edgell and W. Legge Symes, "The Wheatstone-Hipp Chronoscope: Its Adjustment, Accuracy, and Control," *British Journal of Psychology* 2 (1906): 58–88.

6.106 To Parents, Leipzig, 29 August 1887

Our life is quiet and pleasant. I try to work three hours, reading for the "Mind" paper, which is on the same subject as that on which I shall lecture next winter. We try to work three hours together making experiments, or getting such into order. While I work alone Jo practices singing and piano. Jo reads to me an hour after dinner, after which we go out. This evening we played whist for ¾ hour. I had given Jo several lessons, but this was the first time she had played—she learned quickly and plays better than most women who have played all their lives.

6.107 To Parents, Leipzig, 7 September 1887

I am very glad to receive, Mama, your letter of 23ʳᵈ—it is all the more welcome for being longer than usual. You speak of money matter and then afterwards of my marriage. [1] *The two taken together give me considerable cause for thought. I should like best to be married next Summer—beside the natural reasons which make a year a long enough engagement there are two special reasons in our case. The one is our homes being so far apart, the other that I am not sure that she will be quite*

[1] In her letter of 23 August Elizabeth Cattell expressed concern about the future of the Warren Foundry and its dividends.

happy in hers. She having no mother makes a great deal of different. Then her older sister, although right nice, is queer. She spends all her time in church work and in looking after poor people, not attending at all to the house,—the servants run things. Mr. Owen being away a great deal, there are no regular meals etc. Then Mr. Owen himself, while he is very fond of his daughter, is somewhat thoughtless and selfish, and does not understand her very well. He has, for example, told her that he expects her now to pay for her clothes, music lessons, travelling expenses etc. with her own money—$200!

On the other hand we cannot marry unless we have money to live on. I suppose we could get on with very little the first year, but after that we ought to have $2000. I could perhaps earn a little by writing or private teaching, but it would not be well for me to do much of this. The best chance would seem to be to teach in one of the smaller colleges about Phila. or to spend half the year at some college further away—like Lehigh Univ.[2] But this is only a chance.

If we cannot afford to marry next summer, I am inclined to think that it would be best for me to return to Cambridge—at all events if the authorities of the Univ of Penna. would let me. If I stop there seven more terms after Xmas I am almost certain to get a fellowship.[3] This would be an honor and great help to me in future advancement—and would give me $750 for six years. If the Univ. of Penna. would give me $500–$1000 with the $500 we have it would be all right. If I could get a fellowship at Cambridge with $900 from Bailey[4] and $300 I might earn, I should have perhaps enough to pay for term expenses—and if I did not travel the vacations need not cost so very much. I should regret more than I can tell being two more years away from you, but I should at least be home from Jan 1st to Apr. 15th, and should hope that you would come over in the summer.

I simply write what is passing through my mind, and shall be glad if you will tell me what you think.

[2]JMC and his father had both mentioned the possibility of JMC lecturing at Lehigh (see 5.124 and 5.127)

[3]See 6.49.
[4]See 6.48.

6.108 To Parents, Leipzig, 10 September 1887

I went to Borsdorf, five miles out of Leipzig this afternoon, in order to see Lieb-knecht, the leader of the Social-Democratic party in Germany.[1] I spent a couple of hours with him right pleasantly; he was banished 13 years, has been put in prison a number of times, and is now not allowed to live in Leipzig—all for having views about the same as mine. He does not at all want to use force or disturb the present

[1]On Wilhelm Liebknecht see 5.36.

order of things—only advocates certain measures which nearly every one in England and America believes in—secrecy of the ballot, education of the lower classes, limitation of the hours of labor, etc.

He says hundreds of members of his party are in prison without trial, on the charge of conspiracy—it being called conspiracy to attempt to overthrow the present government. The party cast 700 000 votes in the last election, and is gaining rapidly.

6.109 To Parents, Leipzig, 15 September 1887

We had a long day together starting for Dresden before nine and not getting back until after two. I regret the expense but otherwise it has been a perfect day. We got to Dresden at 11, and stayed until 3 in the Gallery—I wonder when I shall see the Sistine Madonna again—perhaps never. We walked about a bit, had dinner and went to the theatre at 6. The Götterdämmerung was spendidly given—it is one of the greatest creations of genius. I imagine The "Ring des Niebelungen" is greater than the Parthenon, or the Sistine Chapel or Hamlet. I wish you could hear one of Wagner's later operas.

6.110 To Parents, Leipzig, 18 September 1887

I am glad, Papa, to receive your letter of two weeks ago.[1] I tried to count up on your birthday whether you were 59 or 60 years old, but could not decide. I hoped it was 59 and that I should be with you on your 60th birthday.[2] I shall be glad if when I come to be sixty years old, I can look back feeling that I have done as much good in the world as you have.

You speak of plans you make for Jo and me, wondering whether they will come to pass. I wish you would write me what they are. I am myself quite in doubt, but after I get back to Cambridge must a least provisionaly decide whether or not I shall be there longer than the one term.

At first I told Jo that we could not be married for three years; then after I learned that she had a little money and would perhaps be not quite happy at home I thought it would be better to be married next summer.[3] Now I am altogether in doubt.

We are both willing to be economical, but there are certain things to be got with money which it were a pity to be without; and circumstances might arise making more money than we should have almost necessary—if, for example, there should be a child and Jo should be ill. We should be thankful for as nice a wedding present as you can afford to give us (out of income) but cannot accept money from you to live on.

[1] Not found.
[2] William Cattell was sixty years old on 30 August 1887.

[3] See 6.99 and 6.107.

But even apart from money matters there are certain advantages in waiting. It would probably be better for Jo's health to be married at 24 than at 22, and both of us would get more fit for the responsibilities and cares before us. We are very happy now—we shall doubtless be equally happy afterwards—but then this peculiar sort of happiness cannot be recalled. Both of us have chances to study and learn which afterwards will be less. I fear you would dread my stopping two years longer at Cambridge. I too should deeply regret being separated from you. I should, how-ever, hope to spend both summers with you. The advantages of stopping for longer at Cambridge for my career as a psychologist are immense.

But please write to me, won't you, just what you think.

6.111 To Parents, Leipzig, 20 September 1887

Prof. Wundt is back from his vacation trip, and I called on him and Mrs. Wundt this morning. They asked us to take supper with them tomorrow. I am glad to see Prof. Wundt once more. My work is in the direct line of his, and just now especially I am writing an account of his laboratory and shall write reviews of his "Psychologie."[1] This latter is now ¾ printed, and I got today the advance sheets which are useful to me.

I met today an English friend Gerrans, fellow of Worcester College, Oxford.[2]

[1] On JMC's article about Wundt's laboratory see 6.100. The review appeared in *Mind* 13 (1888): 435–439.

[2] On Henry T. Gerrans see 5.75.

6.112 To Parents, Leipzig, 25 September 1887

(Two paragraphs of gossip have been omitted.)

I have not worked much during the summer. When I was here in June however I read some and wrote a long review for Mind.[1] Now I have fixed up some experi-ments and read and written some.[2] I am glad to have been here to look after my apparatus, which otherwise would have gone wrong.[3]

I have had to draw $100 on the new letter of credit. I have paid a bill amounting to $104 for books and music, and $75 to the machinist! I owe the latter more, he not knowing what his bill would be. I have three pieces of apparatus, two of them complicated and expensive. I have also, you know, paid $25 for drawings. I think the university should ultimately take the apparatus and drawings.[4]

Good-bye from Leipzig.

[1] See 6.91.
[2] See 6.97 and 6.104.

[3] See 6.105.
[4] See 6.97 and 6.105.

6.113 To Parents, Cambridge, 29 September 1887

(Two paragraphs of gossip have been omitted.)

Dr. MacAlister came back this afternoon. He is much pleased with your hospitality[1]—his friend has also written to the same effect—I am glad you were able to show him some attention as he has been very kind to me—it was chiefly through him that I was appointed fellow commoner.

You forwarded me a letter from Joe Lieper dated Dec. 24th '86![2] I have also received a note from Jastrow who was fellow at Baltimore after me, and a wedding invitation from Gould who was fellow in history when I was there.[3]

[1] Donald MacAlister (see p. 215) visited the United States during the summer of 1887 with a friend, probably Francis Gore Wallace, a noncollegiate student at Cambridge who would later practice medicine at St. Thomas' Hospital in London (Venn, *Alumni Cantabrigiensis*, vol. 6, p. 325). JMC twice wrote asking his parents to entertain his friends during their visit (31 July 1887 and 25 August 1887). On 26 September 1887 Elizabeth Cattell wrote to her son that "D[r] MacAlister . . . dined with us last Tuesday. We invited them to stay with us, but they were invited to a supper at D[r] Osler's to meet D[r] Pepper, and expected to start early Wednesday morning for Washington." (William Osler was a distinguished physician then at the University of Pennsylvania.)
[2] On Leiper see 5.22.
[3] On Jastrow see 2.41 and 6.10. (Jastrow was never a fellow at Johns Hopkins.) On Elgin R. L. Gould see 2.9.

6.114 To Parents, Cambridge, 1 October 1887

I have been working pretty hard this morning, as I must write my article for "Mind" this month and have more time now than after the lectures begin.[1]

Keynes, the university lecturer on moral sciences and secretary of the local lecture board, called on me this afternoon, and I called on Venn and Miss Stuart.[2] Dr. Venn, as you perhaps know, has written important works on logic.

Jo writes every day and very nice letters—she comes to England two weeks from today, when I shall of course meet her.

[1] See 6.100.
[2] John Neville Keynes was a fellow of Pembroke College and a distinguished scholar in political economy and logic (*IESS* 8: 376–377). On John Venn see 6.27.

6.115 To Parents, Cambridge, 2 October 1887

I have received a check for $15 for my review in "Mind"; it is about the first money I have earned by the "sweat of my brow."

I have agreed to read a short paper at the Aristotelian Society with several others on the question "Is mind synonomous with consciousness.¹ The other four are the leading members of the society all being fellows of Oxford.²

I took a walk around the "Madaly Ground" before dinner—I often do this on Sunday—it is about eight miles.³ On Sunday, you know, we dine at four. This evening there was service for the college servants—they look very well when they are dressed up, and there are a lot of them.

¹This symposium (*Proceedings of the Aristotelian Society* 1 [1888–1891]: 5–33, 75–76) was held on 5 December 1887, and JMC did not take part in it; see 6.137.
²Those who did take part in the symposium were Hodgson (see introduction to section 6), Ritchie (6.13), Bosanquet (6.60), and Alexander (6.13). Ritchie and Alexander were fellows at Oxford colleges in 1887, but Hodgson and Bosanquet were former fellows by that date. Stout (6.36) also took part in the program, perhaps substituting for JMC.
³Madingley was a "favourite point for the 'constitutionals' of university men" (Baedeker, *Great Britain*, p. 447).

6.116 To Parents, Cambridge, 5 October 1887

I have been busied this afternoon trying to find a place for a psychological laboratory. All the buildings are very crowded. Some of the colleges are rich, but the university itself is poor, and finds it expensive to house the laboratories and museums which have grown so rapidly during the past few years.¹ I suppose, however, we shall be able to get something. I dine with Ward tomorrow to talk it over.²

The days pass so quickly that I don't accomplish all I should wish. I fear my poor lectures will not get written this term. But I keep working at the subject so hope I shall be able to write them with ease.

¹See introduction to section 6.
²A day earlier Ward had written the following to JMC: "I saw Prof Thomson this morning. He tells me now that he has no room at the Cavendish which he could set apart for psychophysical work this term. It is the usual crowded term of the three and next term he thinks *there might be* a room which he could give up to us *temporarily*. It is, I fear, pretty clear that if a beginning is to be made it must be in some college—Trin. or Joh.—& not in the University." Ward was referring to Joseph John Thomson, fellow of Trinity and from 1884 Cavendish Professor of Experimental Physics, and to the Cavendish Laboratory of Experimental Physics, completed in 1873. (See *DSB* 13: 362–372; Tanner, *Historical Register*, pp. 232, 240; Romualdas Sviedrys, "The Rise of Physical Science at Victorian Cambridge," *Historical Studies in the Physical Sciences* 2 [1970]: 127–151.)

6.117 To Parents, Cambridge, 8 October 1887

This is Jo's birthday; she is, you know, 22 years old. She sent me today a letter from her Uncle, her mother's brother; she has one Aunt, her father's sister, but no other near relatives.[1] Her uncle is, I suppose, a good man; he studied here and is a barrister—which means more than a lawyer in America. He paints, I believe, well. He has a nice house on the Thames. His history is sad. He married a woman who to judge from her picture was very beautiful; she was a fine artist. She died soon after her marriage. The aunt is unmarried.

Yesterday I rowed, and today I played foot-ball, but not violently nor for long. I am trying to take good care of my health, have been taking plenty of exercise and eating and sleeping regularly.

Yesterday I saw the professor of physics, and have made arrangements, to start a laboratory in the building for physics. I shall set up apparatus to make original research, and look after any who wish to study the subject.[2]

[1] For more on Josephine Owen's family see 6.87.

[2] See 6.116, and Michael M. Sokal, "Psychology at Victorian Cambridge—The Unofficial Laboratory of 1887–1888," *Proceedings of the American Philosophical Society* 116 (1972): 145–147.

6.118 To Parents, Cambridge, 12 October 1887

Lectures began today—I look to hear quite a number, as it may be my last chance. Today I heard Ward on advanced and Stout on elementary psychology—advanced psychology is experimental psychology or psycho-physics. Tomorrow I shall hear Keynes on logic and probably some political economy lectures.[1] I may also hear lectures on mathematical physics.

I have been at the laboratory as much as possible—my apparatus came from Germany today, but not in very good order.[2] Fortunately there is perhaps the best apparatus maker in England at Cambridge—though expensive.[3]

[1] On John N. Keynes see 6.114.
[2] See 6.97 and 6.105.

[3] The Cambridge Scientific Instrument Company, owned by Horace Darwin (see 6.56).

6.119 To Parents, Cambridge, 13 October 1887

(This letter was misdated "Thurs. Oct. 14th.")

I am going to London tomorrow—there are several things I want to do. I take lunch with Prof. Robertson[1]—but of course I go chiefly in order to meet Jo.[2] I shall beside go to see her every second or at most third week.

[1] George Croom Robertson (see 5.111).

[2] On 11 October JMC had written to his parents that his fiancée would be returning to England on the 15th.

*Here, I am as much as possible in the physical laboratory. They have been ex-
tremely kind—Prof Thomson lets me work in his room, I may use university apparatus, of which there is a fine collection, there is a boy to wait on me etc.*

*I have been in Love's rooms tonight[3]—he having a brother 'up.' He is one of the
mathematical fellows.*

[3]On A. E. H. Love see 6.36.

6.120 To Parents, Cambridge, 17 October 1887

(One paragraph of gossip has been omitted.)

*I have in the end decided to hear lectures on psychology only this term. It is best to
think about and work on this chiefly, owing to the things I am writing and the
lectures I must deliver. Then I want to make some progress with experiments. I am
now looking on it as not unlikely that I shall have a further chance to hear these
lectures. You do not write advising me. Jo is very sweet and good about the matter,
only wanting to do what will be best for me. I of course in turn think chiefly of her.*

6.121 From Elizabeth Cattell, Philadelphia, 18 October 1887

(Several paragraphs of gossip have been omitted.)

*You seem to think we ought to have told you more about Bryn Mawr. I think the
letter I wrote you telling you, your Papa had called upon the President of the
College (I have forgotten his name)[1] must have miscarried.[2] Your Papa of course did
all he could, but the President wanted evidently to see you, and talk with you
himself, but I think from what your Papa said they will have you lecture there.[3] Your
Papa received a note from D[r] Pepper yesterday in which he says. "Referring to our
conversation last Spring about starting a laboratory, I write to say that I am prepared to join in getting up a small fund for that purpose. I had a letter from your
son asking continued leave until the New Year.[4] Of course I willingly consented, but
if he is to get any apparatus on the other side the time is drawing close." This looks
hopeful. Of course your Papa will call upon him immediately upon his return,[5] and
write to you.*

[1]James E. Rhoads (see 5.154, 6.48, and 6.65).
[2]It did not (see Elizabeth Cattell to JMC, 12
September 1887).
[3]JMC's contacts with Henry M. Thomas in
Baltimore (see 2.9) and with Mary Costelloe
in London (see 6.59) may have led either or

both of them to speak highly of him to their
sister and cousin Martha Carey Thomas (see
5.154), who was dean at Bryn Mawr.
[4]See 6.78.
[5]William Cattell was traveling through the
western United States.

6.122 To Parents, Cambridge, 19 October 1887

I received the sheets of Wundt's "Psychologie" as they are printed. In the part concerned with my work I am mentioned quite often—20 times I should think.[1] Indeed I think more account is taken of my work than of any one else's whatever.[2] I am to review the book for Mind.[3] I have copies of the other review,[4] which was more carefully written than the one for the Independent.[5] I shall send you several.

I am working tolerably hard—about four hours in the morning, three in the afternoon and two in the evening—but the laboratory is not especially trying and is a change from the rest.

[1] Probably an exaggeration, though JMC's work was well cited in Wundt's *Grundzüge der physiologischen Psychologie*, 3rd edition [Leipzig: Engelmann, 1887]; see for example vol. 2, pp. 308–311.
[2] Others cited include Hall (p. 314), Tischer (p. 309), and Kraeplin (p. 315). On Tischer and Kraeplin see 5.137.
[3] See 6.111.
[4] Of Ladd's *Elements of Physiological Psychology*, in *Mind* (see 6.76).
[5] See 6.64.

6.123 To Parents, Cambridge, 22 October 1887

(One paragraph on the weather has been omitted.)

I took a long walk at noon, but came back to go to the laboratory. There I am getting my apparatus into good order, They are very obliging giving me what apparatus I want, batteries etc. for all of which I had to pay at Leipzig.

6.124 To Parents, Cambridge, 24 October 1887

(One paragraph of gossip has been omitted.)

I am going to "coach" two "Girton girls" in psychology.[1] I think it will be a good thing for me to do this as one learns best by teaching. I receive 15sh. a lesson—the regular charge is 10-6, but at Girton, which is two miles away, they give enough in addition to pay for cab-fare.

[1] See 6.32. JMC wrote to his parents on 26 October that one of these unidentified young women had passed the moral sciences tripos the preceding year and the other expected to try soon.

6.125 To Parents, Cambridge, 25 October 1887

(Several paragraphs of gossip have been omitted.)

I have finished my paper on the Leipzig Laboratory, which I suppose will appear in the Jan. Mind.[1] It must still be copied, but I feel somewhat relieved, having been working at it off and on for two months. Still it does not make much matter, as work accumulates faster than it can be done.

[1] See 6.100 and 6.105.

6.126 To Parents, Cambridge, 27 October 1887

Mr. Mason, the "President" of this college is going to send you some papers or a book or something.[1] He lectures on Hebrew, but I imagine is not a great scholar. He offered to give me his paper, and I thanked him saying I would send them to you who knew about such matters, Papa. So he is going to send them himself, and you will have to answer and thank him. You had better write before you read them—you can also thank him for his kindness to me—he is a funny old man, excessively polite, but rather nice and very "good". He is not head of the college[2]— that is a Dr. Taylor[3]—the "master" of a college is considered a very high position & he receives $10000 has a beautiful house, and nothing whatever to do.[4]

I received this evening, Papa, a second copy of the Los Angeles paper,[5] but have not had a letter from you for a long time.

I have done nothing in particular today, was in my room this morning and at the laboratory in the afternoon. I am going to town on Saturday to see Jo.

[1] Peter Hamnett Mason had graduated from St. John's in 1849, was elected Fellow in 1854, and served as president from 1882 through 1902 (Venn, *Alumni Cantabrigiensis*, vol. 4, p. 352.

[2] "The President or Vice-Master . . . presides over the High Table in the Hall, acts as the Master's deputy in his absence, and sometimes discharges other functions" (Tanner, *Historical Register*, p. 12).

[3] Charles Taylor (*DNB*, 1901–1911: 480–482), a graduate of St. John's (1862) and a college

lecturer in theology, had played a major role in the reform of the college throughout the 1870s. He was elected master in 1881 and served until his death. In 1888–1889 he was also vice-chancellor of the university. (Miller, *Portrait of a College*, pp. 95–98.)

[4] This statement exemplifies JMC's opinions of college and university administrators other than his father. See postscript.

[5] William Cattell had been visiting California to raise funds for the Board of Relief (see 6.121).

6.127 To Parents, Cambridge, 2 November 1887

I have felt something relieved today having sent off my paper and got things in order at the laboratory. Yesterday I had the two Girton students there to see experiments on sound and music—there are only two others (beside them & me) in Mr.

Ward's lectures in advanced psychology. I do not think I shall ask the others to come to the laboratory as the term will so soon be over. It will be good for me to go over the experiments which can be made in psychology—I can do them again at home if there are students and apparatus.

6.128 From Elizabeth Cattell, Philadelphia, 8 November 1887

(Several paragraphs of gossip have been omitted.)

Papa did not see D[r] Pepper to day, but he sent word for him to call tomorrow morning. Your Papa called to see Mr. Fullerton this afternoon, he had a very nice talk with him. He told him <u>confidentially</u> that he intended giving up his position at Bryn Mawr,[1] where he had been teaching Philosophy, that he had told the President Mr Rhodes that you would be a good man he thought for the place,[2] he receives $1,000, but seems to think he would be willing to give more. Prof. Fullerton resigns the position at the end of the College year. He is living in West Phil[a], but he says the going & coming takes up too much of his time. Your Papa will write by next mail, and tell you what D[r] Pepper says about the apparatus.[3]

[1] See also 6.143.
[2] See 6.121 and 6.130.
[3] See 6.121 and 6.129.

6.129 From William Cattell, Philadelphia, 9 November 1887

(Two paragraphs of gossip have been omitted.)

I called by appointment on Dr. Pepper this morning: had hoped the Dr. w[d] be prepared with some definite proposition <u>anent</u> the apparatus, but it looks as if he wanted to <u>interest</u> me in raising this money!—of course in view of my natural interest in your success at the University—after our conference however, he said there w[d] be a meeting of "the Committee" within two or at the farthest three days; and he "hoped" to authorize you to expend five hundred Dollars for such apparatus as you w[d] judge to be most necessary for yr immediate work:[1] and this would be independent of the amount he hoped that I w[d] raise for the same object![2]

Upon the whole, I think it w[d] be well for me to subscribe 50 or perhaps 100 Dollars on condition that another 500 shall be secured for yr use. But you may as well regard the $500 as appropriated and give yr orders at once:—or of course "turn in" with the bills, apparatus you have on hand already paid for by yrself.[3] If we secure the other $500 you c[d] order by mail and finish it all up when you return to Europe next year.

[1] But see 6.132.
[2] See 6.176.
[3] See 6.121.

Yr mother wrote you last night about my interview with Prof. Fullerton who has shown himself very friendly to you. It looks as if from next fall you might count upon an income from yr salary at Bryn Mawr & at the university & yr own private means of two thousand Dollars & upon this you ought to live. Times have changed, of course, but for the first 6 years of <u>our</u> married life we lived on less than that!

6.130 To Parents, Cambridge, 10 November 1887

I received today a letter from the president of Bryn Mawr College, asking me to give a course of 18 or 19 lectures next winter and offering me $150. I have replied accepting, indeed I think it is rather an important opening, as between it and the Univ. of Penna. my whole time may ultimately be taken up with a reasonable salary. I should much rather lecture at Bryn Mawr than at N.Y. or elsewhere.[1]

[1]From 9 January 1888 through 29 March 1888, JMC lectured twice a week at Bryn Mawr to 13 students, one of the larger classes at the college. His twenty lectures on "physiological psychology" treated "the sense organs in connection with the sensations and perceptions received through them; also the localization of the brain functions . . . . time, space, the association of ideas, consciousness and memory [and] Mental time and other phenomena. . . ." (James E. Rhoads, *The President's Report to the Board of Trustees, Bryn Mawr College, 1887–1888*, pp, 23, 33).

6.131. To Parents, Cambridge, 15 November 1887

I have applied to the council of the university to let me count last year before I was matriculated. It came before them yesterday and though not yet passed I understand they are disposed to allow it. In this case I should look to try to get further leave of absense from Phila. next year, and get the degree here in '89, and hope to be married that summer. If things went favorably I should then for a while carry on work at both universities.

I am pleased to get tonight a letter from the editor of the "Nineteenth Century."[1] *I sent him a second MS. long ago, asking if he preferred it to the first.*[2] *He replied to Tyrol and I did not get his letter. Now he writes inter al. "I shall lose no time in sending you proofs of <u>both</u> your extremely interesting papers, which I shall hope to publish at the earliest possible date."*[3] *I shall be well paid for these, but that is the least of it; it is a great satisfaction and will be of lasting use to me to have papers in the "Nineteenth Century" which has an immense circulation and prestige.*

[1]James Thomas Knowles (see 6.77).
[2]The "second MS." was "The Time It Takes to Think," a short and popular abstract of his reaction-time experiments. Knowles had asked JMC to expand his first paper, "The Way We Read" (see 6.77).

[3]"The Way We Read" was never published, but "The Time It Takes to Think" appeared the following month in *The Nineteenth Century* 22 (1887): 827–830.

6.132 From Elizabeth Cattell, Philadelphia, 15 November 1887

(Several paragraphs of gossip have been omitted.)

Your Papa, and I have talked over again and again your remaining in England with view to the fellowship, but we think in every way it is undesirable. We were disappointed upon receiving Dr Peppers note, only appropriating $250 for apparatus, for he told your Papa he thought they would be willing to give you $500 but Harry says he is very apt to create great expectations.

6.133 To Parents, Cambridge, 23 November 1887

(One paragraph of gossip has been omitted.)

Jo and Mr. Owen have been here today and we have been of course together. We have been seeing the ordinary things, which I hope I shall be able to show you next summer. We called besides on a Mrs Homeden, who was Mr. Owen's bridesmaid, but whom he had not seen since her marriage many years ago.[1] Curiously enough they turn out to be MacAlister's best friends and knew all about me.[2] Mr. Owen dined this evening in Hall with Stout, and Jo and I got our own supper, it seemed quite like house-keeping and very nice. Through a complication of Stout's—he is alway making such—I had to invite several men to my rooms this evening—Foxwell, Lernier & Main.[3]

There is scarcely time now to order apparatus, but I may buy some in London or transfer some of mine. I shall, of course, be glad, Papa, if you can without undue effort collect an additional $250, and it may be best to subscribe yourself—though I hope it need not be more than $50.[4]

[1] Annie Harwood Holmden was the translator of several contemporary French and German books on the early Christian Church. Her husband, Samuel N. Holmden, was the superintendent of a Christian mission in the city of Cambridge. (Edith MacAlister, *Sir Donald MacAlister of Tarlbert* [London: Macmillan, 1935], pp. 24, 26, 31; *Pre-1956 Imprints* 252: 13–14.)

[2] The Holmdens regularly lent money to MacAlister when he was short of funds, and supported him emotionally when his mother died (MacAlister, *Sir Donald MacAlister*, p. 31).

[3] On Herbert Foxwell see 6.50. Lernier is unidentified. Philip Thomas Main was a fellow of St. John's from 1863 and, from 1869, superintendent of the college's laboratories. Like Sidgwick, he played a major role in women's education at Cambridge, lecturing on chemistry at Girton from 1873 and serving as treasurer of Newnham for eight years. (Venn, *Alumni Cantabrigiensis*, vol. 4, p. 292.)

[4] See 6.129.

6.134 To Parents, Cambridge, 24 November 1887

(This letter was misdated "Thurs. Nov. 23." One paragraph of gossip has been omitted.)

It is very nice having Jo here, and today she has not interfered at all with my work—on the contrary she has been helping me. She was with me in the laboratory this afternoon, where I had demonstrations, and this evening we have been working for three hours arranging association experiments.

6.135 To Parents, Cambridge, 30 November 1887

Jo and Mr. Owen have come again today, but have been mostly at the dentist's. Tonight we had a college feast. Stout invited Mr. Owen, and I stayed with Jo and we had supper together. I was at the laboratory this afternoon and later Jo came and helped me.

My paper on "The Time it takes to Think" is in the current number of "The Nineteenth Century."[1] I have no reprints—I think you can get copies for 25 cts. I shall be glad, Papa, if you will give my papers to people, especially at the Univ.

[1] See 6.131 and 6.77.

6.136 To Parents, Cambridge, 3 December 1887

I have written a number of letters today including one to Dr. Pepper telling him the subjects on which I propose to lecture.[1] I wish they were written. I shall not even have written by the time I return home the review of "Wundt", which must be done by Feb. 1st.[2]

My name has been a good deal in the papers in connection with the paper before the Aristotelian Society[3] & the article in the Nineteenth Century.[4] It might, perhaps,

[1] As lecturer on psychophysics at the University of Pennsylvania, JMC was to give ten lectures between 23 January and 26 March 1888. The titles were The Use of Experiment in Psychology; The Measurement of Mental Time; The Measurement of Mental Intensity; Perceptions of Sound; Perceptions of Light; Perceptions other than those of Sound and Light; Data used in the formation of our ideas of Space and Time; Experiments on Attention, Memory, and the Association of Ideas; Feeling and Volition; and The Correlation between Mind and Matter. (*The Pennsylvanian* 3 [1887–1888]: 239; *Report of the Provost of the University of Pennsylvania for the Two Years Ending October 1, 1889*, pp. 29–30.)

[2] See 6.111.

[3] "The Psychological Laboratory of Leipsic," which JMC presented on 21 November 1887 and which was later published in *Mind* (see 6.100).

[4] See 6.135.

be worth while to get something—say from the Athenaeum, copied into American papers.[5]

[5] The London *Athenaeum*, "a superior literary journal" (Baedeker, *London*, p. 17), carried at least three notes of JMC's Aristotelian Society talk: a four-line notice among the front-page advertisements in the 19 November issue, a ten-line abstract under the heading "Scien-tific Gossip" in the same issue (p. 679); and a 33-line report of the presentation in the 26 November issue (p. 718). The abstract and the report cited JMC's affiliation with the University of Pennsylvania.

6.137 To Parents, Cambridge, 6 December 1887

I was in town last night and did not write. When I attend a society meeting I do not get to Blackheath until after eleven and cannot well write then.[1]

Last night I went on to attend the meeting of the Aristotelian Society. It was a written debate by the leading members[2]*—I was at first going to take part in it, but retired when I agreed to read a paper.*[3]

I spent the night at Mr. Owen's—he himself being, however, in Brussels. Then I came back to Cambridge this morning, as I had an engagement to lunch with Dr. Venn.[4] *I go to town again on Thurs. in order to attend a meeting of the Neurological Society, where I may myself speak, as the proof of the paper on "Inhibition" has been sent me, with this in view.*[5] *I dislike to spend so much money in travelling, but otherwise it is pleasant and profitable.*

[1] Blackheath: The Owen's suburb.
[2] This symposium was entitled "Is Mind Synonomous with Consciousness?" (see 6.115).
[3] The paper on Wundt's laboratory that JMC had presented in November (see 6.100 and 6.115).
[4] John Venn (see 6.27).
[5] The paper, by Charles Arthur Mercier, longtime physician for mental diseases at Charing Cross Hospital, London (*Who Was Who* [Br.], 1916–1928, p. 724) was eventually published in *Brain* 11 (1889): 361–386. Among the other discussants of this paper was John Hughlings Jackson (see 5.116), whose comments along with JMC's were published in *Brain* 11 (1889): 386–405.

6.138 To Parents, Cambridge, 13 December 1887

I am glad to get your letter, Mama, and to learn that Papa seems nearly well.[1]

As to my plans, they are rather vague. I think you scarcely understand the great advantage it would be for me to stop here. The people I meet here are head and shoulders above those I should see at home, and the able men in America are widely scattered, whereas here they are all collected about London,[2] *Cambridge & Oxford being almost suburbs. If I teach enough at Univ of Penna, and Bryn Mawr to earn a living I should have no time whatever for original work. A little later that would make less matter—after my reputation was established and I was fully abreast with current philosophy & science—I might then devote five years to teaching—and afterward go back to writing and research. I think I might do good as a teacher and*

[1] The letter mentioned is 6.132.

[2] See 6.22, 6.27 and 6.73.

educator, but there are others who could do it better than I, whereas I seem well
fitted for the work I have begun. I do not mean to be selfish—if I were wholly
selfish I think I should kill myself,[3] as in my past life I have suffered more than I
have enjoyed and imagine it will be the same in the future. I want to do the most &
best I can for the world, and most of all for those I love. When I received the
appointment I accepted it and intended to come home for your sake. I had then no
definite idea of marrying. Since I am going to marry an English girl I must remem-
ber that her family is in England. Jo has said no word and is ready to leave all for
me, but I must consider her and to some extent her family, as well as you. Of
course it may in the end be best for Jo to go with me at once and permanently to
America; on the other hand it may not be best for you. So I shall settle nothing until
I see you and my opening at home. I want, however, to return here until Xmas next,
when, if things go very well, we might be married. So far as I have a further plan, it
would be to lecture here during the Autumn term and at home during the winter
and spring terms. I shall probably be asked to lecture here next Autumn,[4] and want
to prepare a text-book with Ward.[5]

There is every reason for staying here except for being apart from you and Harry
and the expense. If I get ultimately a fellowship the money spent will be paid back
to me and in a welcome way $700–$800 a year for six years.

[3] See 1.16.   [5] See 6.55.
[4] JMC was indeed asked to lecture at Cam-
bridge during fall 1888 (see 6.141).

James Cattell left for America sometime after 20 December 1887 and arrived
in Philadelphia on New Year's Day 1888. He spent the next three months
teaching at Bryn Mawr College and lecturing at the University of Pennsylvania. The lectures were apparently viewed as successful by William Pepper
and others at the university—including the student editors of *The Pennsylvanian*, who urged their classmates to "avail themselves of the opportunity of
becoming acquainted with one of the most interesting branches of philosophical science" (3 [1887–1888]: 209)—and would soon lead to Cattell's
appointment as professor.

Around 1 April, Cattell left America to return to England. He arrived on
8 April, and went to the Owen home in Blackheath before returning to
Cambridge.

6.139. To Parents, Blackheath, 9 April 1888

I have been in town again this evening—this time without Jo. I called at Prof.
Robertson's and spent an hour with him—he, poor man, was in bed.[1] I have "Mind"

[1] See 6.16.

and Prof. Ladd's reply to me—I don't see any use in replying to a review—I certainly shall not rereply to him—though I imagine I am right & Prof. Robertson fortunately agrees.[2]

Then this evening I was at a meeting of the Aristotelian Society.[3] This morning I went with Jo to the gymnasium but was not allowed to stay.

The weather is bad.

[2]JMC's review of Ladd's *Elements of Physiological Psychology* in *Mind* (see 6.76), though not especially critical, did raise some questions about Ladd's interactionist approach to the mind-body problem, and while avoiding dogma presented evidence in favor of the physicalistic perspective, which JMC had taken since early 1886 (see 5.123 and 5.142). These comments set off quite a debate. The review was published in October 1887 (*Mind* 12 [no. 48]: 583–589) and Ladd's reply in April 1888 (13 [no. 50]: 308–312). Stout replied to Ladd's reply in July 1888 (13 [no. 51]: 467–468; see also 6.145) and Ladd replied to Stout's comments in October 1888 (13 [no. 52]: 527–529). The issue was the mind-body problem, and it is clear that Cattell felt that debate on the question was futile.

[3]The focus of this meeting was the philosophy of Heraclitus (*Proceedings of the Aristotelian Society* 1 [1888–1891]: 83–84), which JMC had studied with G.S. Morris at Johns Hopkins (see 2.4).

6.140 To Parents, Cambridge, 12 April 1888

(Two paragraphs of gossip have been omitted.)

I don't know just what I shall do this term—I must write that review of Wundt first[1]—after that there is much to choose from. I shall, besides, take a good deal of out-of-door exercise, and shall be some in London.

I left Jo well. I think I did not tell you that she sends her love and best thank for the book and the "Melange."[2] She also sends her thanks through you to Jo for the candy.[3]

We did not say much about our marriage—it rests as before. She was willing to wait two years if it seemed best, and she is willing to leave her home for me when I wish. I spoke to her father, though I mentioned no special time. He would rather that I should have a larger income, but I think would not object to our marriage under the circumstances. He intends to write to you, Papa.

[1]See 6.111 and 6.136.
[2]The Lafayette College yearbook.

[3]On Josephine Cattell Fithian Hitchcock see 1.1.

6.141 To Parents, Cambridge, 24 April 1888

I have today seen Profs Foster and Sidgwick—I lunched with the latter—and have fully discussed and carefully considered the prospects here. I think it would be well for me to stay here until Xmas. I should be permitted, or asked, to lecture during the Autumn term, and it would always be a considerable satisfaction and advantage to have lectured at Cambridge.[1] Then it would help me to give my lectures under different conditions, and with the aid of things here, and further association with the men here and in London would be of great value. Then there is a chance that I might be given an honorary M.A. in which case I should be eligible for a fellowship in the college.[2]

So I think I shall plan to work here (with due vacations) through the summer and autumn and to be married the beginning of December. I should then return home, and should not come back here unless I am given a paid lectureship for the following autumn. I think the prestige of having lectured here, would help me as much <u>at home</u>, as being on the ground (with nothing in particular to do) during the autumn <u>term</u>, and the chance of the fellowship etc is worth effort.

In this case I think your coming over would be even more welcome and satisfactory than if I were to be married in June. I could spend a month with you either in England or on the Continent (Wildungen).[3]

[1] See 6.138.
[2] See 6.49.

[3] A German spa at which the elder Cattells had stayed before (see 4.2).

6.142 To Parents, Cambridge, 25 April 1888

I trust the plans I have made meet with your approval. I understood, Papa, that you thought it best for me to lecture here, if it could be arranged. Putting our marriage in December is a compromise between completing the course here and being married in June. It has been a hard matter to decide and has troubled me continually— and even now things may be upset.

I called today at Prof. Lumby's which includes a long walk.[1] Mon., Wed. & Fri. I have two lectures Venn & Stout, and I think this will be all. I have heard Sidgwick's and Ward's lectures.

[1] John Rawson Lumby (DNB 22: 983–984) was a fellow at St. Catherine's College, Cambridge, and Norrisian Professor of Divinity.

6.143 To Parents, Cambridge, 3 May 1888

(One paragraph of gossip has been omitted.)

I have received a long letter from Fullerton. He thinks they have been "acting dishonorably at Bryn Mawr" towards him.[1] I cannot well tell you about it as he writes 8 pages, and there would be much more to explain. They say he agreed to give lectures, which it seems he did not. He has definitely quarrelled with them, and will not lecture there next year.

[1] See also 6.128. The details of Fullerton's experiences are unclear. He had taught philosophy and psychology at Bryn Mawr from 1885, basing his lectures on Lotze's *Outlines* (the book founded upon the lectures JMC had heard at Göttingen in 1880–1881). A month after Fullerton, Woodrow Wilson also resigned his teaching position at Bryn Mawr, citing the salary he was to be paid at Wesleyan University, the lack of research assistance at Bryn Mawr, and the difficulty he had in teaching women (Arthur S. Link, ed., *The Papers of Woodrow Wilson*, vol. 5, *1885–1888* [Princeton, N.J.: Princeton University Press, 1968], pp. 743–747).

6.144 To Parents, Cambridge, 6 May 1888

It occurs to me that, perhaps, I should now have something like a "will," before it did not matter as my things would have gone to Harry anyhow. Should I die before my marriage I should like Harry to have my books at home, and you the pictures and fancy things. I should give Stout my scientific & philosophical books here and my papers—to edit any unfinished work I any have. I should want you to give Jo my other things and my money, and to have her visit you.

6.145 To Parents, Cambridge, 8 May 1888

I have written my review of Wundt, but am not especially well pleased with it. I do not criticize at all, but simply give an account of the contents of the book. I have another week & must try to improve it, if possible.[1] Stout has written a note criticizing Ladd, which is better than if I had replied to him.[2]

I had Green the son of a friend of Mr. Owen to lunch and to play tennis this afternoon.[3] He is at school yet but has a scholarship in King's College.

[1] See 6.11. As published ("Critical Notices," *Mind* 13 [1888]: 435–439), the review was not particularly critical. Perhaps partly in hope of avoiding a controversy similar to that caused by his review of Ladd's *Elements of Physiological Psychology* (see 6.139), JMC even noted with approval Wundt's approach to the mind-body problem.

[2] See 6.139.
[3] Walford Davis Green, the son of Walford Green (a former president of the Wesleyan Congress), who after graduating from King's served as a barrister-at-law and wrote on legal history (*Who Was Who* [Br.], 1941–1950, p. 465).

6.146 To Parents, Cambridge, 11 May 1888

(This letter was misdated "Fri. May 10th.")

I have been out nearly every evening this week. This evening I attended a meeting of the Moral Sciences Club at which Prof. Sidgwick read a paper. Tomorrow I dine at the Vice Chancellor's—the highest officer of the university.[1]

This afternoon I played tennis for a little while and called on Tanner a fellow of this college who has just been married.[2] His new wife seems very nice.

[1]Charles Edward Searle, Master of Pembroke College from 1880 until his death in 1902, was vice-chancellor of the university for the 1887–1888 academic year (Venn, *Alumni Cantabrigiensis*, vol. 5, p. 454). The Chancellor was usually a political figure or member of the royal family whose favor was curried by the appointment; the vice-chancellor was the highest academic officer.

[2]Joseph Robson Tanner (*DNB*, 1931–1940: 846–847) was a historian and fellow of St.

John's from 1886 until his death in 1931. He was to become editor of the Cambridge Medieval History and the Historical Register of the university, and later he tutored the boy who was to become King George V in English constitutional history (Robert Lacey, *Majesty: Elizabeth II and the House of Windsor* [New York: Harcourt Brace Jovanovich, 1977], p. 14). Mrs. Tanner, the former Charlotte Maria Larkman, helped make their home "a center of genial influence."

6.147 To Parents, Cambridge, 14 May 1888

It is not much that I have to say as one day is so much like another. I work in the morning and play tennis or something of the sort in the afternoon. Stout and I have begun to translate Höffding's Psychologie, we think it is better than any in English, but I don't know that we shall finish our translation.[1] The translating itself is excellent training, however.

I called on Prof. Marshall (political economy) this afternoon—he is a very interesting man.[2]

[1]Harold Höffding (*EoP* 4: 48–49) was professor of philosophy at the University of Copenhagen from 1883 through 1915, where he taught, among others, Alfred Lehmann, JMC's companion in Wundt's laboratory (see 5.114). His 1882 Danish text, *Psykologi i Omrids pa Grundlag af Erfaring*, had been recently translated into German by T. Bendixen (*Psychologie in Umrissen auf Grundlage der Erfahrung* [Leipzig: Fues, 1887]). JMC reviewed this translation as one of several "recent books on physical psychology" (*Brain* 11 [1889]: 263–266) and called it "an excellent introduction to psychology, clear and complete." As JMC spoke no Danish, he must

have been translating this German translation. JMC did not finish this translation, just as he had not completed his projected translation of one of Lotze's books (see 2.63 and 3.1). An English translation of this book was published in 1891 as *Outlines of Psychology*.

[2]Alfred Marshall (*DNB*, 1922–1930: 562–564) was a graduate of St. John's College who had grown interested in the "moral sciences" under the influence of Venn and Sidgwick. Professor at Cambridge from 1885 to 1908, he was best known for his work in economic theory.

6.148 To Parents, Cambridge, 27 May 1888

(One paragraph of gossip has been omitted.)

I hear that Dr. Hall has been given the Presidency of a college[1]—I don't know which—and will leave Baltimore, and understand that Mr. Ward has been offered the professorship.[2] I don't know who they can get—positions are rare, but good men for them still rarer.

[1] Hall became President of the new Clark University in Worcester, Massachusetts (see also 6.153 and Ross, *Hall*, pp. 186, 204).

[2] This probably was not true. Ward had been considered earlier for the chair that Hall eventually filled at Johns Hopkins (see 6.2).

6.149 To Parents, Cambridge, 1 June 1888

I have a distinguished guest tonight—Dr. Bain.[1] I met him at the station and brought him to the college, and had Prof. Sidgwick and Mr. Keynes to dine with him. Then he read a paper in my room before the moral sciences club.

I also had company to lunch, A. Ward, McTaggart,[2] and Stout and we played tennis afterwards.

There are a great many strangers here towards the close of this term, and much entertaining. It is of course rather expensive. Dinner itself, fortunately, does not cost much (62 cts), not so much as lunch—but in both cases wine and cigars must be added.

Summer has begun with fine weather. Cambridge is very lovely in Spring and early Summer.

I have not yet decided what to do when term closes. I shall go to Greenwich for a few days, and shall be here mostly during July & August—but I should like some life in the open air and should besides enjoy travelling as it may be my last chance.

[1] On Alexander Bain see 5.89.
[2] John McTaggart Ellis McTaggart, an undergraduate at Trinity College in 1888, was to become a fellow of Trinity, College Lecturer in moral science, and an important British analytic philosopher (*EoP* 5: 229–231; *DNB*, 1922–1930: 550–551).

6.150 To Parents, Cambridge, 6 June 1888

I have received notice from Prof. Barker (Sec'y) that I have been elected member of the American Philosophical Society[1]—I shall accept but trust there are no annual dues. Prof. Barker writes kindly, "We have quite missed you. Your lectures were highly appreciated" etc.[2]

Full term ends tomorrow, but I shall stay here until Thurs. of next week. The boat races begin on Fri. and there are concerts, balls etc.

I have not yet made up my plans for the summer.

[1] The oldest existing learned society in the United States, "The American Philosophical Society Held at Philadelphia for Promoting Useful Knowledge" traced its origins to a group founded by Benjamin Franklin in 1743. Today primarily an honorary society of scientists and scholars, the American Philosophical Society was in the 1880s an active and selective organization. However, the Philadelphia scientific community was represented disproportionately. (Edwin G. Conklin, "A Brief History of the American Philosophical Society," 1975 *Yearbook of the American Philosophical Society*, pp. 37–63.) George Frederick Barker (*DAB* 1: 601–602) was professor of physics at the University of Pennsylvania.
[2] See 6.136.

6.151 To Parents, Cambridge, 8 June 1888

(This letter was misdated "Wed. June 7th." One paragraph of gossip has been omitted.)

I took Mrs. Keynes and her little boy and a young lady guest out boating this afternoon.[1] This evening we have been playing whist in my room.

[1] "Mrs. Keynes" is Florence Ada Brown Keynes, the wife of John Neville Keynes (see 6.114). Later she would serve as mayor of Cambridge. The "little boy" is John Maynard Keynes, later a fellow of King's College and a distinguished economist (*DSB* 7: 316–319).

6.152 To Parents, Cambridge, 13 June 1888

(This letter was misdated "Wed. June 12.")

This is my last letter from Cambridge for a time. It has been a fairly good term, but the time has passed quickly and I do not seem to have accomplished much. I have written book reviews for Mind[1] and Brain[2] and a proposed first chapter of a Practi-

[1] See 6.76 and 6.111.
[2] See 6.80 and 6.147. In addition to the Höffding book, this review also discussed, among other items, the third edition of Wundt's *Grundzüge der physiologischen Psycholo-* gie (see 6.93), Ladd's *Elements of Physiological Psychology* (see 6.64), and Ward's article "Psychology" in the ninth edition of the *Encyclopedia Britannica* (see 6.86).

cal Psychology.[3] Then I have heard lectures and read some. I have besides seen philosophical people here, and have arranged to lecture next term,[4] which latter is not a small matter.

I have drawn £40 more and spent it mostly paying bills. I owed the tailor $85—having gotten two suits and an overcoat. I must spend more than I like on clothes this year, as I bought few last year, and must now get enough for next. Then I owed the grocer circa $15, the shoemaker $10, and the bursar $10 washing $10, gyp $5 and $40 for pictures. My college bill proper is still to pay.

I do not know where I shall go when I leave Greenwich. I look to stay there about a week.

[3] See 6.55. About three weeks earlier, on 24 May 1888, JMC had sent this chapter (on hearing) to Galton for his comments, and in a cover letter had outlined his plans for the rest of the volume as follows: "My plan would be to follow it with the other senses, sight, taste and smell, touch and temperature, organic and motor sensation. Then chapters on the measurement of mental processes—time, intensity and complexity, analagous you see to the units of physics, time, mass and extension. Then some treatment of attention, memory, association of ideas, mental imagery etc." (Galton papers). He also asked Galton to contribute a chapter on physical anthropometry, and the making of anthropometric measurements. See also 6.154.

[4] See 6.165.

6.153 From Elizabeth Cattell, Philadelphia, 15 June 1888

(Several paragraphs of gossip have been omitted.)

Mr. Rolf called yesterday to inquire about you, they go to their Country home next week.[1] He spoke of Mr Hall, said he had been called to found a large University at Wooster Mass,[2] that the University had plenty of money, a gentleman had already given one million and a half of dollars, and that Mr Hall was looking out for Professors, he wondered if it would not be a good place for you.

[1] On Henry Winchester Rolfe see 1.19.   [2] See 6.148.

6.154 To Parents, Cambridge, 12 July 1888

(One paragraph of gossip has been omitted.)

Here I am once more after just four weeks abscence.[1] I hope to do some work, now, between this and the first of September. I must write a paper on the "Association of Ideas", which will be printed as by Mrs. Bryant and me,[2] she having made some of the experiments in her college. Then I want to get forward with the MS of my proposed book.[3] Mr. Galton, with whom I lunched yesterday will contribute the

[1] JMC had spent these four weeks at his fiancée's home outside of London (see 6.152).   [2] See 6.104.
[3] See 6.55 and 6.152.

chapter on anthropology[4]—*he is, you know, the most eminent man living in that department. I may write the work with Stout—but I have not suggested it to him yet.*[5] *Then I want to make some experiments on vision.*[6]

[4] JMC had earlier (7 July 1888) written to Galton the following: "I think a chapter containing directions for making the anthropometric determinations you recommend would be an important part of the proposed work, and if contributed by you yourself would add much to its value. I can understand, however, that you might like to see part of my text before you assent—this I shall try to prepare the summer." (Galton papers.) See also 6.152.

[5] JMC had planned to write this book with Ward (see 6.55), and then tried to talk Galton into being his coauthor (see 6.152 and JMC to Galton, 6 May 1888, 24 May 1888, and 31 May 1888 [Galton papers]) with Ward writing the preface (JMC to Galton, 12 June 1888 [Galton papers]). Galton, however, backed away from this proposal, and JMC looked to Stout (see also 6.156).

[6] See 6.158.

6.155 To Parents, Cambridge, 19 July 1888

(Two paragraphs of gossip have been omitted.)

I enclose a notice on Psychological Laboratories.[1] *The British Medical Journal is as good as the Lancet & has a much larger circulation. It might be well to have it copied in an American paper—one of the Phila. medical papers, for example— they would doubtless take it owing to the references to the Univ. of Penna. If you do not think this worth the while please return the notice to me.*

[1] "Psychological Laboratories," *The British Medical Journal* 7 July 1888: 29–30. This note reviewed the establishment of laboratories in Germany and the United States, abstracted JMC's paper on the Leipzig laboratory (see 6.100), and called for the development of such laboratories in Britain.

6.156 To Parents, Cambridge, 21 June 1888

(One paragraph of gossip has been omitted.)

I am now working with Stout on an Experimental Psychology;[1] there is plenty to do, as it should be finished this year—it will not, however, be very large. I think it will be a good book, and probably successful, though it will scarcely more than pay the expenses of publication.

[1] See 6.55 and 6.154. This book was never completed. But as JMC had written to Galton on 24 May 1888 (Galton papers), "Some such book [a practical text for laboratory instruction] is urgently needed—it does not exist in any language." What probably was the first such laboratory manual in psychology did not appear until the early 1890s, when Edmund Clark Sanford, the first working director of the Clark University Psychological Laboratory, began publishing one in parts in the *American Journal of Psychology* (see 4 [1891–1892]: 141–155, 303–322, 474–490; 5 [1893]: 390–415; 6 [1895]: 593–616; 7 [1896]: 412–424). Sanford himself never completed this manual. He finished only the first part, which was devoted to the simpler mental processes. See Sanford, *A Course in Experimental Psychology*, part 1, *Sensation and Perception* (Boston: Heath, 1894, 1898). Throughout the 1890s JMC continued working on such a laboratory manual, and twice the Columbia College *Bulletin* announced its imminent publication (no. 7, February 1894, p. 37; no. 9, December 1894, p. 31). But this book was never finished, though large sections of its manuscript can be found in the Cattell papers.

6.157 To Parents, Cambridge, 25 July 1888

(Two paragraphs of gossip have been omitted.)

I took a new departure today. I began to work in the machine shops of the university.[1] I wish I had been taught this and carpentry when I was a boy—also gardening. I think this was the chief mistake in my education. I really need the mechanism work to repair my apparatus, but I am too old and busy now to learn it well. In any case it seems to me something we all ought to know. Just now I take it up partly for the exercise—which is most energetic. I worked very hard for four hours this morning. The only objection is the expense, about $20 for 24 times.

[1] "Practical instruction . . . in woodwork and iron work, including pattern-making, fitting, turning, and forging" was given at Cambridge from 1884 (Tanner, *Historical Register*, p. 243). Many early American psychologists, who had to build their own experimental apparatus, also recognized the need for such skills (see Sanford, *A Course in Experimental Psychology*, chap. 9, "Suggestions on Apparatus," pp. 363–419).

6.158 To Parents, Cambridge, 27 July 1888

(A postscript of gossip has been omitted.)

You will perhaps be surprised that I have stopped working in the machine shops after my praise of such work. I wish very much that I had learned when I was young, but I fear now I have not time. It seems best to work at the Cavendish laboratory where I can get results of value, perhaps, and it does not cost me anything.

I was at the Cavendish consequently this morning, and began to get apparatus in order. I want to experiment on vision and have an apparatus of my own for it.[1]

This afternoon I was up the river in a canoe.

[1] JMC published nothing on vision while at Cambridge, but soon after his return to America he began a series of psychophysical experiments in which an observer was asked to distinguish between lights of similar intensities (see JMC and George S. Fullerton, *On the Perception of Small Differences: With Special Reference to the Extent, Force and Time of Movement*, Publications of the University of Pennsylvania, Philosophical Series, no. 2 (1892): 134–135. The apparatus JMC used in these 1889 experiments consisted of a hooded lamp mounted behind an aperture that was regularly closed by a screen mounted on a pendulum bob with a period of one second. When the bob was released, the light shone through the aperture onto a surface for one second before the bob was stopped. This procedure allowed JMC to present his subjects with light stimuli for controlled periods of time, and a slide on which the lamp could be moved closer to or farther from the aperture allowed him to vary the intensity of the stimuli. (See *On the Perception of Small Differences*, pp. 135–138; "A New Machine," *The Pennsylvanian* [University of Pennsylvania weekly student newspaper] 5 [1889–1890]: 17; and figure 15.)

For the rest of the summer of 1888 Cattell continued his experiments at Cambridge, often visiting Jo at her home in Blackheath. The couple also spent some time together, with Jo's younger sister Grace, on a vacation in Bournemouth on the south coast of England. By the middle of September Jo had left Bournemouth for London, and Cattell set out on a walking tour, heading due north from Bournemouth towards Oxford (JMC to parents, 16 September 1888). Meanwhile, William Cattell continued to look after his son's interests in America.

6.159 From William Cattell, Philadelphia, 10 September 1888

(This letter was dated "Mon. 10th." Several paragraphs of gossip have been omitted.)

I enclose you two checks which Dr. Pepper sent me today—one for the amount adv^d by you for apparatus—the other a division of the tuition fees. If you will endorse these payable to my order I can get them cashed & hold the amount subject to yr direction.

6.160 To Parents, Amesbury, 19 September 1888

I have had another good day's walk. The country is very pretty and the weather has been magnificent. I got into Salisbury this morning, and spent a couple of hours in and about the cathedral. Then I walked on to Stonehenge—you know it is a curious circle of stones, built by sunworshipers, all other record of whom is long since perished.[1] This little village is near by, and about as far from a railway as one can get in England.

[1] Baedeker (*Great Britain*, pp. 87–91) starred both the cathedral at Salisbury and Stonehenge.

6.161 To Parents, Pangbourn, 23 September 1888

(This letter was misdated "Sun., Sept. 22nd." One paragraph of gossip has been omitted.)

I have walked on to this place today, but not very far—about 12 miles. On the way I stopped at one of the large English lunatic asylums[1]—I found that I knew the brother of the resident-physician,[2] and he took me through the wards.

[1] The Berkshire County Asylum, in Moulsford, which had been established in 1870 as one of the newer publicly supported county insane asylums, had accommodations for about 500 patients (*Forty-Second Report of the Commissioners in Lunacy to the Lord Chancellor, 1888*, pp. 154–156).

[2] J. Harrington Douty (*Medical Register*, 1908: 514, 1206) was the brother of Edward Henry Douty (*Who Was Who* [Br.], 1897–1916, p. 206), a graduate of King's College who in 1888 was Demonstrator of Anatomy at Cambridge University.

6.162 To Parents, Blackheath, 25 September 1888

We are both working at the experiments on the association of ideas[1]—we have a great mass of papers, and I scarcely know what to make of them. We go to Mrs. Bryant's on Thursday for dinner.

Otherwise we have not done much today, it has been pouring rain mostly, and we only went out for an hour towards evening.

We quite keep house together as Mr. Owen and Grace are away most of the day.

[1] "We": JMC and Josephine Owen. JMC had arrived in Blackheath at the end of his walking tour the day before. On the experiments see 6.75.

6.163 To Parents, Cambridge, 28 September 1888

I came back to Cambridge this evening, and was glad to find letters from you both.[1]
You are very good to us.[2] *I have not at any time intended to apply for the professorship in Australia*[3]—*but it is only on your account that I do not seriously consider it.
Of course I should not in any case be sure to get it, but it is to be awarded by a
committee in England (Nov. 1st), and I am told that I could probably have it. When
I say that I prefer to return to America on your account I mean for our sakes as well
as for your's. I am especially glad to have Jo near or with you.*

*Mr. Owen is willing to have us marry in December, since he knows we are going
home to your house. Jo is going to get her things now. She would like to know what
sort of dresses she will need. It is rather hard as she wants to dress loosely, and
without following the current deformity.*[4]

$1000 seems a large wedding present—almost too large.[5] *I told Jo just recently
that you would give us $500 for a piano—which she thought was very generous.*

*So long as we do not have a house to furnish we do not seem to want anything—
though books and pictures are always welcome. I scarcely know what else to get for
Jo's birthday present. We have not yet spent the $20 you gave her, Mama, when I
came away, but we have arranged what music we shall buy. Jo does not care to
wear jewelry.*

*I can suggest no other wedding-present than books & pictures. We should like to
have editions of Carlyle, Ruskin, Tennyson, Scott, Shelley, Rossetti, Morris and Darwin. Jo has one present—a fair example of such—a jewel-box!*

*In many ways the most acceptable present of all would be to spend part of the
money you give us in a wedding-trip longer than we could otherwise afford.*

[1] From William Cattell, 2 September 1888;
from Elizabeth Cattell, 4 September 1888.
[2] That is, with regard to wedding gifts; see
below.
[3] JMC had apparently mentioned the existence
of such a professorship (see William Cattell
to JMC, 9 September 1888).
[4] About 50 years later the clothing styles of
the 1880s were described as "ugly in line
and drab in color" (Mary Agnes Hamilton,

Newnham: An Informal Biography [London:
Faber and Faber, 1936], p. 139). On 5 November 1888 Elizabeth Cattell wrote to her
son "I did not know how to advise Jo, as I
had no idea of what she expected to spend
on her outfit, but we now dress so much
alike in England and America that whatever
she invests in will be all right."
[5] Elizabeth Cattell had mentioned this sum in
her letter of 4 September 1888.

6.164 To Parents, Cambridge, 3 October 1888

*I keep intending to go up to London, but dont get off. There are three or perhaps
four societies whose meetings I should like to attend—Psychological, Aristotelian,
Neurological and Anthropological. It is of course interesting to hear the papers
which are carefully prepared and keep a year or so ahead of published matter, but
it is perhaps of more importance to meet the people. Tea and coffee is served
before the meeting and a chance for conversation thus given. I shall perhaps spend a
week in London during the vacation which lasts nearly a month. It is possible but not
likely that I shall go to Paris. I wish I could by magic be set at home for Xmas time.*

6.165 To Parents, Cambridge, 13 October 1888

I received this morning papers etc. from Mrs. Bryant, and have my hands tolerably full between this article and my teaching work.[1] *I had one man come this morning who wants to attend my lectures—if he is the only one, he seems at all events to be a good man—he stood first in the London Univ. examination.*[2]

I sent you today several copies of the lecture list.[3] *If you care to make use of them I could get more for 6 cts. each. I shall send one to Fullerton, but not to anyone else.*[4]

[1]On the article mentioned see 6.75.
[2]This student has not been identified.
[3]"List of Lectures proposed by the Special Board of Moral Science, 1888–9." *Cambridge University Reporter*, 8 October 1888, p. 28. For the Michaelmas Term (16 October–December) JMC was listed as lecturing twice a

week on psychophysics, at a fee of £1. 1s. In addition, a note indicated that "Dr. Ward and Mr. Cattell will superintend practical work on this subject at the Cavendish Laboratory. Fee £1. 1.s."
[4]On George S. Fullerton see 2.17.

6.166 To Parents, Cambridge, 15 October 1888

Lectures mostly began today, and I heard Ward and Stout. I am also intending to hear Sidgwick, Keynes and Marshall.[1] *Perhaps it is more than I shall like, but I feel that it is my last chance.*

I have arranged things for tomorrow and think there will be a couple of students at all events.[2]

Today I rowed and intend to on Mon. Wed. and Fri.

[1]On Sidgwick see introduction to chapter 6; Keynes, 6.114; Marshall, 6.147.
[2]JMC's own lectures were to start the next day.

6.167 To Parents, Cambridge, 16 October 1888

I had the laboratory this afternoon—there were five men and two women. I could have had more women, but thought it best to have only those in their 3ʳᵈ year. I have about half the men here in moral sciences, so have no cause to complain. The men are better than any I should be likely to have in America, as they devote themselves entirely to philosophy and are preparing to teach it—They must be on the average 25 years old. I shall lecture on Thurs, as two or three want me to. I shall not have women, and shall lecture in my own room.

The practical work is really more trouble than a lecture, as I must talk continuously for two hours, and arrange things ahead. But I think I can now give a course on psycho-physics without undue worry or work.

6.168 To Parents, Cambridge, 18 October 1888

I was tolerably busy today as I lectured this afternoon and heard three lectures this morning. There were only two (three with Stout) to hear me this afternoon—I was in my own room. There are so very few moral science students here—exclusive of myself Ward only has two and Sidgwick three men. I am not sure that it will be worth while going on on Thurs—still the others do.

Jo is, I think, looking for advice from you, Mama, as to her things.

6.169 From William Cattell, Clarinda, Iowa, 21 October 1888

(Several paragraphs of gossip have been omitted.)

Just before leaving home I wrote to Dr. Ward of the Independent who has always taken a great interest in you.[1] He has more than once said to me that the University of Penn[a] was "notorious" for its policy of getting work done for little or nothing. Dr. Pepper frankly said to me that they had undertaken more than, with their limited means, they c[d] well carry on. This is true of most colleges:—And I am very anxious that you should have more than this one string to yr bow. I enclose you Dr. W.'s reply, and the letter he enclosed from Pres[t] Gates[2]—which were forwarded to me from Philad[a]. Of course I telegraphed Dr. W. at once that you w[d] not be home till Jan[y]—& then wrote him. Altho' what the Pres[t] proposes is out of the question— something may grow from this. At all events I was pleased that Dr. W. moved so promptly in the matter. He is a man of great influence and he knows "the field".

I have no fear but what there will, in time, be an opening for you in the line of yr special studies. Few colleges will have, for some years yet, a Chair of Psycho-Physics—but in a number it will become an important part of Psychology.[3]

[1]On William Hayes Ward see 6.64.
[2]Merrill Edwards Gates was president of Rutgers College (1882–1890) and of Amherst College (1890–1899) (*Who Was Who in America*, 1897–1942, p. 444). The letters mentioned have not been found, but appear to have raised the possibility of JMC teaching at Rutgers.

[3]For William Cattell "Psycho-Physics" meant all experimental psychology, and "Psychology" included both the philosophical and experimental approaches to the study of the mind.

6.170 To Parents, Cambridge, 22 October 1888

I have not written for two days, having been at Cromer, and not alone in the evening. Cromer is our nearest sea-side place and is very pretty, with cliff, hills and woods. The land reaches out into the sea, and the air is peculiarly fresh. It is said to be the only place in England where one can see the sun both rise from, and set in the sea.

Stout went with me, and we discussed philosophy and really worked some, making experiments and writing. I did about as much as I should have done at Cambridge, and had the sea air for two days, with walks and three baths. The only drawback was the expense, and this was not very great, as one can get a return ticket for $2, and wherever I am I must pay for my meals.

I wish I could always live at the sea-shore, or in the mountains.[1]

[1]JMC was to live on a mountaintop most of his life (see postscript).

6.171 From William Cattell, Philadelphia, 28 October 1888

(Several paragraphs of gossip have been omitted.)

On my return, I found a note from Dr. Pepper requesting me to call & see him at my "earliest convenience".[1] I went the next morning;—not without some apprehension of a "hitch" with reference to your university work; But I was soon & happily relieved. He is evidently in earnest about making some arrangements that will give you better facilities for work & make yr connection with the University permanent. He referred again to the highly appreciative way in which Dr. Ferrier spoke of you to him;[2] and said there were two things he w^d purpose to the Trustees if there was a reasonable assurance that you w^d remain permanently at the University (1) that you sh^d receive the title of Professor and be elected for a term of three years instead of receiving an annual appointment as Leturer (2) that a good working Laboratory should be fitted up and appropriated for your exclusive use. I expressed my pleasure at both of these suggestions, but asked him if he expected you to entertain the proposal of a permanent connection with the University upon a salary of $300!—"Of course not" he promptly replied; and then went on to say that, while the University had no funds whatever to vote for yr department, he was not without hopes of securing from certain individuals special subscriptions which w^d raise your salary to one thousand dollars for the next three years. I reminded him that altho' this, under all the circumstances, was a handsome provision for a branch of study that a university "without funds" might well excuse itself from taking up, yet $1000 w^d not afford you a living and I asked whether it would prevent yr delivering a course of lectures elsewhere to supplement this salary.—He replied "that is just what we should like. Your son would be a Professor in the University: his publications would be dated here; the University w^d have credit for his researches &c"— which pleased me very much; as Dr Genth quotes him as saying that he attached no importance to original researches by the Professors.[3] I could hardly believe that Dr Genth rightly understood him—though he puts it in _italics_ by way of emphasis. Certainly if this were Dr Pepper's opinion I should have little hopes of his making any extra or special efforts for your department.

[1]William Cattell had returned from a fundraising trip for the Board of Relief.
[2]On David Ferrier see 5.110.

[3]Frederick Augustus Genth, Jr. was a chemist educated at the University of Pennsylvania, where he taught until 1888 (_Who Was Who in America_, 1897–1942, p. 447).

We had some more pleasant talk about you & yr work and of course I came away much gratified. The Dr. is to notify me in writing of the success he meets in carrying out his plans—the funds for your increased salary being of course to be secured before any thing else is done. I am inclined to think he will succeed—but it is well to remember that he is very <u>sanguine</u>. He has suffered <u>somewhat</u> from the charge of creating expectations that were never realized.[4] People have even given this a very harsh name. But of the sincerity of his intentions I feel quite assured and I shall let you know at once when I hear anything definite (i.e. in writing) from him.

—I wrote the above before dinner;—and when I laid down my pen was on the point of "wondering" what had stirred up the Provost in this matter! Curiously enough, this very afternoon Harry met Prof. Fullerton who spoke of his great desire to have you at the University and the steps he had taken to interest others;—and that he would soon write you anent the whole matter! This is of course the explanation of Dr. Pepper's awakened interest—and of one thing you may be sure Prof. F. is a good & true friend of yours. I shall anxiously await developments—meanwhile I cannot but feel very deeply this exceedingly friendly move on the part of Prof. F. I trust you will never cease to be good friends.

I mail you a copy of Dr. Patton's Inaugural which I have just rec^d—will get another copy for myself.[5] I was sitting on the stage very near to the Dr. during its delivery and when I returned home told yr mother I thought he glanced significantly at me when he said "—it is more important after all to think than to know "the time it takes to think"[6] (see p. 37)—But a friend of mine once told me that he heard Dr Patton discussing with one of the College Professors yr publications in Wundts Journal and that he spoke very highly of them & of you. You will no doubt agree with him as to the importance of the study of "the history of Philosophy". He wd give it the place it has in the German Universities.[7]

I am glad, upon the whole, that you are not at Princeton. Next to Lafayette my heart is there. I have two <u>alma maters</u> in Princeton! But the environment would not suit you I fear. They would probably be sensitive about certain expressions you might use which—clinging to the old paths as I do—would give me no alarm—such as the expression in the reprint from Brain you recently sent me—"biology culminated

[4] See 6.132.

[5] Francis Landey Patton (*DAB* 14: 315–317), former professor at Princeton Theological Seminary, was inaugurated president of Princeton College—a separate institution—on 20 June 1888, succeeding James McCosh (see 4.14). (*The Inauguration of Rev. Francis Landey Patton . . . as President of Princeton College* [New York: Gray Brothers, 1888].)

[6] JMC's paper of this title, which had originally been published in *The Nineteenth Century* (see 6.131), had been reprinted in February 1888 in America (*Popular Science Monthly* 32 [1888]: 448–491).

[7] But see Patton's comment "Better a thousand times for us a roomy American college than a feeble German university," cited in Laurence R. Veysey, *The Emergence of the American University* (Chicago: University of Chicago Press, 1965), p. 52. McCosh had qualms about Patton's ideas and policies, especially as McCosh saw no conflict between religion and the results of German scholarship (see 4.14; Baldwin, *Between Two Wars*, vol. 2, p. 203). But Veysey also wrote that Patton was only "rhetorically" a true reactionary, and stressed that he eventually supported changes in the Princeton curriculum.

with the doctrine of the evolution of species"[8]—*some good people will not hear of evolution in any form. Even Dr. McCosh, who seems to me to hold the true view of evolution has been severely criticized as "going too far" & "conceding too much"!*[9]

[8] JMC made this point in his review "Recent Books on Physical Psychology," *Brain* 11 (1889): 263–266, which had been published in July 1888.

[9] By the mid-1860s McCosh had developed a natural theological view of evolution, "defending Evolution . . . as the method of God's procedure, and [finding] that when so understood it is in no way inconsistent with Scripture" (William Milligan Sloane, ed., *The Life of James McCosh: A Record Chiefly Autobiographical* [New York: Scribner, 1896], pp. 122–124, 234).

6.172 From William Cattell, Philadelphia, 5 November 1888

(Several paragraphs of gossip have been omitted.)

Nothing further from Dr. Pepper. It is very plain now that he was depending upon Prof. F. to raise the $1000—and the Prof. has been disappointed.[1] *He is a very sanguine young man and no doubt led the Provost to believe he had in hand what was only a confident expectation! But, as I wrote you, he is certainly a fast friend of yours & appreciates very highly the work you could do in the University.*[2] *But I am not so hopeful as I was when I first saw Dr Pepper, as to the outcome of this new effort. But your "opportunity" will come—if not in the near future. We must use all the means; and then, while you are doing faithfully & well the present duty, await developments—*

I recall that you said, during yr last visit home you could take some other Professorship if there sh[d] *be no opening in Psychophysics. If there was a vacant chair in Psychology*[3] *or Biology I sh*[d] *advise you to apply for it (if the "environment" were suitable) as yr specialty w*[d] *naturally form a part of such a course and it will be long before any college can afford a chair in Psychoph. with a living salary. But how about a Professorship of Greek or Latin?—in which there might be reasons for you to give a course on Experim*[t] *Psychol*[y]*? Prof. Libbey at Princeton has the chair of Physical Geography—yet I am told that some of his best work is done in a course he gives in Microscopic Anatomy in the department of Biology!*[4]—*subjects wider apart than Ancient Languages & Psychophysics*

[1] That is, George S. Fullerton (see 2.17).
[2] See also 6.129.
[3] That is, from William Cattell's view, philosophical psychology.

[4] William Libbey, Jr. (*Who Was Who in America*, 1897–1942, p. 728), educated at Princeton, was from 1883 until 1923 professor of both physical geography and histology. It might be noted, however, that his father was a trustee.

6.173 To Parents, Cambridge, 10 November 1888

(Two paragraphs of gossip have been omitted.)

I sent today for some catalogues of apparatus, under the circumstances it may be well to get some although at present I have no appropriation from the university.[1] I wish, Papa, you could get $500, as you were proposing last winter.[2] If it were subscribed on condition that there should be a university grant of the same amount I think they would give it, and then I should have enough apparatus for the present.

[1] Some funds had been appropriated for apparatus the previous Fall (see 6.132), but they had long since been spent on the instruments JMC had purchased during the summer of 1887 with the hope that the university would eventually buy them (see 6.97, 6.105, and 6.112).

[2] See 6.129.

6.174 From William Cattell, Philadelphia, 11 November 1888

(One paragraph of gossip has been omitted.)

Dr. Pepper sent me by today's mail your cheque for $300 which I will get changed into a draft the first thing tomorrow morning.[1] It will reach you in the same mail with my letter of yesterday—you can send Dr. Pepper a rec't direct—or if you prefer to send it to me I will hand it to him.

I enclose you also a copy of his letter.[2] I suppose from his not having had you elected to this chair at the last meeting of the Board when it was "created" there must be some rules about nominations laying over, or being referred to a committee. And this w^d explain the contradictory reports in the Papers as to what was done at this meeting—the Telegraph having referred to your having been elected to this "new chair" as Lecturer—& The Ledger that the chair was established & "Nominations" would made in the near future—a rather awkward way of putting it.[3] However the main thing is to secure the position & it seems to me that the departure is in good "shape". You will have the title of Professor in one of the leading Universities of our Country—a suitable Laboratory in the "near future"; and, I hope, a salary of $1000 at the University with leisure to supplement this by a course of lectures in some neighboring Institution. I think Haverford w^d be most likely to offer you such a field and I have been moving (judiciously I trust & without in the least

[1] This was JMC's salary as lecturer on psychophysics for the previous winter (see 6.9 and 6.136). It is interesting to note that at least part of this salary was paid for out of the Seybert Fund for the Study of Spiritualism (see *Annual Report of the Provost of the University of Pennsylvania . . . for the Year Ending October 1, 1887*, pp. 7, 31; see also 2.49).

[2] Pepper's letter has not been found, but its contents may be determined from this letter.
[3] JMC was officially appointed Professor of Psychology on 1 January 1889 (*Report of the Provost . . . for the Two Years Ending October 1, 1889*, pp. 16, 18).

compromising you) in that direction.[4] *The son-in-law of the President of the Board*[5] *is a personal friend of mine & so is also one of the most active members of the Board—and the friend ought, if possible, to be preempted—but more of this anon.*

[4] Haverford had been founded in 1833 as a Quaker college (Rufus M. Jones, *Haverford College, A History and an Interpretation* [New York: Macmillan, 1933]). See also 6.154.

[5] In 1888 the president of the board of Haverford College was Wistar Morris of Overbrook, Pennsylvania (Jones, *Haverford College*, p. 108). His son-in-law has not been identified.

6.175 Parents, Cambridge, 16 November 1888

I am glad to receive letters from you both today—but I am sorry the university arrangement has not been made, I hope they will in any case give me the title & laboratory. I think I should rather for the next three years have as small an income as it is possible to live on, than take a professorship such as Latin. It would indeed scarcely be fair to the college as at first I should not know enough, and later should give it up. Philosophy or psychology would be my subject, and I should prefer it to psychophysics ultimately, though not at present.

Tonight I was at the meeting of the Moral Science Club,[1] *and before I had heard a Mr. Carpenter lecture, whom I had been invited to meet at Lady Wilson's in the afternoon.*[2] *There is mostly something here in the evening. Tonight there was also a concert I should have liked to hear.*

[1] See introduction to section 6, 6.57, and 6.149.

[2] On Lady Wilson see 6.61.

6.176 To Parents, Cambridge, 20 November 1888

I have heard from Fullerton again today with a letter enclosed from one of the Trustees to him, containing the very good news that the professorship has been established, and that I should probably be elected in Jan. It is said "The Provost stated that $1000 for three years would be contributed by friends of the university"—but you of course knew about it before this.

So now I have nearly everything I could wish. You may think it improbable, but I believe we could keep house on $1000 & a house, which we have enough money to buy. "Henry" lived on $300, and I suppose they were never hungry or cold.[1]

[1] This may be a reference to the hero of a popular children's book, *The History and Adventures of Little Henry*, which went through several editions before 1820. In the story a boy named Henry is separated from his wealthy parents and has to live as a beggar until he is reunited with them. (d'Alte A. Welch, *A Bibliography of American Children's Books Printed Prior to 1821* [Worcester, Mass.: American Antiquarian Society, 1972], p. 167.)

I have rather a lot of things to do now—one perplexing thing being deciding as to the purchase of books, pictures, furniture and apparatus.[2]

I am told that $1000 has been collected for apparatus[3]—this seems very liberal—I wish, Papa, you might also have gotten something as you were intending—but in any case do not overwork, if other things seem more pressing.

[2]JMC now had at his disposal what seemed to be a fairly large sum of money, and he used it to order an extensive range of instruments—mostly standard items such as were used in most psychological laboratories of the period, including reaction-time apparatus, kymographs (see 2.41), instruments for the production of standard tones, and models of the brain and the sense organs. (See also George M. Stratton to George H. Howison, 23 June 1895 [Howison papers]; [Hugo Münsterberg,] *Catalogue of The Psychological Laboratory of Harvard University*, 1893.) In May 1890 JMC wrote for a Philadelphia newspaper an account of the apparatus being used in the University of Pennsylvania psychological laboratory, acknowledging a specific donor for each instrument. This note was republished soon afterward in *The Pennsylvanian* 5 (1889–1890): 241. See also figure 15.

[3]See 6.129 and 6.173. In the *Report of the Provost of the University of Pennsylvania for the Two Years Ending October 1, 1889* (p. 170), Pepper indicated that a total of $1,935 had been donated to the University in support of a laboratory of experimental psychology. Among those whose gifts were listed were William Cattell ($100), Fullerton ($50), Pepper ($250), Frederick Fraley (see 2.17; $50), and S. Weir Mitchell (see 3.3, $200).

6.177 To Parents, Cambridge, 22 November 1888

(Three paragraphs of gossip have been omitted.)

I have rather a lot to do just at the end—but the worst thing, finishing the paper on the association of ideas, I can do better at Greenwich then here as Jo prepares the tables and copies out the MS.

6.178 To Parents, Cambridge, 4 December 1888

My stay at Cambridge is now to be counted almost by hours, as I intend to go on Thursday.

Today I concluded my own course here. I had six students, which is as many as Prof. Sidgwick and more than Ward had. I am glad to have lectured here & to have held a position at Leipzig, partly for the experience, partly for the prestige.

I have every reason to think that I shall get on well at the Univ. of Penna, owing to my experience here & at Leipzig, and it was more difficult at home last winter than it is likely to be again.

I do not seem to have made myself quite clear as to money matters, but I dare say the intervening letters have done so. I have

On first letter	*$450*
second "	*500*
draft (Univ.)	*300*
	1250
subtract for photos	*250*
	1000

I look to spend

Bills	*50–100*
Packing	*50–100*
Traveling 1 mo.	*300–400*
Passage home	*150–200*
	550–800

I shall not spend this year more than the $1000 you gave me and next year, I think, will ballance.

6.179 To Parents, Greenwich Park, 6 December 1888

I said good-bye to Cambridge this afternoon. I should like going less if it were not for my marriage. I have been very kindly treated there and there is a powerful fascination about the university, which our American colleges do not possess. But on the other hand there are certain advantages in America—apart from the great one of being with you.

There are arranging a lectureship on psycho-physics at Cambridge which I under-stand was for me.[1] But I could not afford to take it as the salary is very small, $250, the lectureship being nearly always held with fellowships.

Jo is very well. She wrote to you yesterday, and sends her love again today.

Mr. Owen seems much the same. I thought he seemed brighter last time, but Jo says he is altogether different from what he used to be.

[1] A lectureship in physiological and experimental psychology was not established at Cambridge until 10 June 1897, despite JMC's statement here (Tanner, *Historical Register*, p. 124).

6.180 To Parents, Folkestone, 12 December 1888

We were married yesterday. We received your telegram an hour before and sent ours to you immediately after.

There is not much to tell of the ceremony—which was as simple as possible. I walked across the heath, and Jo with Mr. Owen, Edith and Grace drove afterwards. Jo's Aunt & cousin Sara were there, and a few people. The church of England service was used somewhat shortened—we both like Mr. Brash, the clergyman. We drove to the house afterwards, and said good-bye got our baggage, and drove to a station several miles away where we got a train to this place. It is very pretty here and the hotel is good. There was a dense fog at Blackheath yesterday morning, but the weather here has been wonderfully fine.

Mr. Owen was kinder to Jo towards the end, but according to Jo and his sister and cousin he is very different from what he used to be. It was very hard for Grace to have Jo go away, and also a sad thing, for Jo, as she fears Grace will not be happy at home. But she is a nice and good girl and though she may be unhappy, it will not otherwise harm her. Edith seems brighter and happier than when she was at home, and is quite enthusiastic about the Salvation Army.

We intend to travel along the South coast—we are now at the extreme east. It is the pleasantest place in England now in Winter—many of the seaside places being quite warm. Then we shall go north to Liverpool. Mr. Owen intends to come with Grace to see us off. You know we expect to sail on the Servia, Jan. 12.

Dr. Pepper kindly sent a telegram "University sends hearty congratulations." I had two or three more presents sent me from Cambridge and Jo had presents which we will tell you about when we come home.

Jo is quite perfect. She will write tomorrow.

Figure 1. James McKeen Cattell, ca. 1880. From Cattell papers, Library of Congress. This photograph, taken when Cattell was a senior at Lafayette College, was not published until 1900, when it appeared in *Vigintennial Reunion: Class of 1880*. The pin on Cattell's tie represents his membership in Delta Kappa Epsilon fraternity (see document 1.1).

Figure 2. William C. Cattell, ca. 1880. From an article in the 1880 *Melange* (Lafayette College yearbook), celebrating "President Cattell's 25th year at Lafayette."

Figure 3. Francis A. March, ca. 1880. From an article in the 1880 *Melange* celebrating "Dr. March's 25th year at Lafayette."

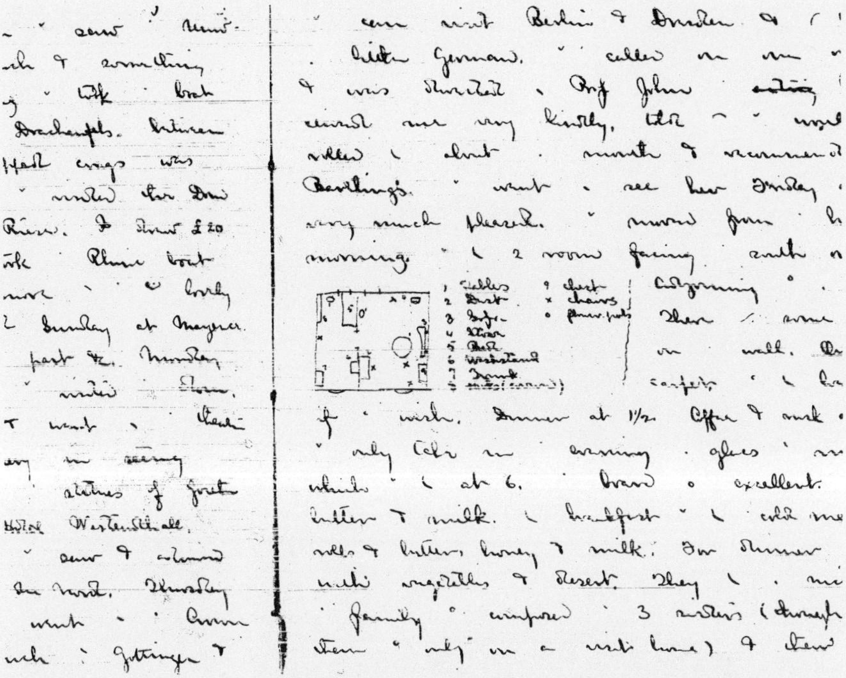

Figure 4. Cattell's sketch of his first room in Germany, 14 September 1880 (see 1.4).

Figure 5. G. Stanley Hall, ca. 1884. From Clark University Archives.

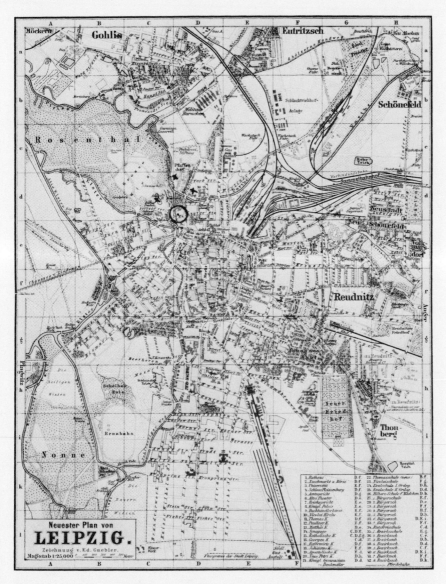

Figure 6. *Neuester Plan von Leipzig*, ca. 1880. Cattell sent this map to his parents on 1 October 1884, circling the site of his new home and marking places of interest (see 5.1).

Figure 7. Wilhelm Wundt, ca. 1886. From Cattell papers, Library of Congress. Wundt presented this photograph (taken by Georg Brokesch, one of the leading photographers of Leipzig) to Cattell in 1886 in celebration of Cattell's completion of the Ph.D. requirements.

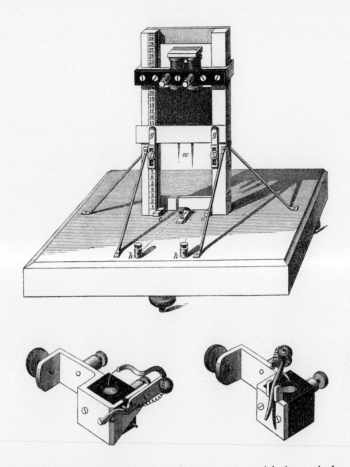

Figure 8. Cattell's modified gravity chronometer, with its switches. From Cattell, "Psychometrische Untersuchungen" (see 5.22 and 5.130). In the first version of this instrument (see 3.4), the breaking of a current in the electromagnet at its top allowed the screen to fall (thus revealing to the subject a stimulus and eliciting a reaction) and also started the Hipp chronoscope. With the introduction of these switches (which were mounted on the columns of the chronometer) and of several electric contacts (for example, *h*, Cattell made it possible to regulate and measure how long a stimulus would be presented. See figure 10.

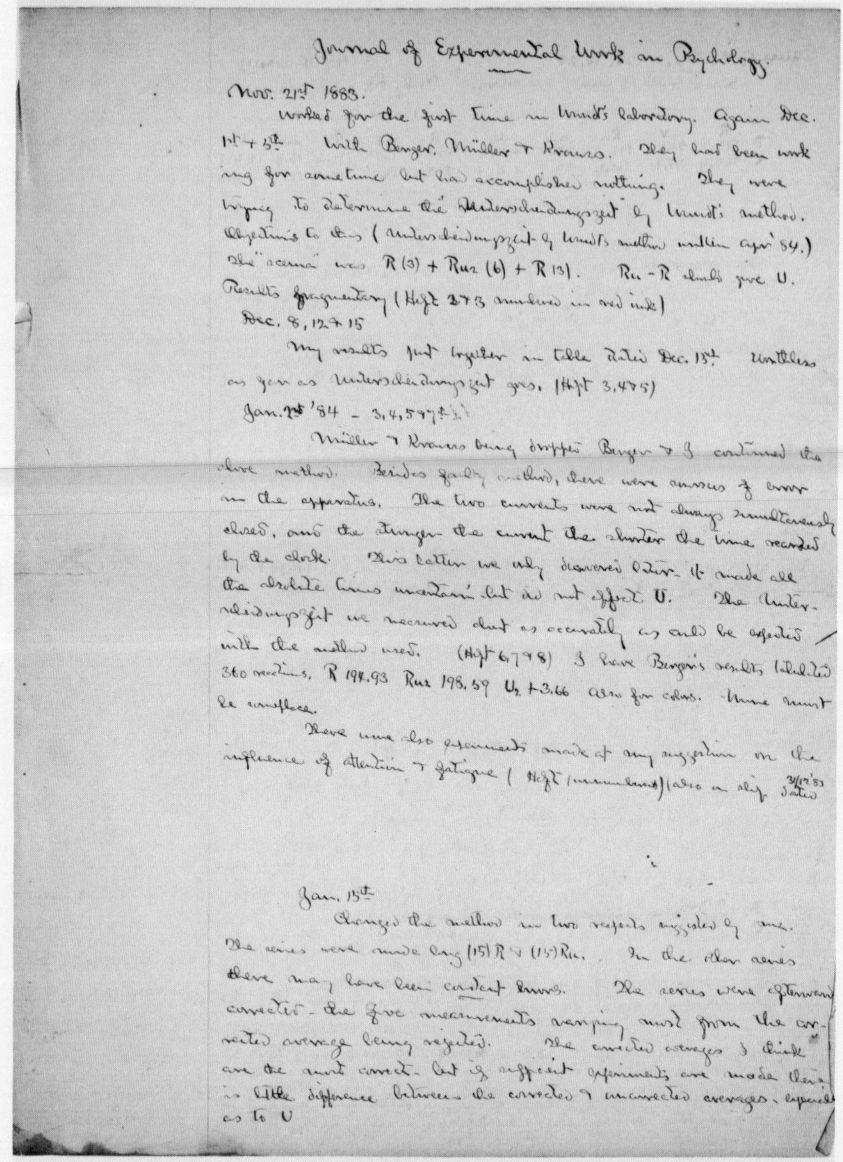

Journal of Experimental Work in Psychology.

Nov. 21st 1883.

Worked for the first time in Wundt's laboratory. Again Dec. 1st + 3rd. With Berger, Müller + Krauss. They had been working for sometime but had accomplished nothing. They were trying to determine the Unterschiedungszeit by Wundt's method. Objections to this (Unterschiedungszeit by Wundt's method written Apr' 84.) The "scema" was R (3) + Rux (6) + R (3). Ru − R should give U. Results fragmentary (Heft 373 numbered in red ink)

Dec. 8, 12 + 15

My results put together in table dated Dec. 15th. Worthless as far as Unterschiedungszeit goes. (Heft 3,4,7,5)

Jan. 2nd '84 – 3,4,5,9,7...

Müller + Krauss being dropped Berger + I continued the old method. Besides faulty method, there were sources of error in the apparatus. The two currents were not always simultaneously closed, and the stronger the current the shorter the time recorded by the clock. This latter we only discovered later. It made all the absolute times uncertain but did not effect U. The Unterschiedungszeit we received out as accurately as could be expected with the method used. (Heft 6,7,9,8) I have Berger's results tabulated 360 reactions, R 194.93 Rux 198.59 U_2 + 3.66 also for colors. These must be somewhere.

There were also experiments made at my suggestion on the influence of attention + fatigue (Heft (numbered)) (also in slip dated 3/12/83)

Jan. 15th

Changed the method in two respects suggested by me. The series were made long (15) R + (15) Ru. In the other series there may have been constant errors. The series were afterward corrected – the five measurements varying most from the corrected average being rejected. The corrected averages I think are the most correct – but if sufficient experiments are made there is little difference between the corrected + uncorrected averages – especially as to U

Figure 9. The first page of Cattell's abstract of his laboratory work during the 1883–1884 academic year (see 3.12).

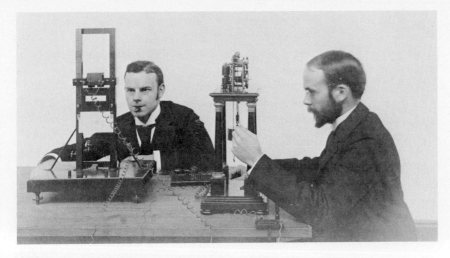

Figure 10. A reaction-time experiment, 1893. From Clark University Archives. This photograph was one of many exhibited by the Clark University
department of psychology at the 1893 Columbian Exposition in Chicago.
The method is this: When the screen of the gravity chronometer in front of
the subject (left) is allowed to fall, a word is revealed and the hands of the
Hipp chronoscope begin to revolve. When the word is read aloud the subject
opens the lip key, thus stopping the chronoscope. See 5.80.

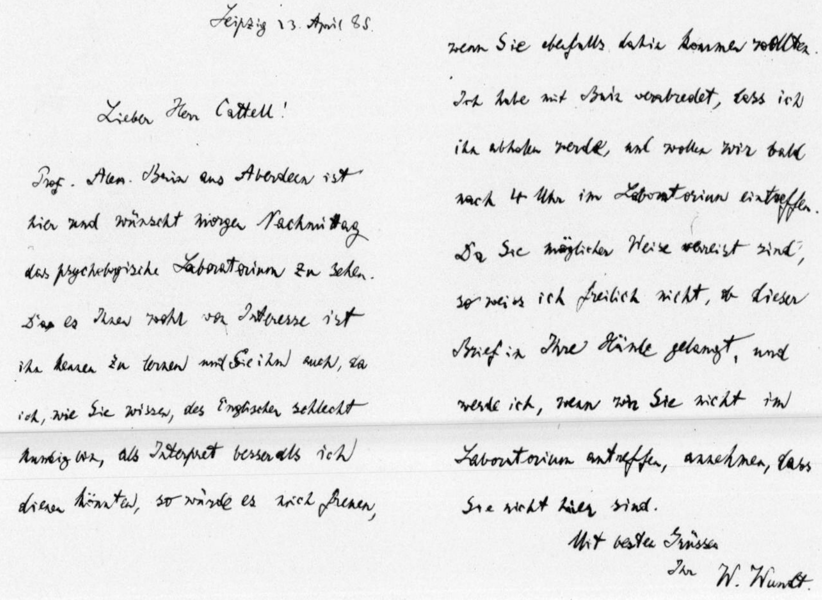

Leipzig 23. April 85.

Lieber Herr Cattell!

Prof. Alex. Bain aus Aberdeen ist hier und wünscht morgen Nachmittag das psychologische Laboratorium zu sehen. Da es Ihnen wohl von Interesse ist ihn kennen zu lernen und Sie ihm auch, da ich, wie Sie wissen, des Englischen schlecht kundig bin, als Interpret besser als ich dienen könnten, so würde es mich freuen, wenn Sie ebenfalls dabei kommen wollten. Ich habe mit Bain verabredet, dass ich ihn abholen werde, und wollen wir bald nach 4 Uhr im Laboratorium eintreffen. Da Sie möglicher Weise verreist sind, so weiss ich freilich nicht, ob dieser Brief in Ihre Hände gelangt, und werde ich, wenn wir Sie nicht im Laboratorium antreffen, annehmen, dass Sie nicht hier sind.

Mit besten Grüssen

Ihr W. Wundt

Figure 11. Letter from Wilhelm Wundt to Cattell, dated 23 April 1885. This note, which well illustrates why Wundt took up the use of a typewriter, was enclosed by Cattell in one of his letters to his parents (see 5.89, which includes a translation).

Figure 12. Francis Galton. From "Psychology in America," Cattell's 1929 address as president of the Ninth International Congress of Psychology.

Figure 13. Academic costume, St. John's College, Cambridge. From the editor's collection. Cattell wore the robes of a Fellow-Commoner while at Cambridge (see 6.5).

Figure 14. Josephine Owen, 1887. From Cattell papers, Library of Congress. Cattell had this photograph taken after his engagement by the photographer who had prepared Wundt's formal portrait, and sent it to his parents in July 1887 (see 6.91).

Figure 15. The experimental psychology laboratory at the University of Pennsylvania, showing apparatus for measuring reaction times. From Samuel W. Fernberger, "The First Psychological Laboratory at the University of Pennsylvania," *Psychological Bulletin* 25 (1928): 445. Illustrated here are some of the instruments that Cattell purchased for the university, including a Hipp chronoscope (left) and a gravity chronometer (second from right). Cattell used other apparatus seen here in later experiments at Pennsylvania. See 6.158.

MR. FRANCIS GALTON'S ANTHROPOMETRIC LABORATORY.

The Laboratory communicates with the Western Gallery containing the Scientific Collections of the South Kensington Museum. Admission to the Gallery is free. It is entered either from Queen's Gate or from Exhibition Road.

Date of Measurement.	Initials.	Birthday. Day. Month.		Eye Color.	Sex.	Single, Married, or Widowed ?	Page of Register.
11 August 88	J McK	25	5	6	Grey	M	Single	626

Head length, maximum from root of nose.		Head breadth maximum.		Height standing, less heels of shoes.		Span of arms from opposite finger tips.		Weight in ordinary clothing.	Strength of squeeze. Right hand.	Left hand.	Breathing capacity.	Keenness of Eyesight. Distance of reading diamond numerals. Right eye. Left eye.		Snellen's type read at 20 feet.	Color Sense.
Inch.	Tenths.	Inch.	Tenths.	Inch.	Tenths.	Inch.	Tenths.	lbs.	lbs.	lbs.	Cubic inches.	Inches.		Inches. No. of Type	? Normal.
7	7	5	8½	66	7	68	9	144	80	82	238	16		12 D18	Yes

Height sitting above seat of chair.		Height of top of knee, when sitting, less heels.		Length of elbow to finger tip left arm.		Length of middle finger of left hand.		Keenness of hearing.	Highest audible note.	Reaction time. To sight.	To sound.	Judgment of Eye. Error in dividing a line of 10 inches in half	in thrds	Error in degrees, estimating an angle of 90°	60°
Inch.	Tenths.	Inch.	Tenths.	Inch.	Tenths.	Inch.	Tenths.	? Normal.	Vibrations per second.	Hundredths of a second.	Hundredths of a second.	Per cent.	Per cent.	90°	60°
34	8	21	1	17	7	4	3	Yes	19,000	30	20	0	3	1	10

One page of the Register is assigned to each person measured, in which his measurements at successive periods are entered in successive lines. No names appear on the Register. The measurements that are entered are those marked with an asterisk (*). Copies of the entries can be obtained through application of the persons measured, or by their representatives, under such conditions and restrictions as may be fixed from time to time.

Figure 16. Cattell's record at Galton's anthropometric laboratory. From Cattell papers, Library of Congress.

Figure 17. James McKeen Cattell, 1888. From Cattell papers, Library of Congress. Cattell had this photograph taken while on his honeymoon in Torquay on the southern coast of England.

Postscript

Cattell at Columbia University

James McKeen Cattell stayed at the University of Pennsylvania for less than three years, despite his father's hopes. As Cattell had noted in letter 6.148, good psychologists were rare in the late 1880s and 1890s; however, American colleges were growing—some were becoming universities—and they were looking for representatives of the new and exciting approach to philosophy. In New York, Columbia College was undergoing just such a change, and by 1890 Dean Nicholas Murray Butler of the Philosophical Faculty wrote that Columbia hoped "to secure within a few months not only a specialist in Experimental Psychology, but a well-arranged laboratory and a fair stock of apparatus."[1] By the fall of 1890 Cattell was lecturing regularly at Columbia while keeping his position at Pennsylvania; a year later, he assumed the position of professor of experimental psychology at Columbia College at a salary of $2,500—twice the highest salary he had been paid at Pennsylvania.[2]

Columbia was to be the site of Cattell's greatest triumphs and disappointments. He established there a department of psychology that was clearly one of the best in the United States. By 1917, when Cattell left, the department had grown to be one of the largest in the country and had granted graduate degrees to many distinguished students.[3] Among the members of its graduate faculty were Edward L. Thorndike and Robert S. Woodworth (both of whom had earned their Ph.D.s under Cattell's direction), Harry L. Hollingworth, and Cattell's acquaintance from Europe

[1] Butler, "Psychology at Columbia College," *American Journal of Psychology* 3 (1890): 277–278.

[2] Seth Low to JMC, 18 December 1890; JMC to Low, 20 December 1890 (Cattell collection, Columbia University).

[3] Robert S. Woodworth, *The Columbia University Psychological Laboratory* (New York: Columbia University, 1942).

Charles A. Strong (see 6.72).[4] Cattell arranged for Strong's father-in-law, John D. Rockefeller, to endow a chair in experimental psychology.[5] He also played an important role in building up Columbia's other departments. He helped arrange the appointment of the respected anthropologist Franz Boas,[6] and saw to it that his students studied with Boas where appropriate. He also worked hard to develop philosophy at Columbia, working with Butler and James H. Hyslop (see 6.10) and recruiting F. J. E. Woodbridge[7] and his old friend George S. Fullerton (see 2.17). In 1904 Cattell arranged for his friend John Dewey (see introduction to section 2) to come to Columbia as professor of philosophy.[8]

But throughout Cattell's years of organizational success other things were happening. His major scientific effort of the 1890s—his program of anthropometric mental testing—had failed completely by 1901. And his relations with the administration at Columbia worsened steadily from year to year. In 1902 Nicholas Murray Butler, the man who had recruited Cattell, became president of the university.[9] Butler was a major figure in the development of American higher education, and it was during his forty-three-year presidency that Columbia became a leading university. But he was dictatorial, and clearly identified more with Columbia's trustees than with its faculty.[10] By 1902 Cattell had assumed a leading role in the movement to increase the participation of faculty members in university governance. This position brought him into direct opposition to administrators and trustees, whom he considered shallow businessmen totally ignorant of scientific and scholarly concerns. In the journals he edited Cattell began publishing articles advocating faculty participation.[11] Many of these articles, and a long series of

[4]On Thorndike see Geraldine Joncich, *The Sane Positivist: A Biography of Edward Lee Thorndike* (Middletown, Conn.: Wesleyan University Press, 1968). On Woodworth see *IESS* 16: 561–564. On Hollingworth see *Who Was Who in America*, 1951–1960, p. 410. Though few of JMC's colleagues were officially members of the faculty of Columbia College (several were officially connected with Teachers College and Barnard College), all had some standing with the Columbia faculty of philosophy and were thus really members of the university's graduate faculty in psychology. (See *A History of the Faculty of Philosophy, Columbia University* [New York: Columbia University Press, 1957].)

[5]JMC to Rockefeller, 3 February 1899 and 8 February 1899 (Cattell papers). See also Alan Nevins, *John D. Rockefeller: The Heroic Age of American Enterprise* (New York: Scribner, 1940), vol. 2, pp. 173, 220, 457.

[6]See Ronald P. Rohner, ed., *The Ethnography of Franz Boas: Letters and Diaries of Franz Boas*

Written on the Northwest Coast from 1886 to 1932 (Chicago: University of Chicago Press, 1969), pp. 241, 311.

[7]JMC to Seth Low, 16 April 1900 (Cattell collection); *EoP* 8: 345–346.

[8]Jane M. Dewey, "Biography of John Dewey," in Paul Arthur Schillp, ed., *The Philosophy of John Dewey* (Evanston, Ill.: Northwestern University Press, 1939), pp. 3–45; George Dykhuizen, *The Life and Mind of John Dewey* (Carbondale: Southern Illinois University Press, 1973).

[9]See *DAB* Supplement 4: 133–138; Laurence R. Veysey, *The Emergence of the American University* (Chicago: University of Chicago Press, 1965), pp. 394–395, 426–428.

[10]See Veysey, *American University*, pp. 306–307.

[11]For example, "Concerning the American University," *Popular Science Monthly* 61 (1902): 170–182; "University Control," *Science* 23 (1906): 475–477.

letters from other university professors supporting Cattell's position, were published in 1913 as *University Control*, a book often cited as a landmark in the development of American academic freedom.[12] The book also played a major role in stimulating Thorstein Veblen (see 2.57) to write *The Higher Learning in America: A Memorandum on the Conduct of Universities by Business Men*, the most significant analysis of this issue published in the United States before 1920.[13]

Cattell and Butler could hardly have avoided clashing. Butler's administrative style brought out in Cattell what his mother had referred to as his "McKeen nature," and by 1905 the two were battling in public. Cattell called Butler an autocrat and other unflattering things. But perhaps Cattell's most serious charge was the one implied in an anecdote he told about one of his daughters:

I once incited one of my children to call her doll Mr. President, on the esoteric grounds that he would lie in any position in which he was placed.[14]

Cattell's charges against Butler were similar to those he had made against Hall and Gilman after the renewal of his fellowship at Johns Hopkins had been denied. Such was Cattell's reputation that in 1913 a book about Columbia bragged that his presence showed how committed the institution was to academic freedom.[15] But this commitment was not absolute. In one of the most serious academic freedom cases of the World War I period, Cattell was dismissed from Columbia in 1917.[16] The immediate reason for this dismissal was a question Cattell had raised about American conscription policy during the war,[17] but it was clear to all that his activities were just the "last straw" and that he was actually fired because of his continual prodding of Butler. He eventually collected $42,000 from Columbia, officially in settlement of his pension but probably, at least in part, in payment for damages the dismissal had caused him.[18]

[12]JMC, *University Control* (New York: Science Press, 1913); see Veysey, *American University*, pp. 392–397.

[13]New York: B. W. Huebsch, 1918.

[14]JMC apparently first used this anecdote in "Academic Slavery," *School and Society* 6 (1917), 421–426; see *Man of Science*, vol. 2, pp. 346–353.

[15]Frederick P. Keppel, *Columbia* (New York: Oxford University Press, 1914), p. 160.

[16]This case has been treated exhaustively, and quite well, in Carol Singer Gruber's "Academic Freedom at Columbia University, 1917–1918: The Case of James McKeen Cattell," *AAUP Bulletin* 58 (1972): 297–305,

and *Mars and Minerva: World War I and the Uses of the Higher Learning in America* (Baton Rouge: Louisiana State University Press, 1975).

[17]JMC had written to members of Congress (on his own stationery, which gave Columbia University as his return address) questioning the policy of sending draftees to fight in Europe. Soon afterwards, Butler received letters from several congressmen charging the university with harboring traitors. (See Gruber, "Academic Freedom at Columbia," p. 301.)

[18]Ibid., passim.

Cattell was almost destroyed by this incident. For many months he stayed at home, writing and publishing bitter pamphlets with such titles as "Memories of My Last Days at Columbia" and "What the Trustees of Columbia Have Done"[19] and seeing only such old friends as Dewey, Boas, and Thorndike.[20] Cattell eventually recovered and continued his career as a scientific editor, which had begun almost twenty-five years earlier. But, as might be expected, he was bitter about his Columbia experiences for the rest of his life.[21]

Cattell as Psychologist

Cattell continued his experiments in America, but the work he had done earlier with Galton and Wundt had begun to have many ramifications even before he left Europe. For example, as has been suggested earlier (see 5.17), Wundt attempted to get around Cattell's objections to the methodology he had used in his reaction-time experiments. In doing so, Wundt distinguished between sensory reactions (those in which the observer's attention was directed towards the stimulus) and muscular reactions (those in which the observer's attention was directed towards the reaction). He claimed that sensory reactions involved apperception whereas muscular reactions did not, and that the reactions Cattell measured were muscular rather than sensory.[22] Later, in part as a result of additional experiments by several of his students,[23] Wundt generalized this hypothesis and wrote of reactions with apperception as "complete" and those without apperception as "incomplete."[24] Cattell performed several experiments in America following the

[19]These pamphlets were published privately in 1918 out of JMC's home, and were headed "Confidential Statement for Members of the American Association of University Professors" or "Confidential Statement for Members of the Faculty Club."

[20]See for example the files of correspondence with these three men in the Cattell papers. See also JMC to Boas, 23 October 1918, 30 October 1918, and especially 1 October 1919: "During the past two years I have stayed pretty continuously at home and sometimes for a month or two at a time." (Boas papers, American Philosophical Society Library.)

[21]Cattell "Autobiography," passim.

[22]For a discussion of this point, including translations into English from the relevant

German documents by Wundt and others, see Solomon Diamond, *The Roots of Psychology: A Sourcebook in the History of Ideas* (New York: Basic Books, 1974), pp. 694–713. See also Robert S. Woodworth, *Experimental Psychology* (New York: Holt, 1938), pp. 306–308.

[23]See, for example, Ludwig Lange, "Neue Experimente über den Vorgung der einfachen Reaction auf Sinneseindrücke," *Philosophische Studien* 4 (1888): 479–510; Götz Martius, "Über die muskuläre Reaction und die Aufmerksamkeit," ibid. 6 (1888): 167–216.

[24]Wundt, *Grundzüge der physiologischen Psychologie*, 3rd edition (Leipzig: Engelmann, 1887), vol. 2, pp. 261–364; "Zur Beurtheilung der zusammengesetzten Reactionen," *Philosophische Studien* 10 (1894): 485–498.

procedures suggested by Wundt's students, and found that the attention of his subjects did not affect their reaction times in any way.[25] He then left this question for others to debate; it formed the basis for what has been termed the reaction-times controversy of the 1890s.[26] The importance of this episode can be overstated, but in many ways it helped delineate two schools of psychology that soon were to compete for the allegiance of psychologists: structuralism and functionalism.[27] Structuralism (which was derived from but not identical to Wundt's work and was in America most closely identified with Edward Bradford Titchener of Cornell)[28] concentrated on what its adherents felt were studies of the way in which the human mind was structured, and accepted Wundt's distinction between complete and incomplete reactions.[29] Functionalism (never as formally defined as structuralism; derived loosely from a Darwinian concern for the ways different individuals functioned in the world) argued that reaction times were determined by many factors other than the direction of the observer's attention.[30] It is too much to say that this distinction was derived directly from Cattell's questioning of Wundt's methodology, but clearly Cattell's challenges of 1883–1884 helped to shape some of the background out of which a functional approach to psychology emerged.

Cattell's more narrowly defined psychophysical experiments (see 5.136 and 6.158) also influenced the development of psychology in America. At Pennsylvania, Cattell was again to challenge several of the assumptions and techniques of classical Wundtian psychophysics—for example, the idea that through introspection an experimental subject could be taught to examine his own sensations and that the sensitivity of the human mind in observing sensations of different types could therefore be determined. Cattell, from his physicalist perspective, questioned that anything like sensation—a purely mental quality—existed. From a series of long and detailed experiments with George S. Fullerton,[31] Cattell argued that even if it is possible to

[25] JMC, "Aufmerksamkeit und Reaktion," *Philosophische Studien* 8 (1893): 403–406; tr. by Robert S. Woodworth as "Attention and Reaction," in *Man of Science*, vol. 1, pp. 252–255.

[26] See David L. Krantz, "The Baldwin-Titchener Controversy: A Case Study in the Functioning and Malfunctioning of Schools," in David L. Krantz, ed., *Schools of Psychology: A Symposium* (New York: Appleton-Century-Crofts, 1969), pp. 1–19.

[27] See, for example, Edna Heidbreder, *Seven Psychologies* (New York: Appleton-Century-Crofts, 1933). It is, of course, quite clear that the importance of schools for the development of American psychology has been greatly exaggerated, and that the differences in their approaches were minor compared with the perspectives they shared.

[28] See Arthur L. Blumenthal, "Wilhelm Wundt and Early American Psychology: A Clash of Cultures," in R. W. Rieber and Kurt Salzinger, eds., *The Roots of American Psychology: Historical Influences and Implications for the Future* (New York: New York Academy of Sciences [*Annals*, vol. 291], 1977), pp. 13–20.

[29] Titchener, "Simple Reactions," *Mind*, new series, 4 (1895): 74–81.

[30] Livingston Farrand, "Note on Reaction Types," *Psychological Review* 4 (1897): 297–298.

[31] JMC and George Stuart Fullerton, *On the Perception of Small Differences; With Special Reference to the Extent, Force and Time of Movement* (Publications of the University of Pennsylvania, Philosophical Series, no. 2, 1892).

introspect it is impossible to quantify sensations.[32] And, as a follower of Comtean ideas, he held that unquantifiable results were of little value. He therefore set out to show that quantifiable (and hence more meaningful) results could be obtained if the stimulus causing the sensation and the subject's error in observing the stimulus were the only data measured[33]— both physical quantities determinable without introspection by an experimenter bearing the same relation to the subject as in a reaction-time experiment.

Not only was Cattell attacking the premise of introspection; he was also shifting the attention of the experimenter to the behavior of the subject. As in his earlier reaction-time experiments, Cattell had the subject's actions observed by a psychologist who carefully controlled the conditions of the experiment. He claimed in a major public address in 1904 that "most of the research work that has been done by me or in my laboratory is nearly as independent of introspection as work in physics or in zoology."[34] This statement has led several writers concerned with the history of psychology to call Cattell a founder of behaviorism, the dominant American psychological perspective of the twentieth century.[35] It is oversimplifying to advance such a thesis on the basis of a rhetorical statement; clearly behaviorism as a self-conscious school owes more to animal experimentation than to any work with humans.[36] But the work behind Cattell's statement—both in psychophysics and in what he called psychometry—shows that he was concerned with behavioral data as early as the 1880s. In fact, in a passage presented as part of the 1904 address but not published, he cited his early work (see 3.12):

When I was a student in the Leipzig laboratory, attempts were being made to measure the time of perception by letting the subject react as soon as he knew from introspection that an object had been perceived . . . . I attempted to continue these experiments, but, feeling no confidence in the validity of my introspection in such a case, took up strictly objective methods in which a movement followed a stimulus without the slightest dependence on introspection.[37]

[32] Ibid., p. 21. See also 6.29. JMC's criticisms of this aspect of classical psychophysical methodology was quite similar to those set forth by William James at about the same time: "Our feeling of pink is surely not a portion of our feeling of scarlet" (James, *The Principles of Psychology*, [New York: Holt, 1890], vol. 1, p. 546). The question of what different psychologists thought was observed in their psychophysical experiments is analyzed in depth in Edwin G. Boring, "The Stimulus-Error," *American Journal of Psychology* 32 (1921): 449–471.

[33] *On the Perception of Small Differences*, p. 28.

[34] "The Conceptions and Methods of Psychology," *Popular Science Monthly* 66 (1904): 176–186 (presented at the International Congress of Arts and Sciences, St. Louis, Missouri, September 1904).

[35] See, for example, Robert S. Woodworth, *Contemporary Schools of Psychology* (New York: Ronald, 1931), pp. 47–49 (2nd edition, 1948, pp. 72–73; 3rd edition, in collaboration with Mary R. Sheehan, 1964, pp. 114–115).

[36] John C. Burnham, "On the Origins of Behaviorism," *JHBS* 4 (1968): 143–151.

[37] Sheaf of pages, "Omitted from St. Louis address" (Cattell papers).

But Cattell is best known in the history of psychology for developing mental testing. His work in this field in America can be traced directly to the influence of Galton (see introduction to section 6), who had been led to Cattell by his attempt to extend physical anthropometry into investigation of the psychological differences between people. Soon after Cattell returned to America he was writing on mental tests and was developing techniques for what might be termed mental anthropometry. Cattell first used the term "mental test" in an 1890 paper that clearly showed the Galton influence.[38] This article proposed two series of tests (one of ten measurements and one of fifty), both of which show that Cattell's interest was at least as much in physical anthropometry as in mental testing. For example, among the shorter series, to which Cattell devoted most of his attention, were such standard anthropometric measurements as the maximum squeezing strength of each hand and smallest spacing between two points of a compass that were sensed as two distinct points. To be sure, some more psychological measurements were included in this series—each subject was asked to bisect a line of a given length, and to judge a ten-second interval—but the focus was clearly anthropometric.[39] There is no evidence that either of Cattell's series of tests was used on any individual or group, and Cattell apparently devoted most of his scientific interest during the early 1890s to the psychophysical experiments discussed above. But Cattell did not lose interest in his tests during this period; he wrote at least one popular article on the subject, and he regularly reviewed books and articles on anthropometry.[40]

Meanwhile, Cattell moved from Pennsylvania to Columbia and made plans to carry out a large-scale program of anthropometric mental testing.[41] In September 1894 he began an examination of 100 students, mostly freshmen at Columbia College's School of Arts and School of Mines. He repeated his tests each fall for the next several years on members of the freshmen class of each school, and also examined several of these classes in their senior years.[42] In all, it was a most ambitious program of testing, at least as extensive as Galton's in London (see introduction to section 6) or as the

[38] JMC, "Mental Tests and Measurements," *Mind* 15 (1890): 373–381.
[39] Ibid., pp. 373–375.
[40] See, for example, JMC, "Tests of the Senses and Faculties," *Educational Review* 5 (1893): 257–265; "Psychological Literature: Anthropometry," *Psychological Review* 2 (1895): 510–511.
[41] JMC to Seth Low, 30 January 1893, 27 September 1894 (Cattell collection).
[42] Articles on this testing program were regularly published in Columbia publications—for example, the *Bulletin* (no. 10 [March 1895]: 21–22; no. 14 [July 1896]: 26) and the *Quarterly* 1 [1898–1899]: 285–287)—and JMC spoke on them before meetings of the New York Schoolmasters' Club, the New York Academy of Sciences, the American Psychological Association, and the American Association for the Advancement of Science (see, for example, "New York Academy of Sciences," *Science* 1 [1895]: 727). The most comprehensive paper on the program was by JMC and Livingston Farrand, a junior colleague who helped him carry out the tests: "Physical and Mental Measurements of the Students of Columbia University," *Psychological Review* 3 (1896): 618–648.

other American testing program of the 1890s: the series of tests developed by Joseph Jastrow of the University of Wisconsin (see 2.41) and given at the 1893 Chicago Columbian Exposition.[43]

Cattell's series of tests was not a polished and carefully defined program of psychological examination. As he noted more than once, he did not know exactly what such a program of tests should measure, or even what the series of tests he used actually did measure. As an American Baconian, Cattell concluded that "the best way to obtain the knowledge we need is to make the tests, and determine from the results what value they have."[44] Indeed, Cattell explicitly designed his program to focus more on "measurements of the body and of the senses" than on "the higher mental processes." As he concluded, "we are concerned here with anthropometric work," and whatever limits the tests had were clearly the results of Cattell's own experimental design.[45]

By the late 1890s Cattell and his colleagues had collected masses of data on the breathing capacity, keenness of sight, and reaction times of hundreds of students at Columbia. Cattell's lead was followed by other psychologists in the mid-1890s, and similar though much smaller testing programs were established at such universities as Yale and Wisconsin.[46] But other psychologists soon began to criticize Cattell's tests. In December 1895 the American Psychological Association organized a Committee on Physical and Mental Tests charged "to consider the feasibility of cooperation among the various psychological laboratories." Two years later, this committee's report to the Association discussed at length what the focus of mental tests should be.[47] Most of the members of the committee, including Cattell, Jastrow, and Edmund C. Sanford of Clark University, urged that tests concentrate on "the senses and the motor capacities"—the areas on which Cattell's program focused—rather than on "the more complex mental processes." But one member, James Mark Baldwin of Princeton, argued against such a program of testing, criticized what he felt was too great a stress on anthropometric measurements, and urged that the tests be given "as psychological a character as possible."[48]

[43] See 2.41, 6.33; Jastrow, "Experimental Psychology at the World's Fair," in JMC, ed., *Proceedings of the American Psychological Association, 1892–1893* (New York: Macmillan, 1894), p. 8; reprinted as "APA's First Publication," *American Psychologist* 28 (1973): 277–292.

[44] JMC to Seth Low, 30 January 1893 (Cattell collection).

[45] "Physical and Mental Measurements," pp. 622–623.

[46] See, for example, Joseph Jastrow, "Statistical Study of Mental Development," *Transactions of the Illinois Society for Child Study* 2 (1897): 100–108; Edward W. Scripture, "Tests of Mental Ability as Exhibited in Fencing," *Studies from the Yale Psychological Laboratory* 2 (1894): 122–124.

[47] James M. Baldwin, JMC, and Joseph Jastrow, "Physical and Mental Tests," *Psychological Review* 5 (1898): 172–179.

[48] Ibid., pp. 174–176.

These criticisms of Cattell's anthropometric tests were mild, but they were only the first wave. In 1898 Stella Emily Sharp, a student of E. B. Titchener at Cornell, published a doctoral dissertation on "individual psychology"[49] in which she compared Cattell's tests with those being developed at the same time in France by Alfred Binet and his colleagues,[50] which were explicitly concerned with what Baldwin had referred to as "the more complex mental processes."[51] Sharp argued that the qualities measured by Cattell's tests were totally independent of one another and not aspects of one or another "fundamental process," that these tests were not useful in the study of any mental or physical trait beyond the narrowly defined sensory or motor capacity being measured, and that an anthropometric testing program could not in itself tell anything about the psychology of the people tested, either individually or as a group.[52] By 1901 one of Cattell's own students, Clark Wissler, was examining his teacher's program of tests, using mathematical techniques beyond Cattell's understanding. Cattell knew that mathematics had never been his forte, but he also knew that Galton had developed important mathematical methods that allowed the measurement of how closely two sets of data were correlated.[53] At Cattell's urging, then, Wissler began working closely with Boas, who with his Ph.D. in physics was much more mathematically sophisticated than any of the other members of the group around Cattell.[54] With Boas's help, Wissler applied Galton's correlation techniques to the results of Cattell's "physical and mental tests."[55]

Wissler found almost no correlation between the results of any one set of Cattell's tests and those of any other. For example, one of Cattell's tests had called for the striking out of ten A's distributed randomly in a 10-by-10 array of letters. In comparing the result of this test with the results of the reaction-time test for the hundreds of individuals who took both tests, Wissler found a correlation of -0.05, so that "an individual with a quick reaction-time is no more likely to be quick in marking out A's than one with a slow reaction-time."[56] He also measured the correlations between the test results and the subjects' academic performance, and the correlations among various sets of grades. Wissler found no correlation between a student's overall academic performance and his scores on the various tests, and determined that a student's performance in any one class correlated better with his standing than did the results of any of Cattell's tests.[57]

[49] Stella Emily Sharp, "Individual Psychology: A Study in Psychological Method," *American Journal of Psychology* 10 (1898): 329–391.

[50] Alfred Binet and Victor Henri, "La Psychologie individuelle," *L'Année psychologique* 2 (1895): 411–465.

[51] See Theta Wolf, *Alfred Binet* (Chicago: University of Chicago Press, 1973).

[52] Sharp, "Individual Psychology," pp. 332, 340–342.

[53] See Ruth Schwartz Cowan, "Francis Galton's Statistical Ideas: The Influence of Eugenics," *Isis* 63 (1972): 509–528.

[54] Rohner, *The Ethnography of Franz Boas*, pp. 208–209.

[55] Clark Wissler, "The Correlation of Mental and Physical Tests," *Psychological Review Monograph Supplements* 3 (1901), no. 6.

[56] Ibid., pp. 22–34.

[57] Ibid., pp. 34–36.

After Wissler's dissertation, Cattell's testing program was effectively dead and his career as an active researcher in experimental psychology was over, although in a sense he continued his interest in the differences between individuals through his statistical studies of eminence,[58] and he eventually founded The Psychological Corporation.[59] But after 1900 Cattell's career was primarily that of a scientific editor and what might be called an entrepreneur of science.[60] In 1894, with Baldwin, he had founded *Psychological Review*, which he co-edited and owned jointly for ten years. Later that year he purchased *Science*, the national weekly founded in 1880 by Thomas A. Edison.[61] *Science*—which had published articles (see 6.33) praising Cattell's psychological work—had ceased publication in 1894 when its backers (notably Alexander Graham Bell) had withdrawn their financial support. Cattell rushed to its rescue, resuscitated it with the help of an editorial committee composed of leading members of the American scientific community, and turned its fortunes around.[62] Where Edison and Bell had failed, Cattell succeeded.

For the next fifty years, until his death in 1944, James McKeen Cattell devoted most of his energies to *Science*. In 1900 it became the official journal of the American Association for the Advancement of Science,[63] and within ten years it clearly had become the most important general scientific periodical in the United States. His editorship gave Cattell an important position within the AAAS and within the American scientific community at large. Especially after his dismissal from Columbia University, he was not loath to use this position to foster his own goals. (But whereas William Cattell had been able to help secure a fellowship and a faculty position for his son, James Cattell was probably less successful in promoting his own goals.) It is clear that James Cattell was opposed to the way the American scientific community was developing in the twentieth century. He felt strongly that

---

[58] For example, "A Statistical Study of Eminent Men," *Popular Science Monthly* 62 (1903): 359–377; "Homo Scientificus Americanus," *Science* 17 (1903): 561–570; "Statistics of American Psychologists," *American Journal of Psychology* 14 (1903): 310–328.

[59] JMC, "The Psychological Corporation," *Annals of the American Academy of Political and Social Science* 110 (1923): 165–171.

[60] This aspect of JMC's career is reviewed well in *DSB* 3: 130–131.

[61] An overview of the history of *Science* is found in Frank Luther Mott, *A History of American Magazines*, vol. 4, *1885–1905* (Cambridge, Mass.: Belknap Press of Harvard University Press, 1957), pp. 307–308. See also "Thomas A. Edison and the Founding of *Science*: 1880," *Science* 105 (1947): 142–148.

[62] [Gardiner G. Hubbard], "Science," *Science* 1 (1895): 352–353. See also Sally Gregory Kohlstedt, "*Science*: The Struggle for Survival, 1880 to 1894," *Science* 209 (1980): 33–42.

[63] JMC, "The Journal *Science* and the American Association for the Advancement of Science," *Science* 64 (1926): 342–347. See also Michael M. Sokal, "*Science* and James McKeen Cattell, 1894 to 1945," *Science* 209 (1980): 43–52.

too much power was concentrated in the elitist National Academy of Sciences and in the hands of the private philanthropic foundations that were underwriting more and more of the work done by scientists.[64] To him the businessman-trustee of a foundation knew as little about science as any university trustee, and he favored a scheme wherein the scientists would control their own community[65]—primarily through the American Association for the Advancement of Science.

Besides *Science*, Cattell also edited the *Popular Science Monthly* from 1900, when he bought it from William J. Youmans. In 1915 he sold the name of this periodical and founded *The Scientific Monthly* to take its place. He edited this journal through the 1930s.[66] From 1907 until his death Cattell owned and edited *The American Naturalist*, a journal that played an important role in the development of genetics in the United States,[67] and from 1915 through 1944 he edited the educational weekly *School and Society*.

The Cattell Family

Cattell had always wanted to live in the country, and preferably in the mountains (see 5.1, 5.31, and 6.170). Soon after he was appointed to his position at Columbia he found his perfect home site about fifty miles north of New York City, almost directly across from West Point and about three or four miles inland from the east bank of the Hudson River, near the ruins of a Revolutionary War fort, Fort Defiance, on the top of a fairly high hill. By November 1891 the house was finished, and for many years thereafter the Cattell address would be Fort Defiance, Garrison-on-Hudson, New York.[68]

Life in the Cattell home centered around Josephine Owen Cattell, and her husband never had any reason to change the opinion he expressed to his parents right after his wedding: "Jo is just perfect." James Cattell had derived from Galton a belief in eugenics, and he felt that his "hereditary material" was good. The couple had seven children, born between 1890 and 1904. Josephine Cattell took total charge of the children from the time of their

[64]JMC, "The Organization of Scientific Men," *Scientific Monthly* 14 (1922): 567–577; "The Carnegie Institution," *Science* 16 (1902): 460–469.

[65]Nathan Reingold, "National Aspirations and Local Purposes," *Transactions of the Kansas Academy of Sciences*, 71 (1968): 235–246.

[66]JMC, "The *Scientific Monthly* and the American Association for the Advancement of Science," *Scientific Monthly* 47 (1938): 468–469.

[67]L. C. Dunn, "The American Naturalist in American Biology," *American Naturalist*, 100 (1966): 481–492; "Ave Atque Vale," ibid. 101 (1967): 427–430.

[68]McKeen Cattell, interviews by Michael M. Sokal, 31 October 1970, 1 November 1970, and 22 May 1971, Fort Defiance, Garrison-on-Hudson, New York.

birth. She tutored all seven children at home until they had learned the rudiments of reading and writing; then Cattell had his graduate students tutor them in more advanced subjects.[69]

All seven of the children went on to college, and two had distinguished scientific careers. McKeen, the second child (born November 1891), earned both Ph.D. (in 1920) and M.D. (in 1924) degress from Harvard and was longtime head of the pharmacology department at Cornell Medical College. Psyche, the third child (born August 1893), earned a Harvard Ed.D. in 1927. In 1939, after a successful career in research at the Harvard School of Public Health and at the Psycho-Educational Clinic, she established a practice in clinical psychology for children that became connected with one of the best-known nursery schools in the United States.[70]

Ware Cattell, the sixth child, took over his father's editorship of *The Scientific Monthly* in the late 1930s, and Jaques (pronounced Jack), the seventh, edited *The American Naturalist* until 1956.[71] In fact, the journals were always a family affair—Ware's fiancée visited Fort Defiance one summer and was informed that she would take charge of the book reviews of *Science* or *The Scientific Monthly* after the marriage.[72]

As James Cattell assumed control of more and more journals, his wife gradually took on many of the editorial duties. This editorial role was only acknowledged sporadically,[73] though often it was she alone who kept the journals coming out on time. For example, in 1917–1918, just after his dismissal from Columbia, Cattell was unable to do any work on them, and most of his children were away at college, and he lacked access to graduate students. Josephine Cattell edited all the journals singlehandedly without missing an issue.[74] After James Cattell's death in 1944, his widow and their son Jaques were listed as transitional editors of *Science* for a year;[75] this was the only time Josephine Cattell's contribution was recognized formally. She died in 1948.

[69]Ibid.; Psyche Cattell, interview by Michael M. Sokal, 29 July 1971.

[70]*Who's Who in the East*, 1st edition (1943), p. 569; 16th edition (1977–1978), p. 123.

[71]Dunn, "Ave Atque Vale," pp. 429–430.

[72]Anna D. Burgess to Michael M. Sokal, 23 September 1978.

[73]See, for example, JMC, "The Journal *Science* and the American Association for the Advancement of Science."

[74]McKeen Cattell and Psyche Cattell, interviews by Michael M. Sokal at Lancaster, Pa.

[75]F. R. Moulton, "The Editing of *Science*," *Science* 101 (1945): 8–10.

Index

Aberdeen, University of, faculty, 5.89, 5.91, 6.36
Adams, Henry, 16
Adamson, Robert, 6.26
Adonais (Shelley), 2.46
Albany, N.Y., 4.7
Alcohol, JMC's experiments with, 2.38, 2.39, 93
Alexander, Samuel, 6.13, 6.70, 6.115
American Association for the Advancement of Science, 339–340
American Exchange, London, 1.3
American Exhibition, London, 6.74
American Naturalist, 340, 341
American Philosophical Society, 6.150
American Psychological Association, 10, 5.25, 337
American Society for Psychical Research, 6.52
American students in Germany, 1–6, 9, 11
American Tract Society, 4.2
American Travellers' Guides, 19
Annapolis, Md., 2.58
Anstie, Francis Edmund, 2.24
Anthropological Institute, 5.156, 222, 6.56, 6.164
Anthropology, at Columbia University, 331
Anthropometric laboratory, London, 113, 5.46, 222, 6.56, figure 16

Anthropometric mental tests, 222–223, 331, 336–338
Anthropometry, 5.46, 222–223, 6.56, 6.74, 6.152, 6.154, figure 16, 336–337
Antioch College, 48
Antwerp, 5.111, 5.149, 5.152, 215
Archer-Hind, Richard Dacre, 6.67
Argand burner, 5.4, 5.6
Aristotelian Society for the Systematic Study of Philosophy, 221, 6.115, 6.164
  announcements of, 6.22, 6.136
  JMC addressing, 6.13, 6.23, 6.57, 6.59, 6.76–6.78, 6.136
  meetings of, 6.26, 6.28, 6.34, 6.60, 6.70, 6.74, 6.137, 6.139
Aristotle, 2.4
Armstrong, Andrew Campbell, 2.64, 5.22
Army Medical Museum, 5.46, 5.64, 5.80, 6.37
Arnault de Nobleville, Louis Daniel, 6.47
Arnold, Matthew, 6.77
Art, 2, 25, 27–28, 30, 44
  in Baltimore, 2.24, 2.52
  in Berlin, 1.6, 6.23
  in Dresden, 2–3, 6.3, 6.109
  in London, 1.3, 5.111, 5.155, 5.156, 221, 6.66, 6.70, 6.74, 6.75

Cattell, James McKeen (cont.)
"The Psychological Laboratory at Leipzic" (*Mind* 13 [1888]: 37–51), 5.149, 6.100, 6.105, 6.106, 6.111, 6.114, 6.125, 6.127, 6.136, 6.137, 6.155
"Psychometrische Untersuchungen" (*Philosophische Studien* 3 [1886]: 305–335, 452–492), 5.22, 5.130, 5.136, 5.142, figure 8. *See also* JMC at Leipzig, Ph.D. dissertation
"Recent Books on Physical Psychology" (*Brain* 11 [1889]: 263–266), 6.80, 6.147, 6.152, 6.171
"Recent Psychophysical Research" (unpublished), 6.13, 6.22, 6.23, 6.57, 6.59
review of George Trumbull Ladd, *Elements of Physiological Psychology* (*Mind* 12 [1887]: 587–589), 6.76, 6.91, 6.95, 6.103, 6.112, 6.139, 6.145, 6.152
review of Wilhelm Wundt, *Grundzüge der physiologischen Psychologie*, 3rd edition (*Mind* 13 [1888]: 435–439), 6.111, 6.122, 6.136, 6.140, 6.145, 6.152
"The Time It Takes to See and Name Objects" (*Mind* 11 [1886]: 63–65), 5.122
"The Time It Takes to Think" (*Nineteenth Century* 22 [1887]: 63–65), 5.60, 5.138, 6.57, 6.131, 6.135, 6.136, 6.171
"The Time Taken Up by Cerebral Operations" (*Mind* 11 [1886]: 220–242), 377–392, 524–538, 5.22, 5.116, 5.134, 5.136, 5.142, 5.154, 5.158
"Über die Tragheit der Netzhaut und des Sehcentrums" (*Philosophische Studien* 3 [1885]: 94–127), 5.103, 5.110, 5.111, 5.117
"Über die Zeit der Erkennung und Benennung von Schriftzeichen, Bildern und Farben" (*Philosophische Studien* 2 [1885]: 635–650), 2.41,

2.51, 5.70, 5.71, 5.73, 5.76, 5.95, 5.98, 5.102, 5.105, 5.108
"The Way We Read" (unpublished), 6.76, 6.77, 6.101, 6.131
and Alexander Bain, 5.91, 5.97, 5.155, 5.156, 6.52, 6.159
and Bryn Mawr College, 6.121, 6.128, 6.129, 6.130, 6.138, 291
and Nicholas Murray Butler, 331–332, 341
at Cambridge University
experiments. *See* JMC, experiments of
initial appointment, 215, 6.2
interest in fellowship at St. John's College, 6.11, 6.14, 6.49, 6.54, 6.65, 6.79, 6.89, 6.107, 6.131, 6.132, 6.138, 6.141
interest in machining, 6.157, 6.158
lecturing, 6.54, 6.65, 6.138, 6.141, 6.152, 6.165–6.168, 6.178, 6.179
rowing, 6.31, 6.87, 6.102, 6.117, 6.151, 6.166
social life, 6.1, 6.7, 6.22, 6.27, 6.36, 6.50, 6.56, 6.62, 6.69, 6.73, 6.78, 6.80, 6.102, 6.133, 6.141, 6.149
studying medicine, 6.1, 6.2, 6.4, 6.35, 6.47, 6.67
studying moral sciences, 6.15, 6.16, 6.50, 6.67, 6.78, 6.118, 6.166
studying philosophy, 6.12, 6.47, 6.50, 6.67, 6.118, 6.166
studying psychology, 6.16, 6.35, 6.47, 6.67, 6.76, 6.82, 6.86, 6.95, 6.102, 6.118, 6.120, 6.127, 6.166
value of to JMC, 219, 6.27, 6.73, 6.78, 6.81, 6.94, 6.107, 6.110, 6.138, 6.141, 6.179
on cities, 6.71, 6.170, 341
and Columbia University
colleagues at, 330–331
dismissal from, 332–333, 341
mental testing at. *See* Anthropometric mental tests
psychology at, 6.156, 330–331, 333–338
on death and suicide, 1.15, 1.16, 2.4, 2.12, 5.116, 5.143, 5.157, 6.11,

Flüela Pass, Switzerland, 5.143
Forsyth, Andrew Russell, 6.13, 6.25, 6.50
Fort Defiance, 340
Foster, Michael, 5.156, 6.1, 6.2, 6.4, 6.36, 6.47, 6.49, 6.55, 6.102, 6.141
Foxwell, Herbert Somerton, 6.50, 6.62, 6.67, 6.102, 6.133
Fraley, Frederick, 2.17, 6.6, 6.176
Frankfurt, 26, 5.111
Frege and Company, 5.1, 5.14
Freiburg, University of, 5.56
French language, JMC's study of, 1.10–1.12
Friedenwald, Julius, 5.18
Fullerton, George Stuart, 6.20, 6.65, 6.129, 6.171, 6.172, 6.176, 331
On the Perception of Small Differences (with JMC), 5.136, 6.29, 6.158, figure 15, 334–335
teaching at Bryn Mawr College, 6.128, 6.143
teaching at University of Pennsylvania, 2.17, 2.19
Functionalism, 334
The Functions of the Brain (David Ferrier), 5.110, 6.47
Future of psychology, JMC on, 5.127

Gallia (steamship), 4.2, 4.9, 4.10
Galton, Francis, 3.7, 221–222, figure 12, 336, 340
Anthropometric Laboratory of, 113, 5.46, 222, 6.56, figure 16, 336
correspondence with JMC, 5.144, 5.155, 218, 222, 6.22, 6.74, 6.80, 6.152, 6.154, 6.156
and JMC's proposed textbook, 6.152, 6.154
and John Shaw Billings, 113, 5.46, 222
meeting with JMC, 5.116, 5.156, 6.28, 6.80
Gambetta, Leon Michele, 1.10
Garrett, John Work, 1.17, 1.21, 2.29, 2.50, 2.61, 4.5, 5.99
Garrett, Mary Elizabeth (daughter of John W.), 2.34, 2.50, 2.52

Garrett, Rachel Harrison (Mrs. John W.), 2.34, 2.50, 5.99
Garrett, Robert (son of John W.), 5.99
Garrison-on-Hudson, N.Y., 340
Gascoyne-Cecil, Robert Arthur Talbot (Lord Salisbury), 6.31
Gates, Merrill Edwards, 6.169
Geissler tube, 3.12
Geistinger, Marie Charlotte Caecelia, 2.51
Geneva, N.Y., 4.7
Geneva, Switzerland, 1.12, 1.13, 1.19, 1.21, 2.4
Genth, Frederick Augustus, Jr., 6.171
Gerhard family, 3.1, 3.7, 5.2, 5.4, 5.27
German language
  American study of, 2, 5
  JMC's study of, 1.4, 1.5
German terminology, JMC's use of, 3.3, 98, 3.12, 5.17
German Theological School, 4.6
German universities, 1–7, 4.15. *See also names of individual universities*
  American interest in, 1–6, 18, 6.171
  breadth of interest in, 5, 9
  guides to, 2.18. *See also* Hart, James Morgan
  influence of on Cambridge, 219, 6.4
  JMC's opinion of, 10–11, 5.122, 5.127, 5.139, 6.24
  legal studies at, 3–4, 1.10, 1.19
  medical studies at, 3, 5.153
Germany, JMC touring, 25–26, 27–28, 1.6, 30, 113, 4.13, 116, 5.1, 5.82, 5.111, 5.116
Gerrans, Henry Tresawna, 5.75, 6.111
Gewandhaus (Leipzig), 84, 5.66, 5.68
Glessen, University of, 6
Gilbert and Sullivan, 2.9, 4.3, 5.54
Gildersleeve, Basil Lanneau, 1.17, 1.21, 48, 2.30, 2.55, 80, 3.8, 4.6, 5.99, 5.108
Gildersleeve, Eliza Fisher (Mrs. Basil L.), 2.34, 2.50
Gilman, Alice (daughter of Daniel C.), 2.34, 2.50
Gilman, Daniel Coit, 1, 2.49, 3.8, 4.7, 6.38

Gilman, Daniel Coit (cont.)

Institute (cont.)
JMC working at, 3.3–3.7, 3.9, 94, 96–99, 3.12, 5.16, 5.36, 5.119, 5.131, 5.136, 5.143, 5.145, 6.104, 6.112
Instruments and apparatus. *See also names of individual instruments*
anthropometric, 5.47, 5.66, 222, 6.50
charts, 6.97, 6.112
cost of, 5.9, 5.21, 5.50, 5.64, 5.100, 5.120, 6.37, 6.97, 6.112, 6.118, 6.129, 6.176
electric batteries, 3.12, 5.10, 5.55, 6.123
light sources and meters, 3.9, 3.12, 5.4, 5.6
manufacturers and suppliers of, 5.6, 6.105. *See also* Krille, Carl.
for reaction, 97, 3.12, 5.30, 5.50, 5.65
for stimulation, 2.41, 3.4, 3.10, 97, 3.12, 5.8, 5.17, 5.62, 5.80, 5.100, 6.105, 6.158
for time measurement, 2.41, 3.4, 3.10, 98–99, 3.12, 4.17, 5.8, 5.9, 5.17, 5.62, 5.80, 5.100, 6.37, 6.105
"Intellectual aristocracy," 220–221, 6.31, 6.61
International Health Exhibition, London, 113, 5.46, 5.111, 222
Inventor's Exhibition, London, 5.111
Iowa, University of, 3.1
Irving, Henry (John Henry Brodribb), 1.7
Isle of Wight, 31, 1.7
Italy, JMC touring, 36, 1.14, 44, 4.13, 5.82, 5.120, 207, 208, 5.147

Jackson, John Hughlings, 5.116, 5.121, 6.54, 6.137
Jaenecka, Martha, 6.82
James, William, 16–17, 48, 6.17, 335
Jastrow, Joseph, 2.41, 2.47, 2.57, 2.58, 96, 5.136, 6.10, 6.33, 6.38, 6.113, 337
Jastrow, Morris, Jr., 6.38
Jena, Germany, 5.91
Jesus College, Oxford, 6.13
Joachim, Joseph, 3.4, 3.5
John, Richard Eduard, 1.4

Johns Hopkins University, 4, 2.49
biology at, 2.3, 2.39
expenses at. *See* Student expenses
JMC applying to be Fellow by Courtesy, 80, 4.6
JMC experimenting at, 2.26, 2.27, 2.32, 2.41, 2.45, 2.51, 2.54, 5.17, 5.22, 5.23, 5.57
JMC's application for fellowship, 1.12, 1.13, 1.15, 1.17–1.23
JMC's application for renewal of fellowship, 2.42, 2.54, 2.57–2.61, 80, 2.62
JMC's second application for fellowship, 5.144, 5.148, 5.152
philosophy at, 47–48, 2.2, 2.4, 2.9, 2.42. *See also* Metaphysical Club
physiological laboratory at, 2.2, 2.9, 2.23, 2.25
physiology at, 48–49, 2.2, 2.23–2.25, 2.31, 2.47, 2.54
professorship of philosophy at, 47–48, 4.5, 4.6, 6.2, 6.148
psychological laboratory at, 2.26, 2.41
psychology at, 48, 2.23, 2.24, 2.26, 2.27, 2.31, 2.35, 2.41, 5.18, 5.136, 6.38
societies at. *See* Metaphysical Club; Scientific Association
student life at. *See* Student life
Jones, George, 4.10, 5.46
Jones, John Sparhawk, 2.43
Journal of Physiology, 5.110
Journal of Speculative Philosophy, 49
Joynes, Edward S., 3

Kant, Immanuel, 6.67, 6.95
Kassel, 26, 4.2
Keats, John, 2.28
Keith family, 3.1
Kessell, Quinta Cattell, xviii
Kew Botanical Gardens, 1.3
Keynes, Florence Ada (Mrs. John Neville), 6.151
Keynes, John Maynard, 6.151
Keynes, John Neville, 6.114, 6.118, 6.149, 6.166
Kiel, University of, 6.93

Psychology (cont.)
at Johns Hopkins, 48, 2.23, 2.24,
2.26, 2.27, 2.31, 2.35, 2.41, 5.18,
5.136, 6.38
at Lafayette, 15
in Britain
at Cambridge, 218, 220, 6.16, 6.67,
6.86, 6.102, 6.118, 6.127, 6.155,
6.165–6.168, 6.178, 6.179
in London, 218, 221. *See also* Psychological Club, London; Robertson,
George Croom
in Germany, 6–11, 5.137, 6.47
American interest in, 10–11, 5.137
British interest in, 218, 220, 6.2,
6.13, 6.16, 6.23
at Göttingen, 7–8, 6.39. *See also*
Lotze, Rudolph Hermann
at Leipzig, 9–11, 1.15, 3.2, 3.3, 3.11,
5.137. *See also* Heinze, Max; Wundt,
Wilhelm
"Psychometry," JMC's use of, 3.7, 3.10,
96, 5.17, 5.22, 5.46, 5.130
Psychophysics, 3.12, 5.57, 5.117,
5.136, 6.10, 6.158, 6.169, 6.173,
334–335
Publishers, American, 2.31, 5.22
Punch, 3.7

Queen's University, Belfast, 4.14

Racine, Jean Baptiste, 1.12
Railroads
American, 4.7
British, 221, 6.25, 6.60, 6.170
continental, 19, 3.1, 5.91, 5.147
Raphael, 2–3, 1.6
Rauch, Christian Daniel, 1.6
Rayleigh, Lord (John William Strutt),
220, 6.31
Reaction-time experiments
by JMC
at Johns Hopkins, 2.41, 2.45, 2.48,
2.57, 3.1, 3.4, 3.5, 5.6, 5.17, 5.22,
5.23
at Leipzig, 3.3–3.7, 3.10, 96–99,
3.12, 5.6, 5.17, 5.22, 5.23, 5.47,
5.65, 5.73, 5.76, 5.80, fig. 8, fig. 10

at University of Pennsylvania, 6.105,
333–334, 338
by Frans Cornelius Donders, 96–98
by Sigmund Exner, 5.57
Galton's interest in, 222
by Joseph Jastrow, 6.10
by Wilhelm Wundt, 97–98, 3.12,
5.47, 333–334
Reaction-times controversy, 5.143,
6.93, 333–334
Reichstag, German, 5.125
Reid, Thomas, 15
Religion. *See* Theology and religion
Remsen, Ira, 2.55
Renan, Joseph Ernest, 1.10
Rensselaer Polytechnic Institute, 5.56
Reprinted articles, distribution of by
JMC and William C. Cattell, 5.61,
5.70, 5.99, 5.105, 5.117, 5.124,
5.138, 5.158, 6.41, 6.135
Rhoads, James E., 5.154, 6.121, 6.128,
6.130
Rhodes, Edward Hawksley, 6.13
Rienzi (Wagner), 3.5, 6.87
Der Ring des Niebelungen (Wagner), 6.109
Ritchie, David George, 6.13, 6.28,
6.115
Robertson, Caroline Anna Crompton
(Mrs. G. Croom), 207, 6.32, 6.75
Robertson, George Croom, 207, 218,
6.16, 6.74
as editor of *Mind*, 5.111, 5.130, 6.69,
6.103, 6.139
meeting with JMC, 5.111, 5.116, 5.155,
6.75, 6.80, 6.119, 6.131
and Psychological Club, 221, 6.25,
6.26, 6.52
Robinson, William Andrew, 1.19, 5.22
Rochester Theological Seminary, 6.72
Rockefeller, John D., 331
Rogers, Julia, 2.50, 2.52
Rolfe, Henry Winchester, 1.19, 1.22,
4.7, 5.13, 5.22, 6.153
Romanes, George John, 6.13, 6.22,
6.34
Rome, 44, 207, 5.144
Root, Elihu, 29–30
Rossetti, Dante Gabriel, 6.163